Weatherization and Energy Efficiency Improvement for Existing Homes

An Engineering Approach

MECHANICAL and AEROSPACE ENGINEERING

Frank Kreith & Darrell W. Pepper
Series Editors

RECENTLY PUBLISHED TITLES

The MEMS Handbook, Second Edition (3 volumes), *Edited by Mohamed Gad-el-Hak*

    MEMS: Introduction and Fundamentals

    MEMS: Applications

    MEMS: Design and Fabrication

Multiphase Flow Handbook, *Edited by Clayton T. Crowe*

Nanotechnology: Understanding Small Systems, Second Edition, *Ben Rogers, Sumita Pennathur, and Jesse Adams*

Nuclear Engineering Handbook, *Edited by Kenneth D. Kok*

Optomechatronics: Fusion of Optical and Mechatronic Engineering, *Hyungsuck Cho*

Practical Inverse Analysis in Engineering, *David M. Trujillo and Henry R. Busby*

Pressure Vessels: Design and Practice, *Somnath Chattopadhyay*

Principles of Solid Mechanics, *Rowland Richards, Jr.*

Principles of Sustainable Energy, *Frank Kreith*

Thermodynamics for Engineers, *Kau-Fui Vincent Wong*

Vibration and Shock Handbook, *Edited by Clarence W. de Silva*

Vibration Damping, Control, and Design, *Edited by Clarence W. de Silva*

Viscoelastic Solids, *Roderic S. Lakes*

Weatherization and Energy Efficiency Improvement for Existing Homes: An Engineering Approach, *Moncef Krarti*

Weatherization and Energy Efficiency Improvement for Existing Homes

An Engineering Approach

Moncef Krarti

CRC Press
Taylor & Francis Group
Boca Raton   London   New York

CRC Press is an imprint of the
Taylor & Francis Group, an **informa** business

Cover image courtesy of Hajer Tnani Krarti.

CRC Press
Taylor & Francis Group
6000 Broken Sound Parkway NW, Suite 300
Boca Raton, FL 33487-2742

First issued in paperback 2017

© 2012 by Taylor & Francis Group, LLC
CRC Press is an imprint of Taylor & Francis Group, an Informa business

No claim to original U.S. Government works
Version Date: 20120112

ISBN 13: 978-1-138-07608-2 (pbk)
ISBN 13: 978-1-4398-5128-9 (hbk)

Library of Congress Cataloging-in-Publication Data

Krarti, Moncef.
  Weatherization and energy efficiency improvement for existing homes : an engineering approach / Moncef Krarti.
    pages cm -- (Mechanical engineering series)
    Includes bibliographical references and index.
    ISBN 978-1-4398-5128-9 (hardback)
    1. Dwellings--Energy conservation. I. Title.

TJ163.5.D86K73 2012
696--dc23   2011049468

Visit the Taylor & Francis Web site at
http://www.taylorandfrancis.com

and the CRC Press Web site at
http://www.crcpress.com

Contents

Preface

Buildings account for 40% of the total primary energy consumption in the world. The contribution of residential buildings to the national energy use is significantly higher than that of commercial buildings. Indeed, household energy consumption is rising in several countries due to the desire for larger homes, expectation of better comfort levels, and use of more appliances.

In the last 5 years, significant investments have been being made, especially in the United States and Europe, to improve the energy efficiency of existing residential buildings through weatherization, energy auditing, and retrofitting programs. It is a consensus among all countries that well-trained energy auditors are essential to the success of these building energy efficiency programs. A recent U.S. study has found that most energy audits conducted for weatherization of residential buildings suffer from common deficiencies, including inadequate utility data analysis, limited scope of the evaluated energy efficiency measures, and inaccurate methods used for estimating energy savings and cost-effectiveness of the recommended retrofits. It is the purpose of this book to provide a training guide for energy auditing specific to weatherization programs targeting residential buildings. In particular, the book presents systematic and well-proven engineering analysis methods and techniques to reduce energy use and better implement weatherization programs for residential buildings.

The current book is organized in 12 chapters. The first four chapters provide basic approaches and fundamental engineering principles that are typically required to perform energy audits of residential buildings. Each of the following four chapters addresses a specific building subsystem or energy efficiency technology. The final four chapters provide basic engineering methods for net-zero retrofits of existing homes, approaches to verify and measure actual energy savings attributed to implementation of energy efficiency projects, reporting guidelines of energy audits, and case studies of both walkthrough and standard energy audits. Each chapter includes some worked-out examples that illustrate the use of some simplified analysis methods to evaluate the benefits of energy-efficient measures or technologies. Selected sets of problems and projects are provided at the end of most chapters to serve as review or homework problems for the users of the book. When using this book as a textbook or a reference, the instructor should start from Chapter 1 and proceed through Chapter 12. However, some of the chapters can be skipped or covered lightly, depending on the time constraints and the background of the students. First, general procedures suitable for building energy audits are presented (Chapter 1). Then, analysis methods are briefly provided for economic evaluation of energy efficiency projects (Chapter 2), and energy modeling and simulation of residential buildings (Chapter 3). Screening methods are outlined to assess the energy efficiency of existing homes (Chapter 4). In residential buildings, improvements to building envelope components including adding thermal insulation and reducing air leakage are essential for weatherization programs (Chapter 5). Various approaches and technologies to reduce electrical energy use are also outlined (Chapter 6). Proven measures to improve the energy efficiency of heating and cooling systems of existing residential buildings are discussed in detail (Chapter 7). Cost-effective measures to improve water management inside and outside residential buildings are also presented (Chapter 8). Analysis methods used for the measurement and verification of actual energy savings attributed to energy retrofits of residential buildings are listed with relevant examples (Chapter 9). Renewable energy technologies, optimization techniques, and analysis approaches suitable to perform net-zero retrofits of residential buildings are presented (Chapter 10). Finally, general guidelines to draft reports after completing energy audits are presented (Chapter 11) with specific example reports associated with two case studies (Chapter 12).

A special effort has been made to use metric (SI) units throughout the book. However, in several chapters English (IP) units are also used since they are still the standard set of units used in the United States. Conversion tables between the two unit systems (from English to metric and metric to English units) are provided. Moreover, weather data as well as thermal properties of construction materials and prototypical characteristics of existing U.S. homes are provided in four appendices.

I wish to acknowledge the assistance of several of my students at the University of Colorado at Boulder in providing some case studies and examples for walkthrough, utility data analysis, and detailed energy modeling. The encouragement of Dr. Frank Kreith was highly appreciated throughout the writing process. Finally, I am greatly indebted to my wife, Hajer, and my children for their continued patience and support throughout the preparation of this book.

Moncef Krarti
June 2011

About the Author

Moncef Krarti, PhD, PE, LEED˚AP, professor and director, Building Systems Program, Civil, Environmental, and Architectural Engineering Department at the University of Colorado, has vast experience in designing, testing, and assessing innovative energy efficiency and renewable energy technologies applied to buildings. He also directed several projects in energy management of buildings. In particular, he has conducted over 1,000 energy audits of various residential buildings and weatherization programs. He has published a textbook on energy audits of building systems that is widely used to teach energy audit techniques. Moreover, he has conducted several training workshops and courses in energy analysis of building energy systems using state-of-the art measurement and simulation techniques. In addition to his experience as an international consultant in energy efficiency, Professor Krarti has published over 200 technical journals and handbook chapters in various fields related to energy efficiency and energy conservation. He is active in several professional societies, including ASME, ASHRAE, and ASES. As part of his activities as a professor at the University of Colorado, he has managed the energy management center at the University of Colorado. He has also helped the development of similar energy efficiency centers in other countries, including Brazil, Mexico, and Tunisia. Dr. Krarti has extensive experience in promoting building energy efficiency technologies and policies overseas, including the development of building energy codes and energy efficiency training programs in several countries, including Tunisia, Sri Lanka, Egypt, and collaborative research with over 10 countries in Europe, Africa, Asia, and South America.

1

Energy Audit Procedures

Abstract

This chapter provides an overview of a general energy audit procedure that is suitable for residential buildings. Today, energy auditing is commonly performed by energy service companies to improve the energy efficiency of buildings. Indeed, energy auditing has a vital role for the success of performance contracting projects. There are several types of energy audits that are commonly performed by energy service engineers with various degrees of complexity. This chapter describes briefly the key aspects of a detailed energy audit and provides a comprehensive and systematic approach to identify and recommend cost-effective energy conservation measures for buildings.

1.1 Introduction

In order to obtain the desired energy use and cost savings from any weatherization project implementation, it is essential that the energy audit be well carried out using proven and sound methodology. Without a good approach for the energy audit, it is difficult to achieve the expected energy savings even after good implementation of all the retrofit recommendations for the weatherization project. Several problems have been identified for conducting energy audits of buildings, including residential dwellings (Shapiro, 2011). The common problem areas include:

1. Poor description of building energy systems: For residential buildings, it is important to describe the thermal performance of the shell, including wall and roof insulation level, infiltration rate, and window type. Moreover, a basic description is needed for the electrical systems, including the lighting fixtures, plug loads, and appliances, as well as for the heating and cooling systems, including motor types and control strategies. A walkthrough audit of the building should help the energy auditor identify most, if not all, the basic features of the desired energy systems.

2. Poor analysis of utility data: As discussed in Chapter 4, monthly utility bills can provide crucial information about the energy performance and level of energy inefficiency of buildings, especially residential dwellings. Using simple analysis techniques, the monthly utility data can be used to provide a benchmarking model to assess the effectiveness of various energy efficiency measures.

3. Inadequate economic analysis: Without accurate and reasonable estimates of the cost of installing energy efficiency strategies, the predictions of the cost-effectiveness as well as of the required initial budget for implementing the energy audit recommendations can be wrongly estimated. Thus, there is a large risk that the audit recommendations for energy efficiency improvement will be abandoned and not considered. Moreover, the use of simplified economic analysis methods such as simple payback period without taking into account the value of money and the energy cost escalation may lead to incorrect recommendations. In particular, simple payback analysis does not capture the merit of two energy efficiency measures having the same payback period. Sound economic analysis using life cycle costing should be considered to finalize any energy audit recommendations for energy improvement. Chapter 2 describes in detail various economic analysis methods, including life cycle cost analysis.

4. Limited energy efficiency measures: Most energy auditors for residential buildings focus on building envelope improvement measures and fail to consider simple, yet cost-effective strategies to reduce energy use of lighting systems and HVAC systems. An energy audit should consider all the building energy systems and identify measures that substantially reduce the overall energy use so their cost-effectiveness can be viable. To help the energy auditing process, it is important the energy end uses for the building can be estimated either through utility data analysis or energy modeling techniques, as outlined in Chapter 6.

5. Inadequate energy saving estimation: The cost-effectiveness of energy efficiency measures depends on their potential in reducing energy use and cost. Most energy auditors use simplified calculation methods to predict annual energy use reduction from the recommended improvement strategies. Due to various factors such as inadequate baseline energy use modeling or erroneous assumptions and calculation methods, predictions of energy savings are often overestimated and consequently lead to wrong recommendations. Chapters 5 through 9 provide detailed calculation methods and techniques to ensure adequate estimates of energy savings for various energy efficiency measures suitable for residential buildings.

It is interesting to note that a review study (Shapiro, 2011) found that energy audits for residential buildings are more likely to have problems such as those listed above than energy audits for commercial buildings. The difference in training level is a possible cause for this finding. Indeed, most energy auditors for single-family homes are trained to focus on building envelope features and do not have sufficient expertise in lighting, HVAC, and domestic hot water (DHW) systems.

This chapter describes a general but systematic procedure for energy auditing suitable for residential buildings. Some of the commonly recommended energy conservation measures are briefly discussed. Finally, an overview of basic characteristics of existing US homes is provided.

1.2 Energy Use Associated to Residential Buildings

1.2.1 Background

Buildings account for 40% of the total primary energy consumption in the world. In some countries, buildings consume over 50% of the national energy use, as outlined in Figure 1.1. The energy use by residential buildings is higher than that for commercial buildings in almost all countries, resulting in a rather high contribution of households to the national primary energy use for several countries, as illustrated in Figure 1.2. In fact, the residential energy consumption is rising in all regions of the world due to the desire of larger homes, the expectation of better comfort levels, and the use of more appliances. Worldwide, households consumed about 82 EJ (i.e., 10^{18} joules) in 2005 (IEA, 2009). The growth of household energy consumption from 1990

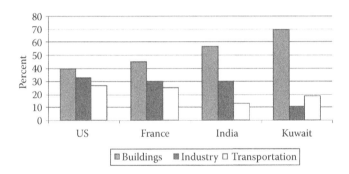

FIGURE 1.1 Energy use by sector for selected countries.

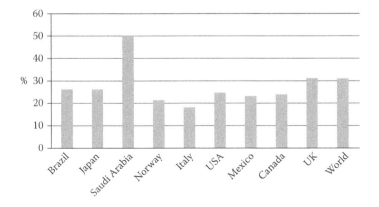

FIGURE 1.2 Percent of residential sectors on national energy use for selected countries.

TABLE 1.1   Worldwide Household Energy End-Use Distribution for 1990 and 2005

Energy End Use	1990	2005
Space heating	58%	53%
Appliances	16%	21%
Hot water heating	17%	16%
Lighting	4%	5%
Cooking	5%	5%

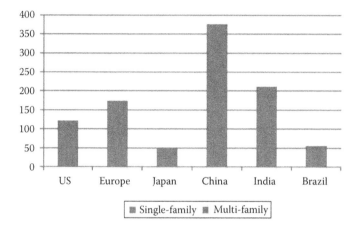

FIGURE 1.3   Number of single-family and multifamily housing units for selected countries.

to 2005 has been 22% for Organization for Economic Cooperation and Development (OECD) countries, compared to 18% for non-OECD countries. As noted in Table 1.1, space heating accounts for over 53% of the total household primary energy use.

Figure 1.3 shows the number of households living in single-family homes and multifamily apartment units for selected countries. Except for China, most households live in single-family homes rather than apartment units. Typically, the energy consumption of a single-family home is larger than that of an apartment unit. The energy end uses for a typical household in selected countries are presented in Figure 1.4. It is clear that in most countries, space heating and domestic water heating use the most energy, except for India, where cooking accounts for most of the energy used by households due to the heavy reliance on biomass with low energy conversion efficiency.

In the following section, a more detailed overview of the energy consumption attributed to residential buildings is provided for selected representatives of a wide range of developed and developing countries. It should be noted that energy efficiency programs for both new and existing buildings have been or are being implemented in almost every country in order to reduce energy consumption associated to the building sector. These energy efficiency programs vary widely in terms of scope and effectiveness. In several countries, the lack of well-trained energy analysts and auditors is the main barrier for improving the energy efficiency of buildings.

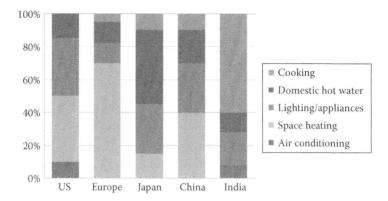

FIGURE 1.4   Average household energy end-use distribution for selected countries.

1.2.2 Energy Performance of Residential Buildings for Selected Countries

In this section, energy performance and energy end uses for existing residential buildings for selected countries are briefly discussed. These countries are representatives of various regions throughout the world.

1.2.2.1 United States

Table 1.2 summarizes the average energy end uses for U.S. households by housing type and number of occupants (DOE, 2005). As noted earlier, space heating accounts for a significant portion of typical U.S. households, especially for units located in small apartment buildings (with two to four units). Moreover, the energy use attributed to appliances represents over 25% of total energy consumed by an average U.S. household. As indicated in Table 1.2, an average apartment unit uses about half the annual energy of a single-family home. All the main energy uses are significantly higher in single-family homes. Individual apartments use considerably less energy than single-family homes owing to their smaller floor area, smaller household size, and lower exterior wall area. In particular, apartments in buildings containing five or more units use about half the heating energy and half the energy for lighting and other appliances of an average single-family home.

In the United States, 45% of multifamily housing stock was built before 1970, and only 14% was built after 1990, with more modern building efficiency. In this section, a more detailed description of the characteristics of U.S. single-family detached homes is presented, depending on their location and vintage.

1.2.2.2 France

Most of the residential buildings in France are detached single-family homes with about 14.5 million units representing over 60% of the housing stock. Table 1.3 summarizes the evolution of some characteristics of the residential buildings in France over the last two

TABLE 1.2 Average Energy End-Use Consumption for a U.S. Household

	U.S. Households (millions)	Energy End Uses (million Btu of consumption per household)					
		All End Uses	Space Heating (Major Fuels)	Air Conditioning	Water Heating	Refrigerators	Other Appliances and Lighting
Total	111.1	94.9	40.5	9.6	19.2	4.6	24.7
			Type of Housing Unit				
Single-family detached	72.1	108.4	44.2	11.0	21.7	5.2	29.3
Single-family attached	7.6	89.3	41.7	6.7	19.0	4.0	20.9
Apartments with 2–4 units	7.8	85.0	48.5	6.3	15.6	3.5	16.3
Apartments with 5+ units	16.7	54.4	25.0	6.6	12.2	3.0	11.8
Mobile homes	6.9	70.4	26.1	9.2	13.3	4.2	21.4
			Household Size				
1 person	30.0	70.7	37.4	6.1	11.7	3.9	14.4
2 persons	34.8	96.4	41.9	10.1	18.5	4.9	24.4
3 persons	18.4	104.1	41.4	10.7	21.7	5.0	28.8
4 persons	15.9	108.4	41.0	11.4	24.2	4.8	31.4
5 persons	7.9	117.1	41.9	13.1	27.2	4.9	34.5
6 or more persons	4.1	123.8	41.7	12.8	33.3	4.9	38.4

Source: DOE, *A Look at Residential Energy Consumption in 1997*, DOE/EIA-0632 (97), U.S. Department of Energy, 2005.

TABLE 1.3 Evolution of Housing Characteristics in France

Characteristic	1988	1992	1996	2002	2006
Housing units built before 1949 (%)	39.1	36.8	35.6	33.2	30.6
Average floor area per housing unit in m²	85	86	88	90	91
Number of occupants per housing unit	2.6	2.5	2.5	2.4	2.3
Housing units with hot water (%)	9.6	6.2	4.1	2.6	1.5

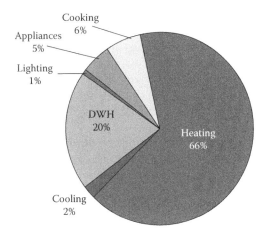

FIGURE 1.5 Energy end uses for the 2005 housing stock in France.

decades. As indicated in Table 1.3, a significant part of the housing stock in France is old, with over 30% built before 1949. In fact, the replacement rate of the housing stock in France is very low (about 0.2% per year), with over 60% of the residential buildings built before 1975. Due to their age, it is estimated that 80% of the current homes need retrofitting to improve their energy efficiency. As shown in Figure 1.5, the energy end uses for the housing stock in France for 2005 are mostly attributed to heating (WBCSD, 2009). In fact, space and water heating account for 86% of the total energy consumption of the housing stock in France.

1.2.2.3 China

The building sector accounts for 30% of the national primary energy use in China, the most populous country in the world. The building sector has been growing steadily over the last decade. Currently, China builds 2 billion m² of new buildings every year. However, the majority of these buildings are not energy efficient. Among 43 billion m² of the total building stock in China, including 10 billion m² of urban housing, only 4% of them have some measures of energy conservation features.

China's household energy consumption jumped from 2.69 EJ in 1980 to 7.49 EJ in 2006, with an annual growth rate of 4.0%. The share of electricity of household energy consumption rose from 4.7% in 1980 to 52.6% in 2006, which has surpassed coal as the most popular fuel type at home since 2000 (Shui et al., 2009). As indicated in Figure 1.4, space heating uses 40% of the total energy consumed in the Chinese residential building stock. It should be noted that energy consumption per household in rural China is high compared to energy used by a household located in an urban area. Indeed, 80% of the energy used by a rural Chinese household consists of wood or agricultural waste, with a low efficiency of conversion.

1.2.2.4 Mexico

Residential energy consumption in Mexico is steadily increasing, particularly as the urban population continues to grow. Similarly, the number of homes in Mexico is also rising; between 1996 and 2006 housing units increased from 20.4 million to 26 million, where nearly 78% were urban homes as of 2006. Urban homes also have greater quantities of equipment, such as refrigerators, washing machines, televisions, etc., and have subsequently higher annual energy consumption per household. Furthermore, it is estimated that the number of housing units will reach nearly 50 million by 2030, emphasizing the importance of implementing energy efficiency measures in both new and existing homes.

Furthermore, energy consumption from the residential sector in Mexico makes up nearly 16% of the nation's total demand. This makes domestic energy the third largest end user in the country, only behind the transportation and industrial sectors, as illustrated in Figure 1.6(a). It is also important to disaggregate total energy consumption into the usable energy sources. Figure 1.6(b) shows the total residential energy consumption by end use, in petajoules, for 2008. This indicates that the highest area of energy consumption is liquid petroleum gas (LPG), followed by wood and electricity. Although the total residential electricity consumption is less than that for LPG and wood, many of the national energy conservation efforts are geared toward reducing electricity use.

Figure 1.7 provides the energy end uses for two prototypical homes in Mexico. When the home is not air conditioned, lighting accounts for over 40% of the total electricity used, as shown in Figure 1.7(a). However, when the home is air conditioned, space cooling is dominant and uses 28% of the total energy used, as indicated in Figure 1.7(b) (Griego et al., 2011). For fuel use (typically LPG), hot water heating accounts for almost 60% of the total home's consumption. Hot water is used for both space heating and domestic applications.

1.2.2.5 Australia

Transport and industry are the main sectors that consume primary energy in Australia. Residential buildings account for about 12% of the national primary energy use, with a total consumption estimated at 400 PJ in 2001 (Dickson et al., 2003). Energy consumption in the residential sector is projected to increase at 2.1% annually until 2020. It is hoped that renewable energy, especially solar hot water systems, would displace about 7 PJ in 2001 (Dickson et al., 2003). Energy consumption in the residential sector is projected to increase at 2.1% annually until 2020. It is hoped that renewable energy, especially solar hot water systems, would displace about 7 PJ of electricity used

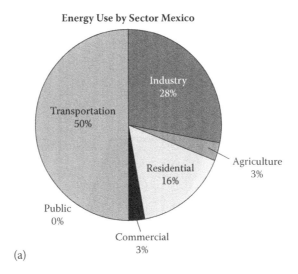

FIGURE 1.6 (a) Total energy consumption by sector in Mexico, 2008 (4,815 PJ). (b) Total national residential energy consumption by source, 2008 (750 PJ).

by households. Electricity and natural gas remain the primary energy resources for households in Australia. Figure 1.8 indicates typical energy end use distributions for an Australian household. Space heating and domestic hot water alone account for more than 80% of the total primary energy consumption for a typical household.

1.2.2.6 Tunisia

The building sector in Tunisia has evolved and will continue to grow in the next decade, as shown in Figure 1.9, illustrating the increase in national energy end use of buildings from 21% in 1992 to 37% in 2020. This growth is mostly attributed to a significant

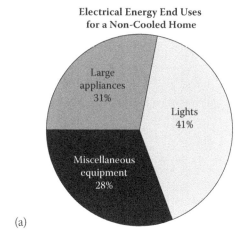

(a)

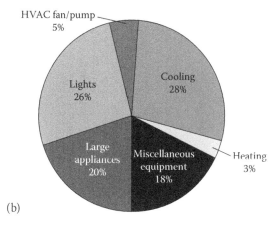

(b)

FIGURE 1.7   Energy end uses for (a) non-air-conditioned home (electricity use only), (b) air-conditioned home (electricity only), and (c) LPG end uses for a typical home in Mexico.

increase in the residential building stock from 1 million in 1972 to over 2.7 million in 2004 due to significant urban population (TEMA, 2010). Based on a national survey, it is estimated that 85% of Tunisians own their homes, and that each housing unit has an average floor area of 1,000 ft² (100 m²). The total primary energy use attributed to the Tunisian residential buildings has been evaluated at 47.0 PJ in 2004, with 64% associated to electricity use. Each housing unit consumes on average 1,153 kWh per year of electricity, of which 39% is attributed to the refrigerator energy use. Most households use natural gas for space heating and cooking use. Figure 1.10 indicates the primary energy end use distribution for a typical housing unit in Tunisia (TEMA, 2010). Space heating, domestic hot water heating, and refrigeration account for almost 80% of the primary energy needs for a typical Tunisian household.

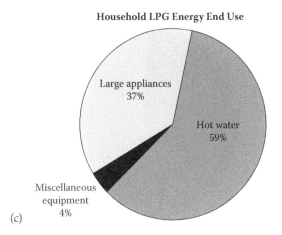

(c)

FIGURE 1.7 *(Continued)*

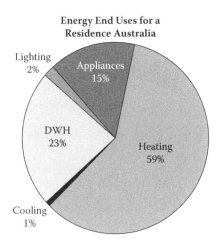

FIGURE 1.8   Energy end uses for a typical home in Australia.

1.3 Types of Energy Audits

The term *energy audit* is widely used and may have different meanings, depending on the energy service companies. Energy auditing of buildings can range from a short walk-through of the facility to a detailed analysis with hourly computer simulation. Generally, four types of energy audits can be distinguished, as briefly described below. A more detailed description of these types of energy audit is provided by Krarti (2010).

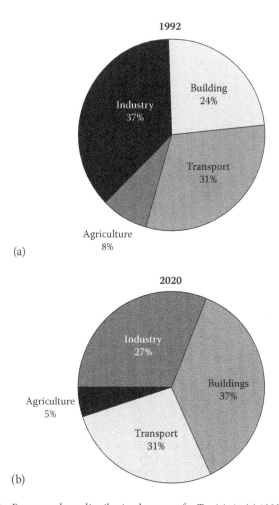

FIGURE 1.9   Energy end use distribution by sector for Tunisia in (a) 1992 and (b) 2020.

1.3.1 Walkthrough Audit

This audit consists of a short on-site visit of the facility to identify areas where simple and inexpensive actions can provide immediate energy use or operating cost savings. Some engineers refer to these types of actions as operating and maintenance (O&M) measures. Examples of O&M measures include setting back heating set-point temperatures, replacing broken windows, and insulating exposed hot water or steam pipes. Table 1.4 illustrates typical findings from a walkthrough audit and potential energy improvement measures that the auditor can recommend. A sample of a walkthrough audit for a residence is provided in Chapter 12.

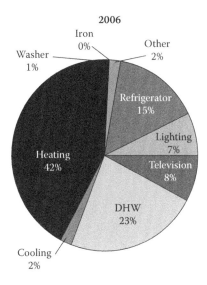

FIGURE 1.10   Energy end uses for a typical home in Tunisia.

1.3.2 Utility Data Analysis

The main purpose of this type of audit is to carefully analyze the energy use of the residential building. Typically, the utility data over several years are evaluated to identify the patterns of energy use, including weather-dependent and -independent loads. Using regression analysis, the auditor can identify the actual thermal performance of the home, often referred to as the energy signature, and assess if a detailed audit is warranted. Screening techniques suitable for evaluating the thermal performance of residential buildings using monthly utility data are introduced in Chapter 3. Figure 1.11 provides an example of regression analysis to analyze the correlation between monthly natural gas usage and outdoor air temperature for a home.

1.3.3 Standard Energy Audit

The standard audit provides a comprehensive energy analysis for the energy systems of the facility. In addition to the activities described for the walkthrough audit and for the utility cost analysis described above, the standard energy audit includes the development of a baseline for the energy use of the facility and the evaluation of the energy savings and the cost-effectiveness of appropriately selected energy conservation measures. The step-by-step approach of the standard energy audit is similar to that of the detailed energy audit, which is described in the following section.

Typically, simplified tools are used in the standard energy audit to develop baseline energy models and to predict the energy savings of energy conservation measures. Among these tools are the degree-day methods and linear regression models (Fels, 1986). In addition, a simple payback analysis is generally performed to determine the

TABLE 1.4 Summary of Walkthrough Audit Findings and Potential Energy Efficiency Measures

	Existing Conditions	Potential Improvements
	Constructions	
Roof	Shingles, no insulation	Add batt or blown-in insulation in the attic
Walls	Siding, 2 × 6 frame construction, batt insulation, gypsum board	
Windows	Clear double-pane glazing Cracks around the frames	Weather stripping and caulking around the window frames
Doors	North façade: one 7 × 3 ft solid wood door	
	Window/Wall Ratio	
North façade	15%	
East façade	16%	
South façade	23%	
West façade	4%	
	Occupancy	
Schedule	2 adults, 24/7	
	2 children, 4 p.m.–8 a.m. M–F, 24/7 S/S	
Lighting fixtures	Some incandescent lamps, mostly CFLs	Replace incandescent with CFLs
Lighting power density	0.75 W/ft²	
Plug load	Most appliances are Energy Star labeled	Use power strip to reduce phantom loads
Equipment power density	0.25 W/ft²	
	HVAC	
Type	Forced-air gas furnace	
Efficiency	80%	Furnace needs to be replaced with high-efficiency furnace
Set point	70°, except 65° from 9 a.m. to 5 p.m.	Lower thermostat setback to 60 to 62°F during the unoccupied hours
	DHW	
Capacity	40 gallons	
Efficiency	85%	

cost-effectiveness of energy conservation measures. Examples of standard audits are provided in Chapter 12.

1.3.4 Detailed Energy Audit

This energy audit is the most comprehensive but also time-consuming. Specifically, the detailed energy audit includes the use of instruments to measure energy use for the whole building or for some energy systems within the building (for instance, by end uses: lighting systems, plug loads, and heating and cooling equipment). In addition, sophisticated

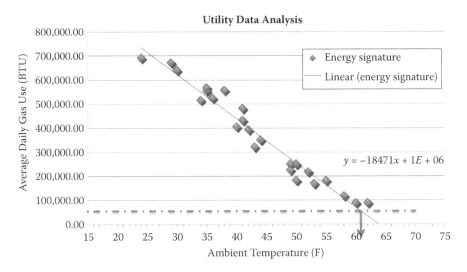

FIGURE 1.11 Utility data analysis for a home using monthly gas usage and outdoor ambient temperature.

computer simulation programs are typically considered for detailed energy audits to evaluate and recommend energy retrofits for the building. Figure 1.12 illustrates three-dimensional views of an audited house and its associated detailed hourly energy simulation model.

The techniques available to perform measurements for an energy audit are diverse. During on-site visits, handheld and clamp-on instruments can be used to determine the variation of some building parameters, such as the indoor air temperature, the luminance level, and the electrical energy use. When long-term measurements are needed, sensors are typically used and connected to a data acquisition system so measured data can be stored and be remotely accessible.

The computer simulation programs used in the detailed energy audit can typically provide the energy end-use distribution. The simulation programs are often based on dynamic thermal performance of the building energy systems and typically require a high level of engineering expertise and training. Moreover, the detailed energy models have to be calibrated to utility data or long-term measurements. Figure 1.13 shows a comparative analysis between monthly utility data and predictions from a calibrated detailed energy simulation model for an audited house. Refer to Chapter 3 for a more detailed discussion of the energy analysis tools that can be used to estimate energy and cost savings attributed to energy conservation measures.

In the detailed energy audit, more rigorous economical evaluation of the energy conservation measures is generally performed. Specifically, the cost-effectiveness of energy retrofits may be determined based on the life cycle cost (LCC) analysis rather than the simple payback period analysis. LCC analysis takes into account a number of economic parameters, such as interest, inflation, and tax rates. Chapter 2 describes some of the basic economic analysis methods that are often used to evaluate energy efficiency projects.

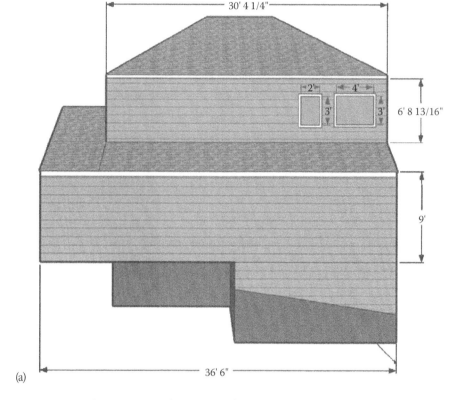

(a)

FIGURE 1.12 Three-dimensional renderings of the actual (a) audited house and (b) its thermal model.

Table 1.5 provides a summary of the energy audit procedure recommended for residential buildings. Energy audits for thermal and electrical systems are separated since they are typically subject to different utility rates.

1.4 Characteristics of Existing U.S. Homes

Energy use for residential housing across the United States is about 8.4% of the total U.S. energy consumption. In recent years, the new housing sector has made notable improvements in energy efficiency due to more stringent codes and standards. Moreover, several programs have been developed to increase the energy efficiency of new residential homes, such as the U.S. Department of Energy (DOE) Building America program (DOE, 2009) and the Energy Star® program (Energy Star, 2009) for homes. However, new homes represent only 8% of the U.S. total housing stock; homes built before 2000 represent 92% (DOE, 2005). To reduce energy use in the residential building sector, older homes need to be retrofitted so their energy performance can be improved.

Retrofit efforts have so far been limited for the residential buildings. The largest effort for retrofitting existing housing includes the Energy Star Home Performance program

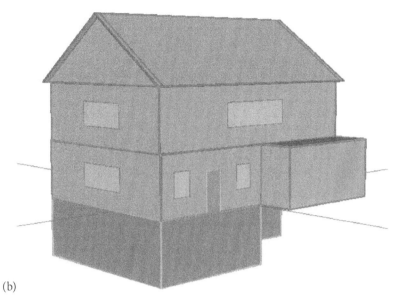

(b)

FIGURE 1.12 *(Continued)*

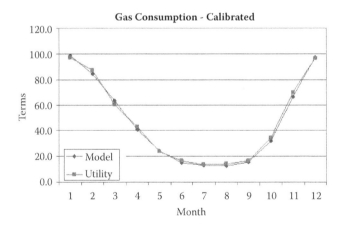

FIGURE 1.13   Comparison between utility data and house energy simulation model predictions.

developed by the USEPA and the USDOE. The Home Performance program offers a whole house approach to improve both energy efficiency and comfort (Energy Star, 2009). This program provides training and technical support to local contractors to perform energy audits and retrofits for interested homeowners. Nonprofit organizations and energy utility companies from several states are currently participating in this program. For example, Xcel Energy (2009) in Colorado offers home energy audits and incentives programs for energy efficiency. Similarly, the Energy Trust of Oregon (2009) offers home energy audits and cash incentives for implementing retrofit projects. In

TABLE 1.5   Summary of Energy Audit Approaches for Residential Buildings

Phase	Thermal Systems	Electric Systems
Utility analysis	Thermal energy use profile (building signature)	Electrical energy use profile (building signature)
	Thermal energy use per unit area	Electrical energy use per unit area
	Thermal energy end uses (heating, DHW)	Electrical energy end use distribution (cooling, lighting, appliances, fans, etc.)
	Fuel type used	
	Weather effect on thermal energy use	Weather effect on electrical energy use
	Utility rate structure	Utility rate structure
On-site survey	Construction materials (thermal resistance type and thickness)	HVAC system type
		Lighting type and density
	HVAC system type	Equipment type and density
	DHW system	Energy use for heating
	Hot water/steam use for heating	Energy use for cooling
	Hot water/steam for cooling	Energy use for lighting
	Hot water/steam for DHW	Energy use for equipment
	Hot water/steam for specific applications (hospitals, swimming pools, etc.)	Energy use for air handling
		Energy use for domestic hot water
Energy use baseline	Review architectural, mechanical, and control drawings	Review architectural, mechanical, electrical, and control drawings
	Develop a base case model (using any baselining method ranging from very simple to more detailed tools)	Develop a base case model (using any baselining method ranging from very simple to more detailed tools)
	Calibrate the base case model (using utility data or metered data)	Calibrate the base case model (using utility data or metered data)
Energy conservation measures	Heat recovery system (heat exchangers)	Energy-efficient lighting
	Efficient heating system (furnaces)	Energy-efficient appliances
	Temperature setback/setup	Energy-efficient motors
	Improved controls	HVAC system retrofit
	HVAC system retrofit	Improved controls
	DHW use reduction	Temperature setup
		Energy-efficient cooling system

order to assess the potential impact of regional and national retrofit programs, identification of the characteristics of existing housing stock is needed. Useful characteristics for homes include building envelope features (i.e., insulation level, glazing type, and air leakage), lighting system, plug load, heating and cooling equipment, age (vintage), location, and annual energy use.

Early efforts to identify U.S. home characteristics include studies by Ristchard et al. (1992), Anderson et al. (1985), Sherman and Dickerhoff (1998), and Huang and Gu (2002). Most of these studies utilize two common databases for the U.S. housing stock, including the U.S. Census Bureau's American Housing Survey (AHS) and the U.S. Department of Energy's (DOE) Residential Energy Consumption Survey (RECS). The two databases are briefly described:

> **AHS database:** Every two years, the U.S. Bureau of the Census collects data via surveys and interviews to develop a U.S. housing stock inventory of the United States. The main purpose of the AHS database is to measure specific changes

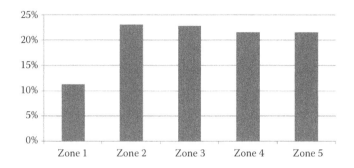

FIGURE 1.14 Distribution of single-family detached houses among climate zones.

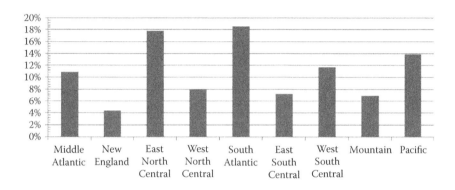

FIGURE 1.15 Distribution of single-family detached houses by region.

in the housing inventory and assess progress toward the goal of providing sustainable living conditions to all Americans.

RECS database: These data are a collection of statistical information covering the characteristics of the U.S. housing units, with emphasis on the consumption of and expenditures for energy in these housing units (DOE, 2005). The RECS survey is typically conducted every three or four years.

Both AHS and RECS databases are based on surveys of a small fraction of existing residential buildings. Indeed, there are over 100 million housing units in the United States, ranging from single-family homes to large apartment buildings. These housing units are distributed throughout the United States and are located in a wide range of climatic zones and regions, as illustrated by Figures 1.14 and 1.15 for single-family homes (DOE, 2005). Figure 1.16 describes the census U.S. regions, while Figure 1.17 shows DOE climate zones based on heating and cooling degree-days. Consequently, characteristics of these homes vary significantly, especially in terms of building envelope features and types of heating and cooling systems. Moreover, the age of the housing units can span over 200 years. In the United States most of the homes are built between 1960 and 2000, as indicated in Figure 1.18.

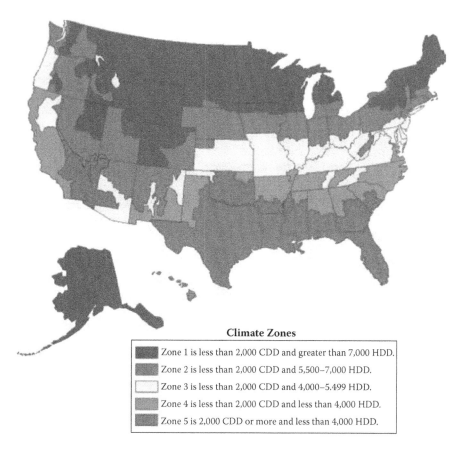

Climate Zones

Zone 1 is less than 2,000 CDD and greater than 7,000 HDD.

Zone 2 is less than 2,000 CDD and 5,500–7,000 HDD.

Zone 3 is less than 2,000 CDD and 4,000–5,499 HDD.

Zone 4 is less than 2,000 CDD and less than 4,000 HDD.

Zone 5 is 2,000 CDD or more and less than 4,000 HDD.

FIGURE 1.16   Department of Energy U.S. climate zones. (From DOE, *A Look at Residential Energy Consumption in 1997*, DOE/EIA-0632 (97), U.S. Department of Energy, 2005.)

Figure 1.19 shows the distribution of the houses per location and vintage. As expected, older houses are more predominant in the Northeast and Midwest, where colonization began and the older settlements are located. As indicated in Figure 1.19, most of the more recent construction activities took place in the South. The West had a slight construction peak during the 1970s and 1980s. Figure 1.20 shows the distribution of the existing U.S. house stock per type and vintage. Single-family residences represent about two-thirds of the U.S. households. As shown in Figure 1.20, only after the 1960s were mobile homes built in significant numbers.

Table 1.6 summarizes typical housing characteristics that are considered in both the AHS and RECS databases. Based on the characteristics, 80 prototypes for homes have been developed to represent regions, climate zones, and vintages (Albertsen et al., 2011). Specifically, 16 cities have been considered to represent both the census regions, as illustrated in Figure 1.17, and DOE climate zones, as outlined in Figure 1.16. For each city, five prototypes have been defined, depending on the vintage, including houses built. Appendix A provides the characteristics of these prototype homes. Table 1.7

FIGURE 1.17 DOE U.S. regions. ((From DOE, *A Look at Residential Energy Consumption in 1997*, DOE/EIA-0632 (97), U.S. Department of Energy, 2005.)

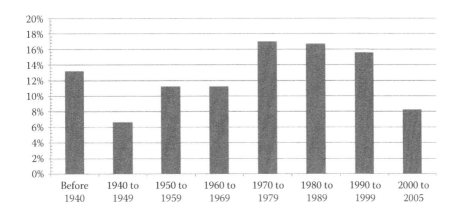

FIGURE 1.18 Distribution of all U.S. housing units by vintage.

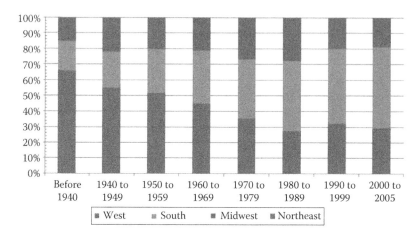

FIGURE 1.19   Distribution of U.S. housing stock by vintage. (From DOE, *A Look at Residential Energy Consumption in 1997*, DOE/EIA-0632 (97), U.S. Department of Energy, 2005.)

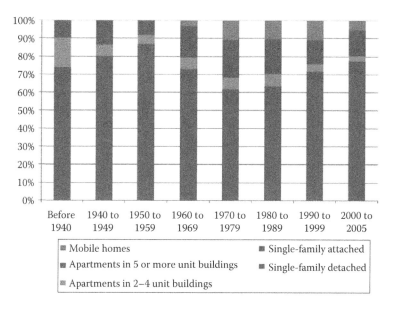

FIGURE 1.20   Distribution of U.S. housing stock by type and vintage. (From DOE, *A Look at Residential Energy Consumption in 1997*, DOE/EIA-0632 (97), U.S. Department of Energy, 2005.)

TABLE 1.6   Basic Characteristics
Surveyed for Each Housing Type

Variable	Variable Range or Level
Single Family (attached and detached)	
Forced-air distribution	Yes
	No
Floor area, m^2 (ft^2)	<149 (1,600)
	149 to 223 (1,600 to 2,399)
	>223 (2,400)
Year built	<1940
	1940–1969
	1970–1989
	1990–1997
Garage or carport	Yes
	No
Foundation type	Basement
	Crawl space
	Slab
Number of stories	1
	2
	≥3
Apartments	
No. units in building	2–4
	5–9
	10–19
	20–39
	>40
Floor area, m^2 (ft^2)	<93 (1,000)
	>93 (1,000)
Year built	<1940
	1940–1969
	1970–1989
	>1990
Central heating system	Yes
	No
Manufactured Homes	
Floor area, m^2 (ft^2)	<149 (1,600)
	>149 (1,600)
Year built	<1940
	1940–1969
	1970–1989
	>1990
Central forced-air system	Yes
	No

TABLE 1.7   Representative Cities and Climate Zones

Census Region	City	Heating Degree-Days	Cooling Degree-Days	Climate Zone
New England	Boston	5,822	888	2
Mid-Atlantic	New York	4,658	1,597	3
East North Central	Chicago	5,906	1,166	2
West North Central	Minneapolis	6,911	1,003	1
	Kansas City	4,638	1,635	3
South Atlantic	Washington	4,607	1,343	3
	Atlanta	2,678	1,843	4
	Miami	131	4,439	5
West South Central	Fort Worth	1,921	3,119	4
	New Orleans	1,150	3,204	5
Mountain	Denver	5,582	925	2
	Albuquerque	3,826	1,562	3
	Phoenix	755	4,714	5
Pacific North	Seattle	4,446	164	3
Pacific South	San Francisco	2,295	71	4
	Los Angeles	1,182	526	4

provides the heating and cooling degree days as well as the climatic zones for selected cities within census regions. Tables 1.8 and 1.9 illustrate the features of 10 prototypes, presenting two distinct climate zones and five vintages.

The characteristics listed for U.S. homes related to various vintages and regions in Appendix A can be used to assess the cost-effectiveness of various retrofit programs at the regional as well as national level. Chapter 9 outlines the most cost-effective energy efficiency measures for various U.S. regions as well as the best packages that have the lowest life cycle costs.

TABLE 1.8 Five Housing Archetypes for New England (Boston)

	Prototype A	Prototype A1	Prototype B	Prototype B1	Prototype C
Year Built	Pre-1950s	Pre-1950s	1950–1979	1950–1979	1980–2000
No. bedrooms	3	3	3	3	3
No. bathrooms	1	1	1	1	2
No. stories	2	2	2	2	2
Floor conditioned area (ft²)	2,090	2,090	1,626	1,626	2,092
Window area (ft²)	233	233	212	212	242
Window-to-wall ratio	0.12	0.12	0.12	0.12	0.12
Glazing layers	1	2	1	2	2
Windows	Wood	Wood	Wood	Wood	Wood
Wall siding type	Wood	Wood	Wood	Wood	Wood
Wall type	Plaster	Plaster	Drywall	Drywall	Drywall
R-values					
Wall	0	7	0	7	13
Ceiling	0	22	22	22	27
Floor	0	0	0	0	0
Foundation type	Basement	Basement	Basement	Basement	Basement
Foundation insulation	None	None	None	None	None
Garage	Detached	Detached	Attached	Attached	Attached
Heating equipment	Natural gas	Natural gas	Natural gas	Natural gas	Natural gas
Furnace efficiency AFUE	80%	80%	80%	80%	80%
Cooling equipment	Window	Window	Window	Window	Window
A/C efficiency SEER	7	7	8	8	10
Cooling set points	76	76	76	76	76
Heating set points	67	67	67	67	67
Infiltration ACH 50 Pa	0.7	0.7	0.7	0.7	0.4
Appliances	Non-ES[a]	Non-ES	Non-ES	Non-ES	Non-ES
Lighting (fluorescent)	14%	14%	14%	14%	14%

[a] Non-Energy Star.

TABLE 1.9   Five Housing Archetypes for Mountain Region (Phoenix)

	Prototype A	Prototype A1	Prototype B	Prototype B1	Prototype C
Year Built	Pre-1950s	Pre-1950s	1950–1979	1950–1979	1980–2000
No. bedrooms	3	3	3	3	3
No. bathrooms	1	1	1	1	2
No. stories	1	1	1	1	1
Floor conditioned area (ft²)	1,823	1,823	1,734	1,734	2,017
Window area (ft²)	159	159	141	141	238
Window-to-wall ratio	0.12	0.12	0.12	0.12	0.12
Glazing layers	1	1	1	1	1
Windows	Wood	Wood	Aluminum	Aluminum	Aluminum
Wall siding type	Wood	Wood	Brick	Brick	Stucco
Wall type	Plaster	Plaster	Drywall	Drywall	Drywall
R-values					
Wall	0	7	0	7	13
Ceiling	0	11	11	11	30
Floor	0	0	0	0	0
Foundation type	Basement	Basement	Slab	Slab	Slab
Foundation insulation	None	None	None	None	None
Garage	Detached	Detached	Attached	Attached	Attached
Heating equipment	Natural gas	Natural gas	Natural gas	Natural gas	Natural gas
Furnace efficiency AFUE	80%	80%	80%	80%	80%
Cooling equipment	Window	Window	Central	Central	Central
A/C efficiency SEER	7	7	8	8	10
Cooling set points	80	80	78	78	78
Heating set points	70	70	70	70	70
Infiltration ACH 50 Pa	0.4	0.3	0.4	0.3	0.1
Appliances	Non-ES[a]	Non-ES	Non-ES	Non-ES	Non-ES
Lighting (fluorescent)	14%	14%	14%	14%	14%

[a] Non-Energy Star.

1.5 Summary

Energy audit of residential buildings encompasses a wide variety of tasks and requires expertise in a number of areas to determine the best energy conservation measures suitable for an existing building. This chapter provided a description of a general but systematic approach to perform energy audits to support weatherization and energy efficiency programs for existing residential buildings. If followed carefully, the approach helps facilitate the process of analyzing a seemingly endless array of alternatives and complex interrelationships between buildings and energy system components associated with energy auditing projects for specific individual dwellings as well as for community scale retrofitting programs involving large numbers of existing residential buildings.

2

Economic Analysis

Abstract

This chapter provides an overview of the basic principles of economic analysis that are essential to determine the cost-effectiveness of various energy conservation measures suitable for retrofitting residential buildings. The purpose of this chapter is to describe various evaluation methods, including life cycle cost (LCC) analysis technique, used to make decisions about retrofitting energy systems. The simple payback period method is also presented, since it is widely used in the preliminary stages of energy audits.

2.1 Introduction

In most applications, initial investments are required to implement energy conservation measures. These initial costs must be generally justified in terms of a reduction in the operating costs (due to energy cost savings). Therefore, most improvements in the efficiency of energy systems have a delayed reward; that is, expenses come at the beginning of a retrofit project, while the benefits are incurred later. For an energy retrofit project to be economically worthwhile, the initial expenses have to be lower than the sum of savings obtained by the reduction in the operating costs over the lifetime of the project.

The lifetime of an energy system retrofit project spans typically over several years. Therefore, it is important to properly compare savings and expenditures of various amounts of money over the lifetime of a project. Indeed, an amount of money at the beginning of a year is worth less at the end of the year and has even less buying power at the end of the second year. Consequently, the amounts of money due to expenditures or savings incurred at different times of a retrofit project cannot be simply added.

In engineering economics, savings and expenditures of amounts of money during a project are typically called *cash flows*. To compare various cash flows over the lifetime of a project, a life cycle cost analysis is typically used. In this chapter, basic concepts of engineering economics are described. First, some of the fundamental principles and parameters of economic analysis are presented. In addition, data are provided to help the reader estimate relevant economic parameters. Then, the general procedure of an economic evaluation of a retrofit project is described. Finally, some of the advantages and disadvantages of the various economic analysis methods are discussed.

2.2 Basic Concepts

There are several economic parameters that affect a decision between various investment alternatives. To perform a sound economic analysis for energy retrofits, it is important that the auditor be (1) familiar with the most important economic parameters, and (2) aware of the basic economic concepts. The parameters and concepts that significantly affect the economic decision making include:

- The time value of money and interest rates, including simple and compounded interest
- Inflation rate and composite interest rate
- Taxes, including sales, local, state, and federal tax charges
- Depreciation rate and salvage value

In the following sections, the above-listed parameters are briefly described to help the reader better understand the life cycle cost analysis procedure, to be discussed later in Section 2.5. A more detailed description of these parameters and common engineering economic analysis approaches is provided by Krarti (2010).

2.2.1 Interest Rate

When money is borrowed to cover part or all the initial cost of a retrofit project, a fee is charged for the use of this borrowed money. This fee is called *interest* (*I*) and the amount of money borrowed is called *principal* (*P*). The amount of the fee depends on the value of the principal and the length of time over which the money is borrowed. The interest charges are typically normalized to be expressed as a percentage of the total amount of money borrowed. This percentage is called the *interest rate* (*i*):

$$i = \frac{I}{P} \tag{2.1}$$

It is clear that an economy with low interest rates encourages money borrowing (for investment in projects or simply for buying goods), while an economy with high interest rates encourages money saving. Therefore, if money has to be borrowed for a retrofit project, the interest rate is a good indicator of whether or not the project can be cost-effective.

For the economic analysis of an energy efficiency project, the interest rate is typically assumed to be constant throughout the lifetime of the project. Therefore, it is common to use an average interest rate when an economic analysis is performed for a retrofit project. Table 2.1 provides historical data for long-term interest rates for selected countries. Example 2.1 illustrates the difference between simple and compounded interest rates.

EXAMPLE 2.1

A homeowner has $2,500 available and has the option to invest this money in either (1) a bank that has an annual interest rate of 5% or (2) buying a new furnace for his home. If he decides to invest all the money in the bank, how much will the homeowner have after 15 years? Compare this amount if simple interest rather than compounded interest had been paid.

TABLE 2.1 Average Long-Term Interest Rates for Selected Countries

Period/Year	France	Germany	Japan	United States
Period				
1961–1973	6.9	7.2	7.0	5.3
1974–1980	11.2	8.1	8.0	8.6
1981–1990	12.0	7.8	6.5	10.3
1990–1995	8.5	7.5	5.1	7.2
Year				
1995	7.5	6.9	3.4	6.6
2000	5.4	5.3	1.7	6.0
2005	3.4	3.4	1.4	4.8
2010	4.1	4.0	2.0	4.4

Source: OECD, Economic Statistics, http://www.ocde.org, 2011.

Solution: If the interest is compounded, with P = $2,500, N = 15, and i = 0.05, the investment will accumulate to the total amount F:

$$F = \$2,500 * (1+0.05)^{15} = \$5,197$$

Thus, the homeowner's original investment will have doubled over the 15-year period.

If simple interest had been paid, the total amount that would have accumulated is slightly less:

$$F = \$2,500 * (1+0.05 * 15) = \$4,375$$

2.2.2 Inflation Rate

Inflation occurs when the cost of goods and services increases from one period to the next. While the interest rate, i, defines the cost of money, the inflation rate, λ, measures the increase in the cost of goods and services. Therefore, the future cost of a commodity, FC, is higher than the present cost, PC, of the same commodity:

$$FC = PC(1+\lambda) \tag{2.2}$$

Table 2.2 presents historical data for inflation rates for selected countries.

If the interest and inflation rates are compounded during the same period, a composite interest rate, θ, can be defined to account for the fact that inflation decreases the buying power of money due to increases in the cost of commodities:

$$\theta = \frac{i-\lambda}{1+\lambda} \tag{2.3}$$

It should be mentioned that theoretically the composite interest rate can be negative. In this case, the value of the money is reduced with time.

TABLE 2.2　Average Inflation Rates for Selected Countries

Period/Year	France	Germany	Japan	United States
Period				
1971–1980	9.8	5.0	8.8	7.1
1981–1990	6.2	2.5	2.1	4.7
1990–1995	1.9	2.8	1.0	2.6
Year				
1995	1.8	1.2	−0.1	2.8
2000	1.8	1.4	−0.5	3.4
2005	1.9	1.9	−0.6	3.4
2010	0.7	0.4	−1.4	1.0

Source: OECD, Economic Statistics, http://www.ocde.org, 2011.

EXAMPLE 2.2

Determine the actual value of the $2,500 investment for the homeowner of Example 2.1 if the economy experiences an annual inflation rate of 3%.

Solution: The composite interest rate can be determined using Equation (2.3):

$$\theta = \frac{i - \lambda}{1 + \lambda} = \frac{0.05 - 0.03}{1 + 0.03} = 0.01942$$

With P = $2,500 and N = 15, the investment will accumulate to the total amount F:

$$F = \$2,500 * (1 + 0.01942)^{15} = \$3,336$$

2.2.3　Tax Rate

In most economies, the interest incurred from an investment is subject to taxation. If this taxation has a rate, t, over a period that coincides with the interest period, then the amount of taxes, T, to be collected from an investment, P, with an interest rate, i, is determined as follows:

$$T = tiP \tag{2.4}$$

Therefore, the composite interest rate defined by Equation (2.3) can be generalized to account for both inflation and tax rates related to present and future values:

$$\theta = \frac{(1 - t)i - \lambda}{1 + \lambda} \tag{2.5}$$

EXAMPLE 2.3

If the homeowner is in the 28% tax bracket, determine the actual value of his $2,500 investment considered in Example 2.1 if the economy experiences an annual inflation rate of 3%.

Solution: The composite interest rate can be determined using Equation (3.15):

$$\theta = \frac{(1-t)i - \lambda}{1+\lambda} = \frac{(1-0.28)*0.05 - 0.03}{1+0.03} = 0.00583$$

With $P = \$2,500$ and $N = 15$, the investment will accumulate to the total amount F:

$$F = \$2,500 * (1+0.00583)^{15} = \$2,728$$

2.2.4 Cash Flows

In evaluating energy efficiency projects, it is important to account for the total cash receipts and disbursements due to an implementation of an energy conservation measure (such as the installation of a new furnace) during the entire lifetime of the project. The difference between the total cash receipts (inflows) and total cash disbursements (outflows) for a given period is called a cash flow.

Over the project lifetime, an accurate accounting of all the cash flows should be performed. For energy efficiency improvement projects, the cash flow accounting can be outlined in a tabular format, as illustrated in Table 2.3 for the cash flows attributed to the installation of a new furnace. In particular, Table 2.3 accounts for the cost related to the initial cost of a new furnace installation (counted as a disbursement for year 0) and

TABLE 2.3 Cash Flows for an Installation of a New Boiler over a Lifetime of 10 Years

End of Year	Total Cash Receipts	Total Cash Disbursements	Total Cash Flows	Comments
0	$0	$4,000	− $4,000	Installation cost of a new furnace
1	$400	$0	+ $400	Net cost savings
2	$380	$0	+ $380	" " "
3	$360	$0	+ $360	" " "
4	$340	$0	+ $340	" " "
5	$330	$0	+ $330	" " "
6	$320	$0	+ $320	" " "
7	$310	$0	+ $310	" " "
8	$305	$0	+ $305	" " "
9	$300	$0	+ $300	" " "
10	$295	$0	+ $295	" " "

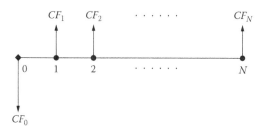

FIGURE 2.1   Typical cash flow diagram.

the cost savings due to the higher energy efficiency of the new boiler (counted as receipts in years 1 through 10). The reduction in the annual receipts is attributed to the aging of the equipment.

Note that the cash flows are positive when they represent inflows (i.e., receipts) and are negative when they are outflows (i.e., disbursements). To better visualize the evolution over time of cash flows, a cash flow diagram can be used as depicted in Figure 2.1. Note that in Figure 2.1, the initial cash flow, $C_0 = -\$4,000$ (disbursement), is represented by a downward-pointing arrow. Meanwhile, the cash flows occurring later, C_1 through C_N with $N = 10$, are receipts and are represented by upward-pointing arrows.

It should be noted again that cash flows cannot be simply added since the value of money changes from one period to the next. In the next section, various factors are defined to correlate cash flows occurring at different periods.

2.3 Compounding Factors

Two types of payment factors are considered in this section. These payment factors are useful in the economical evaluation of various retrofit projects. Without loss of generality, the interest period is assumed to be 1 year. Moreover, a nominal discount rate, d, is used for the analysis described in this chapter. The discount rate is an effective interest rate that includes the effects of several parameters, such as inflation and taxation, as shown by Equation (2.5).

2.3.1 Single Payment

In this case, an initial payment is made to implement a project by borrowing an amount of money, P. If this sum of money earns interest at a discount rate, d, then the value of the payment, P, after N years is a future amount, F. The ratio F/P is often called single-payment compound amount (SPCA) factor. The SPCA factor is a function of i and N and is defined as

$$SPCA(d,N) = F/P = (1+d)^N \tag{2.6}$$

Using the cash flow diagram of Figure 2.1, the single payment represents the case where $C_0 = P$, $C_1 = \ldots = C_{N-1} = 0$, and $C_N = F$ as illustrated in Figure 2.2.

FIGURE 2.2 Cash flow diagram for single payment.

The inverse ratio P/F allows calculation of the value of the cash flow, P, needed to attain a given amount of cash flow F after N years. The ratio P/F is called single-payment present worth (SPPW) factor and is equal to

$$SPPW(d,N) = P/F = (1+d)^{-N} \tag{2.7}$$

Table 2.4 provides the SPPW values for various discount rates, d, and life cycles, N.

2.3.2 Uniform Series Payment

In most energy retrofit projects, energy cost savings are estimated annually after an initial investment to implement one package of energy efficiency measures. It is hoped that during the lifetime of the project, the sum of all the annual energy cost savings can surpass the initial investment.

Consider then an amount of money, P, that represents the initial investment, and a receipt of an amount, A, that is made each year and represents the cost savings due to the retrofit project. To simplify the analysis, the amount A is assumed to be constant for all the years during the lifetime of the project. Therefore, the cash flows can be set to $C_0 = P$, $C_1 = \ldots = C_N = A$, as depicted in Figure 2.3.

It can be shown that the ratio A/P, called uniform series capital recovery (USCR) factor, can be expressed as a function of both d and N:

$$USCR(d,N) = A/P = \frac{d(1+d)^N}{(1+d)^N - 1} = \frac{d}{1-(1+d)^{-N}} \tag{2.8}$$

The uniform series present worth (USPW) factor, needed to calculate the value of P knowing the amount A, is the ratio P/A and can be expressed as follows:

$$USPW(d,N) = P/A = \frac{(1+d)^N - 1}{d(1+d)^N} = \frac{1-(1+d)^{-N}}{d} \tag{2.9}$$

Table 2.5 provides the USPW values for various discount rates, d, and life cycles, N.

TABLE 2.4   SPPW Values for Selected Discount Rates, *d*, and Number of Years, *N*

SPPW	Discount Rate, *d* (%)										
N (years)	0	1	2	3	4	5	6	7	8	9	10
1	1.000	0.990	0.980	0.971	0.962	0.952	0.943	0.935	0.926	0.917	0.909
2	1.000	0.980	0.961	0.943	0.925	0.907	0.890	0.873	0.857	0.842	0.826
3	1.000	0.971	0.942	0.915	0.889	0.864	0.840	0.816	0.794	0.772	0.751
4	1.000	0.961	0.924	0.888	0.855	0.823	0.792	0.763	0.735	0.708	0.683
5	1.000	0.951	0.906	0.863	0.822	0.784	0.747	0.713	0.681	0.650	0.621
6	1.000	0.942	0.888	0.837	0.790	0.746	0.705	0.666	0.630	0.596	0.564
7	1.000	0.933	0.871	0.813	0.760	0.711	0.665	0.623	0.583	0.547	0.513
8	1.000	0.923	0.853	0.789	0.731	0.677	0.627	0.582	0.540	0.502	0.467
9	1.000	0.914	0.837	0.766	0.703	0.645	0.592	0.544	0.500	0.460	0.424
10	1.000	0.905	0.820	0.744	0.676	0.614	0.558	0.508	0.463	0.422	0.386
11	1.000	0.896	0.804	0.722	0.650	0.585	0.527	0.475	0.429	0.388	0.350
12	1.000	0.887	0.788	0.701	0.625	0.557	0.497	0.444	0.397	0.356	0.319
13	1.000	0.879	0.773	0.681	0.601	0.530	0.469	0.415	0.368	0.326	0.290
14	1.000	0.870	0.758	0.661	0.577	0.505	0.442	0.388	0.340	0.299	0.263
15	1.000	0.861	0.743	0.642	0.555	0.481	0.417	0.362	0.315	0.275	0.239
16	1.000	0.853	0.728	0.623	0.534	0.458	0.394	0.339	0.292	0.252	0.218
17	1.000	0.844	0.714	0.605	0.513	0.436	0.371	0.317	0.270	0.231	0.198
18	1.000	0.836	0.700	0.587	0.494	0.416	0.350	0.296	0.250	0.212	0.180
19	1.000	0.828	0.686	0.570	0.475	0.396	0.331	0.277	0.232	0.194	0.164
20	1.000	0.820	0.673	0.554	0.456	0.377	0.312	0.258	0.215	0.178	0.149
21	1.000	0.811	0.660	0.538	0.439	0.359	0.294	0.242	0.199	0.164	0.135
22	1.000	0.803	0.647	0.522	0.422	0.342	0.278	0.226	0.184	0.150	0.123
23	1.000	0.795	0.634	0.507	0.406	0.326	0.262	0.211	0.170	0.138	0.112
24	1.000	0.788	0.622	0.492	0.390	0.310	0.247	0.197	0.158	0.126	0.102
25	1.000	0.780	0.610	0.478	0.375	0.295	0.233	0.184	0.146	0.116	0.092
26	1.000	0.772	0.598	0.464	0.361	0.281	0.220	0.172	0.135	0.106	0.084
27	1.000	0.764	0.586	0.450	0.347	0.268	0.207	0.161	0.125	0.098	0.076
28	1.000	0.757	0.574	0.437	0.333	0.255	0.196	0.150	0.116	0.090	0.069
29	1.000	0.749	0.563	0.424	0.321	0.243	0.185	0.141	0.107	0.082	0.063
30	1.000	0.742	0.552	0.412	0.308	0.231	0.174	0.131	0.099	0.075	0.057

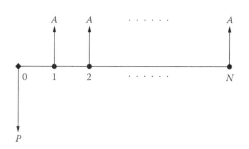

FIGURE 2.3   Cash flow diagram for uniform series payment.

TABLE 2.5   USPW Values for Selected Discount Rates, *d*, and Number of Years, *N*

USPW N (years)	Discount Rate, d (%)										
	0	1	2	3	4	5	6	7	8	9	10
1	1.00	0.99	0.98	0.97	0.96	0.95	0.94	0.93	0.93	0.92	0.91
2	2.00	1.97	1.94	1.91	1.89	1.86	1.83	1.81	1.78	1.76	1.74
3	3.00	2.94	2.88	2.83	2.78	2.72	2.67	2.62	2.58	2.53	2.49
4	4.00	3.90	3.81	3.72	3.63	3.55	3.47	3.39	3.31	3.24	3.17
5	5.00	4.85	4.71	4.58	4.45	4.33	4.21	4.10	3.99	3.89	3.79
6	6.00	5.80	5.60	5.42	5.24	5.08	4.92	4.77	4.62	4.49	4.36
7	7.00	6.73	6.47	6.23	6.00	5.79	5.58	5.39	5.21	5.03	4.87
8	8.00	7.65	7.33	7.02	6.73	6.46	6.21	5.97	5.75	5.53	5.33
9	9.00	8.57	8.16	7.79	7.44	7.11	6.80	6.52	6.25	6.00	5.76
10	10.00	9.47	8.98	8.53	8.11	7.72	7.36	7.02	6.71	6.42	6.14
11	11.00	10.37	9.79	9.25	8.76	8.31	7.89	7.50	7.14	6.81	6.50
12	12.00	11.26	10.58	9.95	9.39	8.86	8.38	7.94	7.54	7.16	6.81
13	13.00	12.13	11.35	10.63	9.99	9.39	8.85	8.36	7.90	7.49	7.10
14	14.00	13.00	12.11	11.30	10.56	9.90	9.29	8.75	8.24	7.79	7.37
15	15.00	13.87	12.85	11.94	11.12	10.38	9.71	9.11	8.56	8.06	7.61
16	16.00	14.72	13.58	12.56	11.65	10.84	10.11	9.45	8.85	8.31	7.82
17	17.00	15.56	14.29	13.17	12.17	11.27	10.48	9.76	9.12	8.54	8.02
18	18.00	16.40	14.99	13.75	12.66	11.69	10.83	10.06	9.37	8.76	8.20
19	19.00	17.23	15.68	14.32	13.13	12.09	11.16	10.34	9.60	8.95	8.36
20	20.00	18.05	16.35	14.88	13.59	12.46	11.47	10.59	9.82	9.13	8.51
21	21.00	18.86	17.01	15.42	14.03	12.82	11.76	10.84	10.02	9.29	8.65
22	22.00	19.66	17.66	15.94	14.45	13.16	12.04	11.06	10.20	9.44	8.77
23	23.00	20.46	18.29	16.44	14.86	13.49	12.30	11.27	10.37	9.58	8.88
24	24.00	21.24	18.91	16.94	15.25	13.80	12.55	11.47	10.53	9.71	8.98
25	25.00	22.02	19.52	17.41	15.62	14.09	12.78	11.65	10.67	9.82	9.08
26	26.00	22.80	20.12	17.88	15.98	14.38	13.00	11.83	10.81	9.93	9.16
27	27.00	23.56	20.71	18.33	16.33	14.64	13.21	11.99	10.94	10.03	9.24
28	28.00	24.32	21.28	18.76	16.66	14.90	13.41	12.14	11.05	10.12	9.31
29	29.00	25.07	21.84	19.19	16.98	15.14	13.59	12.28	11.16	10.20	9.37
30	30.00	25.81	22.40	19.60	17.29	15.37	13.76	12.41	11.26	10.27	9.43

EXAMPLE 2.4

Find the various compounding factors for $N = 10$ years and $d = 5\%$.

Solution: The values of the compounding factors for $d = 0.05$ and $N = 10$ years are summarized below:

Compounding Factor	Equation Used	Value
SPCA	Equation (2.6)	1.629
SPPW	Equation (2.7)	0.613
USCR	Equation (2.8)	0.129
USPW	Equation (2.9)	7.740

2.4 Economic Evaluation Methods

To evaluate the cost-effectiveness of energy retrofit projects, several evaluation tools can be considered. The basic concept of all these tools is to compare among the retrofit alternatives the net cash flow that can be obtained during the lifetime of the project. As discussed earlier, a simple addition of all the cash flows such as those represented in Figure 2.1 is not appropriate. However, by using the compound factors discussed in Section 2.3, "conversion" of the cash flows from one period to another is feasible. This section provides a brief description of the common economic evaluation methods used in engineering projects.

2.4.1 Net Present Worth

Referring to the cash flow diagram of Figure 2.1, the sum of all the present worth of the cash flows can be obtained by using the single payment present worth factor defined in Equation (2.7):

$$NPW = -CF_0 + \sum_{k=1}^{N} CF_k * SPPW(d,k) \tag{2.10}$$

Note that the initial cash flow is negative (a capital cost for the project), while the cash flows for the other years are generally positive (revenues).

In the particular but common case of a project with constant annual revenues (due to energy operating cost savings), $CF_k = A$, the net present worth is reduced to

$$NPW = -CF_0 + A * USPW(d,N) \tag{2.11}$$

For the project to be economically viable, the net present worth has to be positive or at worst zero ($NPW \geq 0$). Obviously, the higher the value of NPW, the more economically sound is the project. The net present worth value method is often called the net savings method since the revenues are often due to the energy cost savings from implementing the project.

2.4.2 Rate of Return

In this method, the first step is to determine the specific value of the discount rate, d', that reduces the net present worth to zero. This specific discount rate is called the rate of return (ROR). Depending on the case, the expression of NPW provided in Equation (2.10) or Equation (2.11) can be used. For instance, in the general case of Equation (2.10), the rate of return, d', is the solution of the following equation:

$$-CF_0 + \sum_{k=1}^{N} CF_k * SPPW(d',k) = 0 \tag{2.12}$$

To solve accurately this equation, any numerical method can be used. However, an approximate value of d' can be obtained by trial and error. This approximate value can be determined by finding the two d-values for which the NPW is slightly negative and slightly positive, and then interpolate linearly between the two values. It is important to remember that a solution for ROR may not exist.

Once the rate of return is obtained for a given alternative of the project, the actual market discount rate or the minimum acceptable rate of return is compared to the ROR value. If the value of ROR is larger ($d' > d$), the project is cost-effective.

2.4.3 Benefit-Cost Ratio

The benefit-cost ratio (BCR) method is also called the savings-to-investment ratio (SIR) and provides a measure of the net benefits (or savings) of the project relative to its net cost. The net values of both benefits (B_k) and costs (C_k) are computed relative to a baseline case. The present worth of all the cash flows is typically used in this method. Therefore, the benefit-cost ratio is computed as follows:

$$BCR = \frac{\sum_{k=0}^{N} B_k * SPPW(d,k)}{\sum_{k=0}^{N} C_k * SPPW(d,k)} \tag{2.13}$$

The alternative option for the project is considered economically viable relative to the base case when the benefit-cost ratio is greater than 1.0 ($BCR > 1.0$).

2.4.4 Discounted and Simple Payback Periods

In this evaluation method, the period Y (expressed typically in years) required to recover an initial investment is determined. Using the cash flow diagram of Figure 2.1, the value of Y is the solution of the following equation:

$$CF_0 = \sum_{k=1}^{Y} CF_k * SPPW(d,k) \tag{2.14}$$

If the payback period Y is less than the lifetime of the project N ($Y < N$), then the project is economically viable. The value of Y obtained using Equation (2.14) is typically called discounted payback period (DPP) since it includes the value of money.

In the vast majority of applications, the time value of money is neglected in the payback period method. In this case, Y is called simple payback period (SPP) and is the solution of the following equation:

$$CF_0 = \sum_{k=1}^{Y} CF_k \tag{2.15}$$

In the case where the annual net savings are constant $(CF_k = A)$, the simple payback period can be easily calculated as the ratio of the initial investment over the annual net savings:

$$Y = \frac{CF_0}{A} \tag{2.16}$$

The values for the simple payback period are shorter than for the discounted payback periods since the undiscounted net savings are greater than their discounted counterparts. Therefore, acceptable values for simple payback periods are typically significantly shorter than the lifetime of the project.

2.4.5 Summary of Economic Analysis Methods

Table 2.6 summarizes the basic characteristics of the economic analysis methods used to evaluate single alternatives of an energy retrofit project. Example 2.2 uses the presented economic analysis methods to assess the cost-effectiveness of a boiler retrofit.

It is important to note that the economic evaluation methods described above provide an indication of whether or not a single alternative of a retrofit project is cost-effective. However, these methods cannot be used or relied on to compare and rank various alternatives for a given retrofit project. The life cycle cost (LCC) analysis method is more appropriate.

TABLE 2.6 Summary of the Basic Criteria for the Various Economic Analysis Methods

Evaluation Method	Equation	Criterion
Net present worth (NPW)	$NPW = -CF_0 + \sum\limits_{k=1}^{N} CF_k * SPPW(d,k)$	NPW > 0
Rate of return (ROR)	$-CF_0 + \sum\limits_{k=1}^{N} CF_k * SPPW(d',k) = 0$	$d' > d$
Benefit-cost ratio (BCR)	$BCR = \dfrac{\sum\limits_{k=0}^{N} B_k * SPPW(d,k)}{\sum\limits_{k=0}^{N} C_k * SPPW(d,k)}$	BCR > 1
Discounted payback period (DPP)	$CF_0 = \sum\limits_{k=1}^{Y} CF_k * SPPW(d,k)$	Y < N
Simple payback period (SPP)	$Y = \dfrac{CF_0}{A}$	Y << N

EXAMPLE 2.5

After finding that the old boiler has an efficiency of only 60% while a new boiler would have an efficiency of 85%, a building owner of Example 2.1 has decided to invest $10,000 in getting a new boiler. Determine whether or not this investment is cost-effective if the lifetime of the boiler is 10 years and the discount rate is 5%. The boiler consumes 5,000 gallons per year at a cost of $1.20 per gallon. An annual maintenance fee of $150 is required for the boiler (independent of its age). Use all five methods summarized in Table 2.6 to perform the economic analysis.

Solution: The baseline case for the economic analysis presented in this example is the case where the boiler is not replaced. Moreover, the salvage value of the boiler is assumed insignificant after 10 years. Therefore, the only annual cash flows (A) after the initial investment on a new boiler are the net savings due to higher boiler efficiency, as calculated below:

$$A = Fuel - Use_{before} * (1 - \eta_{before}/\eta_{after}) * Fuel - cost / gallon$$

Thus,

$$A = 5000 * \left(1 - 0.60/0.85\right) * \$1.20 = \$1,765$$

The cost-effectiveness of replacing the boiler is evaluated as indicated below:

1. *Net present worth.* For this method $CF_0 = \$10,000$ and $CF_1 = \ldots = CF_{10} = A$, $d = 0.05$, and $N = 10$ years. Using Equation (2.11) with USPW = 7.740 (see Example 2.4):

$$NPW = \$3,682$$

   Therefore, the investment in purchasing a new boiler is cost-effective.

2. *Rate of return.* For this method also $CF_0 = \$10,000$ and $CF_1 = \ldots = CF_{10} = A$, while $SPPW(d', k)$ is provided by Equation (2.7). By trial and error, it can be shown that the solution for d' is

$$d' = 12.5\%$$

   Since $d' > d = 5\%$, the investment in replacing the boiler is cost-effective.

3. *Benefit-cost ratio.* In this case, $B_0 = 0$ and $B_1 = \ldots = B_{10} = A$, while $C_0 = \$10,000$ and $C_1 = \ldots = C_{10} = 0$.

   Note that since the maintenance fee is applicable to both the old and new boiler, this cost is not accounted for in this evaluation method (only the benefits and costs relative to the baseline case are considered).
       Using Equation (2.13):

$$BCR = 1.368$$

   Thus, the benefit-cost ratio is greater than unity ($BCR > 1$) and the project of getting a new boiler is economically feasible.

4. *Discounted payback period.* For this method, $CF_0 = \$10,000$ and $CF_1 = \ldots = CF_{10} = A$. Using Equation (2.15), Y can be solved: $Y = 6.9$ year.

   Thus, the discounted payback period is shorter than the lifetime of the project ($Y > N = 10$ years), and therefore replacing the boiler is cost-effective.

5. *Simple payback period.* For this method, $CF_0 = \$10,000$ and $A = \$1,765$. Using Equation (2.16), Y can be easily determined: $Y = 5.7$ years.

   Thus, the simple payback period method indicates that the boiler retrofit project can be cost-effective.

2.5 Life Cycle Cost Analysis Method

The life cycle cost (LCC) analysis method is the most commonly accepted method to assess the economic benefits of energy efficiency projects over their lifetime. Typically, the method is used to evaluate at least two alternatives of a given project (for instance, evaluate two alternatives for the same retrofit project: replace a furnace or consider a ground source heat pump to condition a house). Only one alternative will be selected for implementation based on the economic analysis.

The basic procedure of the LCC method is relatively simple since it seeks to determine the relative cost-effectiveness of various alternatives. For each alternative, including the baseline case, the total cost is estimated over the project lifetime. The cost is commonly determined using one of two approaches: the present worth or the annualized cost estimate. Then, the alternative with the lowest total cost (or LCC) is typically selected.

Using the cash flow diagram of Figure 2.1, the LCC amount for each alternative can be computed by projecting all the costs (including costs of acquisition, installation, maintenance, and operating of the energy systems related to the energy conservation project) to:

1. One single present value amount that can be computed as follows:

$$LCC = \sum_{k=0}^{N} CF_k * SPPW(d,k) \tag{2.17}$$

   This is the most commonly used approach in calculating LCC in energy retrofit projects.

2. Multiple annualized costs over the lifetime of the project:

$$LCC_a = USCR(d,N) * \left[\sum_{k=1}^{N} CF_k * SPPW(d,k) \right] \tag{2.18}$$

Note that the two approaches for calculating the LCC values are equivalent.

In several energy efficiency projects, the annual cash flow remains the same after the initial investment. In these cases, LCC can be estimated based on the initial cost, *IC*, and the annual cost, *AC*, as follows:

$$LCC = IC + USPW(d, N)^* AC \qquad (2.19)$$

EXAMPLE 2.6

The building owner of Example 2.5 has three options to invest his money, as briefly described below:

1. Replace the entire older boiler (including burner) with a more efficient heating system. The old boiler-burner system has an efficiency of only 60%, while a new boiler-burner system has an efficiency of 85%. The cost of this replacement is $10,000.
2. Replace only the burner of the old boiler. This action can increase the efficiency of the boiler-burner system to 66%. The cost of the burner replacement is $2,000.
3. Do nothing and replace neither the boiler nor the burner.

Determine the best economical option for the building owner. Assume that the lifetime of the retrofit project is 10 years and the discount rate is 5%. The boiler consumes 5,000 gallons per year at a cost of $1.20 per gallon. An annual maintenance fee of $150 is required for the boiler (independently of its age). Use the life cycle cost analysis method to determine the best option.

Solution: The total cost of operating the boiler-burner system is considered for the three options. In this analysis, the salvage value of the boiler or burner is neglected. Therefore, the only annual cash flows (A) after the initial investment on a new boiler are the maintenance fee and the net savings due to higher boiler efficiency. To present the calculations for LCC analysis, it is recommended to present the results in a tabular format and proceed as shown below:

Cost Item	Option A	Option B	Option C
Initial Investment			
(a) Replacement cost ($)	10,000	2,000	0
Annual Operating Costs			
(b) Fuel use (gallons)	3,530	4,545	5,000
(c) Fuel cost ($) [$1.2*(b)]	4,236	5,454	1,000
(d) Maintenance fee ($)	150	150	150
(f) Total operating cost ($) [(c) + (d)]	4,386	5,594	6,150
USPW factor			
[d = 5%, N = 10, Equation (3.22)]	7.740	7.740	7.740
Present worth ($)			
[(a) + USPW*(f)]	43,948	45,298	47,601

Therefore, the life cycle cost for option A is the lowest. Thus, it is recommended for the building owner to replace the entire boiler-burner system.

This conclusion is different from that obtained by using the simple payback analysis (indeed, the payback period for option A, relative to the base case C, is SPP(A) = ($10,000)/($1,765) = 5.66 years, while for option B, SPP(B) = ($2,000)/($546) = 3.66 years).

Note: If the discount rate was $d = 10\%$ (which is unusually high for most markets), the USPW would be equal to $USPW = 6.145$ and the life cycle cost for each option will be

LCC(A) = $36,952   LCC(B) = $36,375   LCC(C) = $37,791

Therefore, option B will become the most effective economically and will be the recommended option to the building owner.

2.6 General Procedure for an Economic Evaluation

It is important to remember that the recommendations for energy conservation projects should be based on an economically sound analysis. In particular, the auditor should ask several questions before making the final recommendations, such as:

- Will project savings exceed costs?
- Which design solution will be most cost-effective?
- What project size will minimize overall building costs?
- Which combination of interrelated project options will maximize net savings?
- What priority should be given to various project alternatives if the owner has a limited investment level?

As discussed earlier, the best suitable economic assessment method is the LCC method described in Section 2.5. Before the application of the LCC, several data are needed to perform an appropriate and meaningful economic analysis. To help the auditor collect the required information and apply the LCC method, the following systematic approach in any economic evaluation is proposed:

Step 1: Define the problem that the proposed retrofit project is attempting to address and state the main objective of the project. For instance, a house has an old furnace that does not provide enough heat to condition the entire house. The project is to replace the boiler with main objective to heat all the conditioned spaces within the building.

Step 2: Identify the constraints related to the implementation of the project. These constraints can vary in nature and include financial limitations or space requirements. For instance, the new boiler cannot be gas fired since there is no supply of natural gas near the house.

Step 3: Identify technically sound strategies and alternatives to meet the objective of the project. For instance, three alternatives can be considered for the old boiler replacement: (1) a new boiler with the burner of the old boiler, (2) a new boiler-burner system, or (3) a new boiler-burner system with an automatic air-fuel adjustment control.

Step 4: Select a method of economic evaluation. When there are several alternatives, including the base case (which may consist of the alternative of doing nothing), the LCC method is preferred for energy projects. When a preliminary economic analysis is considered, the simple payback period method can be used. As mentioned earlier, the payback period method is not accurate and should be used with care.

Step 5: Compile data and establish assumptions. The data include the discount rates, energy costs, installation costs, operating costs, and maintenance costs. Some of these data are difficult to acquire, and some assumptions or estimations are needed. For instance, an average discount rate over the life cycle of the project may be assumed based on historical data.

Step 6: Calculate indicators of economic performance. These indicators depend on the economic evaluation method selected. The indicators include simple payback period (SPP), the net present worth (NPW), and the life cycle cost (LCC).

Step 7: Evaluate the alternatives. This evaluation can be performed by simply comparing the values of LCC obtained for various alternatives.

Step 8: Perform sensitivity analysis. Since the economic evaluation performed in step 6 is typically based on some assumed values (for instance, the annual discount rate), it is important to determine whether or not the results of the evaluation performed in step 7 depend on these assumptions. For this purpose, the economic evaluation is repeated for all alternatives using different but plausible assumptions.

Step 9: Take into account unqualified effects. Some of the alternatives may have some effects that cannot be accounted properly in the economic analysis but may be a determining factor in the decision making. For instance, the environmental impacts (emission of pollutants) can be important to disqualify an otherwise economically sound alternative.

Step 10: Make recommendations. The final selection will be based on the findings of the three previous steps (i.e., steps 7 to 9). Typically, the alternative with the lowest LCC value will be recommended.

Once the project for energy retrofit is selected based on an economic analysis, it is important to decide on the financing options to actually carry out the project and implement the measures that allow a reduction in energy cost of operating the facility.

2.7 Summary

Energy auditing and retrofitting of residential buildings encompasses a wide variety of tasks and requires expertise in a number of areas to determine the best energy conservation measures suitable for an existing building. This chapter described a general but systematic approach to perform the economic evaluation of various alternatives of energy retrofit projects. In particular, several analysis methods described in this chapter (such as the net present worth, the benefit-cost ratio, the rate of return, and the payback period) are suitable for evaluating single alternatives but may not be used to rank among alternatives. The life cycle cost (LCC) analysis method is more suitable for evaluating several alternatives and should be used to select the most economical option among possible alternatives for a retrofit project. Under certain market conditions, the simple payback method can lead to erroneous conclusions, and thus should be used only to provide an indication of the cost-effectiveness of an energy retrofit project. For sound economic analysis, the simple payback analysis method should not be used.

Problems

2.1 A homeowner is considering adding a sunspace to his home. The project is carried during 4 years, with $20,000 spent the first year, $15,000 the second year, $10,000 the third year, and $5,000 the fourth year. Determine how much money the homeowner needs to set aside to pay for the sunspace addition project:
   a. If the interest rate is 4% per year, compounded annually.
   b. If the interest rate is 8% per year, compounded annually.

2.2 You want to buy a new home at a cost of $250,000. Determine how much money you should pay every year for 20 years to pay for the house based on equal annual payments:
   a. If the interest rate is 5% per year, compounded annually.
   b. If the interest rate is 10% per year, compounded annually.

2.3 An energy audit of a residential heating system reveals that the boiler-burner efficiency is only 60%. In addition, the energy audit showed that an electric water heater is used. The house uses annually 1,500 gallons of oil at a cost of $1.40 per gallon. The total electric bill averaged $84.50 per month for an average monthly consumption of 818 kWh. Of this total, about 35% was for domestic water heating. It is suggested to the owner of the house to equip the boiler with a tankless domestic water heater, replacing the existing electric water heater. The existing water heater is 12 years old and has no resale value and little expected life. The owner of the house is expected to spend $300 within the coming year to replace the electric water heater. The cost of the tankless heater is $400.

To improve heating efficiency, it is suggested to:
   a. Replace the existing burner with a new one (this burner replacement will improve efficiency to 65%) costing $520.
   b. Replace the entire heating plant with higher efficiency (85%), with a cost of $2,000.

For two discount rates (5 and 10%), provide the LCC analysis of the following options and make the appropriate recommendations:
   • Keep the boiler-burner and replace the electric water heater with a like system.
   • Replace the burner and electric water heater with like systems.
   • Replace the boiler-burner with an efficient boiler and replace the electric heater with a like system.
   • Replace the existing boiler-burner and electric water heater with a new boiler/tankless domestic water heater.

Note: State all the assumptions made in your calculations.

2.4 An energy audit indicates that a home's consumption is 20,000 kWh per year. By using Energy Star appliances and CFLs, a savings of 20% can be achieved for the electrical energy use. The cost for the home retrofit is about $15,000. Assuming that the average energy charge is $0.09 per kWh, is the expenditure justified based on a minimum rate of return of 18% before taxes? Assume a 20-year life cycle and use the present worth, annual cost, and rate of return methods.

2.5 An electrical energy audit indicates that lighting consumes 5,000 kWh per year. By using more efficient lighting fixtures, electrical energy savings of 15% can be achieved. The additional cost for these energy-efficient lighting fixtures is about $900. Assuming that the average energy charge is $0.08 per kWh, is the expenditure justified based on a minimum rate of return (i.e., a discount rate) of 7.0%? Assume a 10-year life cycle and use the present worth, annual cost, and rate of return methods.

2.6 A residential building is heated with a 78% efficient gas-fired furnace (costing $1,600). The annual heating load is estimated at 160 MMBtu. The cost of gas is $6.00/MMBtu. For the discount rate of 5% and 10-year cycle, determine:
   a. The life cycle of the existing heating system.
   b. If it is worth considering a 90% efficient gas-fired furnace that costs $2,800.
   c. The variation of the life cycle cost savings between the two systems (the existing system and the proposed in (b)) vs. the life of the system for life cycles ranging from 3 to 20 years.
   Conclude.

2.7 An air conditioner (A/C) consumes 7,500 kWh annually with an overall efficiency of COP = 3.0. If this A/C is replaced by a more energy efficient A/C (COP = 3.7) at a cost of $3,500, determine:
   a. The simple payback period of the chiller replacement.
   b. If the expected life of the old chiller is 15 years, whether it is cost-effective to replace the chiller in an economy with a discount rate of 7%.

2.8 Two air conditioner systems are proposed to cool a house. Each A/C has a rated capacity of 5 tons and is expected to operate 500 full-load equivalent hours per year. Air conditioner A has a standard efficiency with a COP of 2.8 and costs $1,600, while A/C B is more efficient with a COP of 3.5 at a cost of $2,750. The electricity cost is estimated to be $0.09/kWh. For a lifetime of 20 years and a discount rate of 5%:
   a. Estimate the payback time for using air conditioner B instead of A.
   b. Calculate the life cycle costs for both air conditioners.
   Conclude.
   c. Estimate the rate of return for using air conditioner B instead of A.
   d. Determine the highest cost for air conditioner B at which air conditioner A is more competitive.
   e. Determine the cutoff electricity price at which both air conditioners A and B have the same life cycle cost.
   Assume that the average cost of electricity is $0.08/kWh.

3

Principles of
Thermal Analysis

Abstract

This chapter provides a basic overview of fundamental concepts useful to carry out an energy analysis of buildings, including envelope components and heating and cooling systems. In particular, basic principles of heat transfer, psychrometrics, and energy modeling are introduced. These principles are the basis of screening, evaluating, and analyzing the energy performance of existing building energy systems. Specifically, these principles are used to estimate energy use savings associated with various energy efficiency measures discussed in this book.

3.1 Introduction

To analyze energy consumption and estimate the cost-effectiveness of energy conservation measures, an auditor can use a myriad of calculation methods and simulation tools. The existing energy analysis methods vary widely in complexity and accuracy. To select the appropriate energy analysis method, the auditor should consider several factors, including speed, cost, versatility, reproducibility, sensitivity, accuracy, and ease of use (Sonderegger, 1985). There are hundreds of energy analysis tools and methods that are used worldwide to predict the potential savings of energy conservation measures. In the United States, the DOE provides an up-to-date listing of selected building energy software (DOE, 2009).

In this section, some of the most relevant energy analysis techniques and methods for residential buildings are described. In particular, simplified thermal analysis methods such as the variable-base degree-day and the thermal network techniques are introduced with some applications to estimate thermal performance of existing residential buildings. In addition, detailed energy modeling programs suitable for residential buildings are introduced. The detailed models are typically time-consuming to develop and require some calibration approach to ensure that their predictions match measured and billing data. Some calibration techniques of detailed building energy simulation tools are described in Chapter 10.

First, an overview of fundamental principles of thermal analysis of residential buildings is provided, including heat transfer through building envelope, air infiltration, properties of air, and thermal comfort. Then, simplified energy analysis methods suitable for residential buildings are outlined. Finally, a brief description of various forward and inverse modeling techniques for whole building energy performance analysis is presented. The methods and techniques presented in this chapter are the basis of a wide range of testing and calculation methods to estimate and verify the energy savings for several retrofit measures of residential buildings introduced throughout the chapters of this book.

3.2 Basic Heat Transfer Concepts

Heat transfer from the building envelope can occur through various mechanisms, including conduction, convection, and radiation. In this section, fundamental concepts of heat transfer are briefly reviewed. These concepts and associated metrics are typically used to characterize the thermal performance of various components of the building envelope and are useful to estimate the energy use savings accrued by retrofits of building envelope components.

3.2.1 Heat Transfer Mechanisms

Heat transfer occurring between one surface and a fluid or between two surfaces can be estimated using the concept of R-value as defined in Equation (6.1):

$$\dot{q} = \frac{1}{R} \cdot (T_1 - T_2) \tag{3.1}$$

where $\dot{q}$ is the heat transfer per unit area, T_1 and T_2 are the temperatures of either two surfaces or one surface and a fluid, and R is the R-value of the heat transfer. The expressions for the R-value for selected systems and heat transfer mechanisms typical in buildings are summarized in Table 3.1.

TABLE 3.1   Expression of the R-Value for Selected Heat Transfer Mechanisms

Heat Transfer Mechanism/Medium	R-Value	Comments
Conduction in plane walls	$R_{wall} - \text{value} = \dfrac{L}{k}$	L: thickness of the wall k: thermal conductivity
Convection over a surface	$R_{conv} - \text{value} = \dfrac{1}{h_{conv}}$	h_{conv}: convection heat transfer coefficient
Radiation between two surfaces	$R_{conv} - \text{value} = \dfrac{1}{h_{rad}}$	h_{rad}: radiation heat transfer coefficient $h_{rad} = \varepsilon\sigma(T_1^2 + T_2^2)(T_1 + T_2)$
Conduction in cylinders	$R_{cyl} - \text{value} = \dfrac{r_2 Ln(r_2 / r_1)}{k}$	r_2: outer radius r_1: inner radius L: length k: thermal conductivity

3.2.2 Heat Transfer from Walls and Roofs

In buildings, heat transfer through walls and roofs is dominated by conduction and convection. Typically, one-dimensional heat conduction is considered to be adequate for above-grade building components unless significant thermal bridges exist, such as at wall corners or at slab-on-grade floor edges. Specifically, heat transfer from a homogeneous wall or roof layer, illustrated in Figure 3.1, can be calculated as follows using the Fourier law for conduction heat transfer:

$$\dot{q} = \frac{k}{d}.A.(T_i - T_o) \tag{3.2}$$

where A is the area of the wall, T_i is the inside wall surface temperature, T_o is the outside wall surface temperature, k is the thermal conductivity of the wall, and d is the thickness of the wall.

To simplify estimation of the conduction heat transfer expressed by Equation (3.2), a thermal resistance R-value or a U-value is defined for the layer as shown below:

$$R = \frac{d}{k} = \frac{1}{U} \tag{3.3}$$

The concept of thermal resistance can be extended to convection heat transfer that occurs at the outer or inner surfaces of the building envelope:

$$R_{conv} = \frac{1}{h} \tag{3.3b}$$

where h is the convective heat transfer coefficient of the surface.

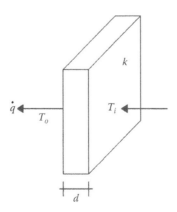

FIGURE 3.1 Conduction heat transfer through one-layer wall.

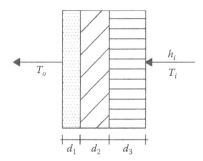

FIGURE 3.2   Heat transfer from a multilayered wall.

In buildings, a wall or a roof consists of several layers of homogeneous materials, as illustrated in Figure 3.2:

Heat transfer from a multilayered wall or roof can be estimated by calculating first its overall R-value:

$$R_T = \sum_{j=1}^{N_L} R_j \qquad (3.4)$$

where R_j is the R-value of each homogeneous layer part of the construction of the wall or roof assembly (it includes the R-value due to convection at both inner and outer surfaces of the wall or roof, obtained by Equation (3.3)), and N_L is the number of layers (including the convection boundary layers) that are part of the wall or roof assembly. For instance, in the wall assembly presented in Figure 3.2, $N_L = 5$ (three conductive layers and two convective layers).

The overall U-value of the wall or roof can be defined simply as the inverse of the overall R-value:

$$U_T = \frac{1}{R_T} \qquad (3.5)$$

It should be noted that practitioners usually prefer to use R-values rather than U-values since the U-values are small, especially when insulation is added to the wall or roof assembly. For doors and windows, the use of U-values is more common since these components have low R-values. Appendix B provides the thermal properties of common materials used in building construction. These thermal properties are the basis of calculating the R-values and U-values of building envelope components, as illustrated in Example 3.1.

EXAMPLE 3.1

Problem: Estimate the overall *R*-value and *U*-value of a 2 × 4 wood frame wall construction typically used in U.S. residences. The wall assembly is made up of ½ in. gypsum board, 2 × 4 stud wall with R-11 fiberglass insulation, ½ in. plywood sheathing, and hardboard siding.

Solution: The table shows the calculation details of the *U*-value and *R*-value for a wood frame wall that consists of the following layers from inside to outside. The various properties of materials and the convective heat transfer coefficients are provided in Appendix B.

Calculation of a Wall *R*-Value and *U*-Value

Layer	*R*-Value between Framing	*R*-Value at Framing
Inside convection, still air, nonreflective vertical wall	0.68	0.68
Gypsum, ½"	0.45	0.45
Stud, 2 × 4 nominal		4.38
Insulation, 3.5"	11	
Plywood, ½"	0.62	0.62
Siding	0.67	0.67
Outside convection, 15 mph	0.17	0.17
Path *R*-value	13.59	6.97
Path *U* = 1/(*R*-value)	0.0736	0.1434
Path area ratio	0.85	0.15

Effective *U*:   0.08407 Btu/hr.ft².°F
Effective *R*-value: 11.89

3.2.3 R-Value Estimation Using Field Testing Method

Under steady-state conditions, it is possible to estimate the *R*-value of the wall by measuring three temperatures as indicated in Figure 3.3: the indoor air temperature, T_o, the outdoor air temperature, T_i, and the indoor surface temperature, $T_{i,s}$.

Since the heat flux through the wall can be estimated using the equivalent thermal network shown in Figure 3.3,

$$\dot{q} = \frac{A}{R_T} \cdot (T_i - T_o) = Ah_i(T_i - T_{i,s}) \tag{3.6}$$

Thus, the *R*-value of the wall, R_T, can be expressed as follows as a function of the indoor heat convection coefficient, h_i, and the temperatures, T_i, T_o, and $T_{i,s}$:

$$R_T = \frac{1}{h_i} \frac{(T_i - T_{i,s})}{(T_i - T_o)} \tag{3.7}$$

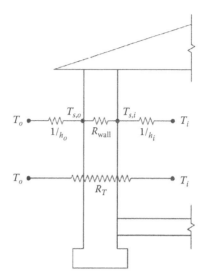

FIGURE 3.3   Temperature measurements for estimating wall R-value.

Alternatively, the *U*-value of the wall can be obtained using the temperature measurements, T_i, T_o, and $T_{i,s}$:

$$U_T = h_i \frac{(T_i - T_o)}{(T_i - T_{i,s})} \tag{3.8}$$

The field testing for estimating the *R*-value or *U*-value of the wall should be conducted when there is a large difference between the indoor and outdoor temperatures and when there is very little solar radiation hitting the wall. Thus, it is recommended to perform the temperature measurements in a cold night. An infrared (IR) camera can be used to measure surface temperature, as shown in Figure 3.4. The measurements for Figure 3.4 were performed on a winter night when the outdoor air temperature was 46.4°F, while the indoor air temperature was 70.2°F. The indoor surface temperature as indicated in Figure 3.4 was 76.6°F. Using an indoor convection heat transfer *R*-value on a vertical wall of 0.68 hr.ft².°F/Btu (refer to the data provided in Appendix B), the *R*-value of the exterior was estimated to be 5.56 hr.ft².°F/Btu. This value compares well with the calculated *R*-value of 5.62 as detailed in Table 3.2.

Table 3.3 summarizes the percent error in estimating the wall *R*-value using temperature measurements for various indoor-outdoor temperature differences ($T_i - T_o$). It is clear from Table 3.3, that the larger the difference between indoor and outdoor air temperatures, the more accurate is the estimation of the wall *R*-value using the field test illustrated in Figure 3.3. In particular, to achieve an accuracy of 5% or less in estimating the wall *R*-value, an indoor-outdoor temperature of at least 4°C or 7°F is required, especially when the wall is poorly insulated.

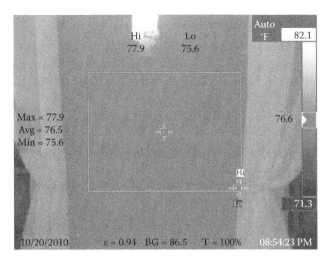

FIGURE 3.4 Thermal image for estimating a wall for *R*-value.

TABLE 3.2 *R*-value Calculation for the Exterior Wall Construction

Material	Thickness (ft)	Conductivity (Btu/h-ft-°F)	*R*-Value (h-ft-°F/Btu)
Outside convection			0.17
Face brick, 4 in.	0.333	0.7576	0.44
Concrete (140 lb), 8 in.	0.667	0.7576	0.88
Vertical air layer	—	—	3
Gypsum board, ½ in.	0.042	0.0926	0.45
Inside convection			0.68
Total	1.042	—	5.62

TABLE 3.3 Percent Error for *R*-Value Estimation Based on Temperature Measurements

$[T_i - T_o]$ (°F)	$[T_i - T_o]$ (°C)	$R = 2.5$	$R = 5$	$R = 10$
40	22.2	0.98%	0.54%	0.28%
35	19.4	1.12%	0.61%	0.32%
30	16.7	1.30%	0.72%	0.38%
25	13.9	1.56%	0.86%	0.45%
20	11.1	1.95%	1.07%	0.56%
15	8.3	2.60%	1.43%	0.75%
10	5.6	3.90%	2.15%	1.13%
5	2.8	7.81%	4.29%	2.25%
4	2.2	9.76%	5.37%	2.81%
3	1.7	13.01%	7.16%	3.75%
2	1.1	19.52%	10.74%	5.63%
1	0.6	39.04%	21.47%	11.25%
0.5	0.3	78.08%	42.95%	22.50%
0.1	0.1	390.39%	214.73%	112.50%

3.2.4 Air Infiltration Heat Loss/Gain

Air can flow in and out of the building envelope through leaks. This process is often referred to as air infiltration and exfiltration. Thus, air infiltration (and air exfiltration) is rather an uncontrolled flow of air, unlike ventilation (and exhaust), for which air is moved by mechanical systems. Generally, air infiltration can occur in all building types but is more important for smaller buildings, such as detached residential buildings. In larger buildings, air infiltration is typically less significant for two reasons:

1. The volume over envelope surface area (from which air leakage occurs) is small for larger buildings.
2. The indoor pressure is generally maintained higher than outdoor pressure by mechanical systems in larger buildings.

Typically, air infiltration is considered significant for low-rise buildings and can affect energy use, thermal comfort, and especially structural damages through rusting and rotting of the building envelope materials due to the humidity transported by infiltrating/exfiltrating air. Without direct measurement, it is difficult to estimate the leakage airflow through the building envelope. There are two basic measurement techniques that allow estimation of the air infiltration characteristics for a building. These measurement techniques include fan pressurization/depressurization and tracer gas.

Fan pressurization/depressurization techniques are commonly known as blower door tests and can be used to estimate the volumetric airflow rate variation with the pressure difference between outdoors and indoors of a building. Several pressure differential values are typically considered and a correlation is found in the form of

$$\dot{V} = C.\Delta P^n \tag{3.9}$$

where C and n are correlation coefficients determined by fitting the measured data of pressure differentials and air volumetric rates. It should be noted that the lower limit of n is 0.5, when the leaks are concentrated and provide direct paths for air infiltration (such as the case of large cracks and holes). The upper limit of n is 1.0, when the leaks are diffuse and provide long paths for air infiltration (such as the case of small holes in a wall where air has to travel the entire length or width of the wall to reach indoors).

Using the correlation of Equation (3.9), an effective leakage area (ELA) can be determined as follows:

$$ELA = \dot{V}_{ref} \cdot \sqrt{\rho \big/ 2.\Delta P} \tag{3.10a}$$

When English units are used, the effective leakage area (in inches squared) can be estimated using a modified Equation (3.10a) as follows:

$$ELA = 0.186\dot{V}_{ref} \cdot \sqrt{\rho \big/ 2.\Delta P} \tag{3.10b}$$

where $\dot{V}_{ref}$ is the reference volume air rate through the building at a reference pressure difference (between indoors and outdoors) of typically 4 Pa and is obtained by extrapolation from Equation (3.9). The ELA provides an estimate of the equivalent area of holes in the building envelope through which air leaks can occur.

To determine the building air infiltration rate under normal climatic conditions (due to wind and temperatures effects), the LBL infiltration model developed by Sherman and Grimsrud (1980) is commonly used:

$$\dot{V} = ELA. \left(f_s . \Delta T + f_w . v_w^2 \right)^{1/2} \tag{3.11}$$

where DT is the indoor-outdoor temperature difference, v_w is the period average wind speed, and f_s is the stack coefficient (Table 3.4 provides the crack coefficients for three levels of building height), and f_w is the wind coefficient. Table 3.5 lists the wind coefficients for various shielding classes and building heights.

TABLE 3.4 Stack Coefficient, f_s

	IP Units			SI Units		
	House Height (Stories)			House Height (Stories)		
	One	Two	Three	One	Two	Three
Stack coefficient	0.0150	0.0299	0.0449	0.000139	0.000278	0.000417

Source: ASHRAE, *Handbook of Fundamentals*, American Society of Heating, Refrigerating, and Air-Conditioning Engineers, Atlanta, GA, 2009.
Note: IP units for f_s: (ft³/min)²/in.⁴×°F. SI units for f_s: (L/s)²/cm⁴×°C.

TABLE 3.5 Wind Coefficient, f_w

	IP Units			SI Units		
	House Height (Stories)			House Height (Stories)		
Shielding Class[a]	One	Two	Three	One	Two	Three
1	0.0119	0.0157	0.0184	0.000319	0.000420	0.000494
2	0.0092	0.0121	0.0143	0.000246	0.000325	0.000382
3	0.0065	0.0086	0.0101	0.000174	0.000231	0.000271
4	0.0039	0.0051	0.0060	0.000104	0.000137	0.000161
5	0.0012	0.0016	0.0018	0.000032	0.000042	0.000049

Source: ASHRAE, 2009.
Note: IP units for f_w: (ft³/min)²/in.⁴×mph. SI units for f_w: (L/s)²/cm⁴×(m/s)².

[a] Description of shielding classes:

1. No obstructions or local shielding.
2. Light local shielding: Few obstructions, few trees, or small shed.
3. Moderate local shielding: Some obstructions within two-house height, thick hedge, solid fence, or one neighboring house.
4. Heavy shielding: Obstructions around most of perimeter, buildings or trees within 30 ft (10 m) in most directions; typical suburban shielding.
5. Very heavy shielding: Large obstructions surrounding perimeter within two house heights; typical downtown shielding.

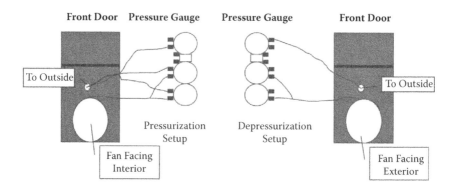

FIGURE 3.5   Typical blower door setup for both pressurization and depressurization tests.

Blower door tests are still being used to find and repair leaks in low-rise buildings. Typically, the leaks are found by holding a smoke source and watching where the smoke exits the house. Several weather stripping methods are available to reduce air infiltration through building envelopes, including caulking, weather stripping, landscaping around the building to reduce the wind effects, and installing air barriers to tighten the building envelope.

It should be mentioned, however, that the blower door technique cannot be used to determine accurately the amount of fresh air supplied to the building through either infiltration or ventilation. For this purpose, it is recommended to use the tracer gas techniques described below.

In a typical blower test, the house should be first prepared. In particular, windows are closed, interior doors that are normally open are kept open, and the fireplace ash is cleaned. The main entrance door is generally used to place the blower fan to either introduce air (for pressurization test) or extract air (for depressurization test). The airflow rate is generally measured using a pressure gauge attached to the blower setup. The pressure gauge should be first checked to make sure that it reads zero with the fan set off. An additional pressure gauge is used to measure the differential in pressure between the inside and outside of the house. Figure 3.5 shows the setup for both the depressurization and pressurization tests. Example 6.2 illustrates how the results of blower door tests can be used to determine the infiltration rate in a house. The results and the analysis presented in Example 3.2 are based on actual tests performed for one single-family home.

EXAMPLE 3.2

A blower door test has been performed in a house located in Boulder, Colorado. The results of the pressurization and depressurization tests are summarized in Table 3.6 based on the results of the blower door. Determine the leakage areas and ACHs for both pressurization and depressurization tests.

Solution: To determine the air leakage characteristics of a house using blower door tests, the following procedure is used:

TABLE 3.6 Summary of the Blower Door Results
for Both Pressurization and Depressurization Tests

	Pressurization Test			Depressurization Test	
	Pressure Pa	Flow cfm		Pressure Pa	Flow cfm
1	82.0	2,345	1	104.6	2,073
2	75.7	2,241	2	99.6	2,026
3	69.8	2,114	3	94.3	1,949
4	64.4	2,006	4	90.5	1,910
5	58.8	1,908	5	84.8	1,836
6	55.3	1,866	6	77.7	1,739
7	48.8	1,778	7	69.4	1,624
8	45.4	1,666	8	61.0	1,487
9	39.2	1,503	9	56.1	1,368
	31.6	1,335		47.0	1,209
	24.6	1,167		40.1	1,075
	20.9	1,073		30.7	919
	16.4	918		−23.3	748
				19.6	639
				16.5	583
				10.1	414
				6.1	294

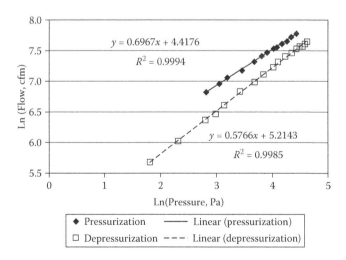

FIGURE 3.6 Flow rate as a function of the pressure difference in a log-log scale.

1. First, the data consisting of pressure differential (ΔP [Pa]) and airflow rate ($\dot{V}$ [CFM]) presented in Figure 3.6 are plotted in a log-log scale, as illustrated in Figure 3.6.

TABLE 3.7   Regression Coefficients and Estimation of Equivalent Leakage Area

	n	(From linear regressions) Ln(C)	C	Flow at 4 Pa	Equiv. Leakage Area
Pressurization	0.697	4.418	82.9	217.8 cfm	56.4 in.^{2}
Depressurization	0.577	5.214	183.9	409.0 cfm	105.9 in.^{2}
					Average = 81.1 in.^{2}
Density of air	0.062 lbm/ft^3		at 5,000 ft elevation		
Reference ΔP	4 Pa				
	0.016 in. H_2O				

TABLE 3.8   Calculation of Winter, Summer, and Annual Average Air Infiltration

Leakage area	81.1 in.^{2}		
Stack coeff.	0.0313 (ft^3/min)^2/(in.$^{4*\circ}$F)	From Table 3.4	
Wind coeff.	0.0039 (ft^3/min)^2/(in.^{4*}(mph))	For shielding class 4, from Table 3.5	
House volume	12,800 ft^3		

For Boulder:	Winter	Summer	Annual
T_avg_outdoor (°F)	31.3	70.2	50.5
T_avg_indoor (°F)	65.0	70.2	70.0
Avg. wind speed (mph)	7.8	8.1	8.3
Annual infiltration (cfm)	92.3	40.8	76.2
Annual ACH	0.43	0.19	0.36

Note: These calculations are based on the average ELA of the pressurization and depressurization tests.

2. Then, a linear regression analysis is used to determine the coefficients C and n of Equation (3.9):

$$\dot{V} = C.\Delta P^n$$

The regression coefficients for both the pressurization and depressurization tests are summarized in Table 3.7. Based on these regression coefficients, the infiltration rate under a normal pressure differential (ΔP_{ref} = 4 Pa) can be calculated. Finally, the leakage areas for both pressurization and depressurization tests can be obtained using Equation (3.10b):

$$ELA = 0.186 \dot{V}_{ref} \cdot \sqrt{\rho / 2.\Delta P}$$

For Boulder, Colorado, the air density should be adjusted for altitude and is found to be ρ = 0.062 lbm/ft^3.

3. The seasonal average leakage airflow rate expressed in air change per hour (ACH) can be determined using the LBL model expressed by Equation (3.11). Table 3.8 summarizes the calculation results and the ACH values for the winter, summer, and annual average infiltration rates.

It should be noted that ASHRAE recommends a leakage of 0.35 ACH for proper ventilation of a house with minimum heat loss.

Tracer gas techniques are commonly used to measure the ventilation rates in large buildings such as high-rise apartment buildings. By monitoring the injection and concentration of a tracer gas (a gas that is inert, safe, and mixes well with air), the exchange of air through the building can be estimated. For instance, in the decay method, the injection of tracer gas is performed for a short time and then stopped. The concentration of the decaying tracer gas is then monitored over time. The ventilation is measured by the air change rate (ACH) within the building and is determined from time variation of the tracer gas concentration:

$$c(t) = c_O.e^{-ACH.t} \tag{3.12}$$

or

$$ACH = \frac{\dot{V}}{V_{bldg}} = \frac{1}{t}.Ln\left[\frac{c_O}{c(t)}\right] \tag{3.13}$$

where c_o is the initial concentration of the tracer gas, and V_{bldg} is the volume of the building.

3.2.5 Coupling between Air Infiltration and Heat Conduction

The thermal load of air infiltration is rather difficult to assess. It has been traditionally thought that the sensible thermal load due to infiltration air is simply calculated as follows:

$$Q_{inf} = \rho.c_{p,a}.\dot{V}_{inf}.(T_i - T_o) = \dot{m}_{inf}.c_{p,a}.(T_i - T_o) \tag{3.14}$$

Equation (3.14) assumes that infiltrating air enters the condition space (kept at temperature T_i) and has the same temperature as the outdoor temperature, T_o. However, various recent studies showed that actually infiltrating air can warm up through the building envelope before entering a condition space. This heat exchange occurs especially when the air leakage happens in a diffuse manner (through long airflow channels inside the building envelope). As a consequence, the actual thermal load due to air infiltration is lower than that determined by Equation (3.14) by a fraction that depends on the heat exchange rate between air infiltration and heat conduction through the building

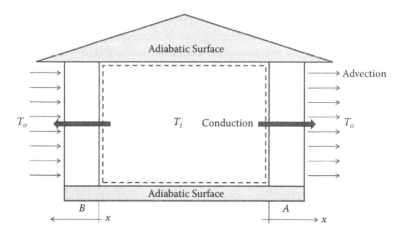

FIGURE 3.7   Simplified house model for thermal coupling between air infiltration and heat conduction through walls.

envelope. In particular, several analytical studies have indicated that the infiltration thermal load can be reduced by a heat recovery factor, *f*, as shown in Equation (3.15):

$$Q_{inf} = (1-f).\dot{m}_{inf}.c_{p,a}.(T_i - T_o) \tag{3.15}$$

When the building envelope is not airtight, it can be assumed that the airflow by infiltration occurs through direct and short paths from outdoors to the indoors without significant heat recovery, and thus Equation (3.15) can be used to estimate the thermal load due to air infiltration. However, when the air infiltration is diffuse with long paths, a significant heat recovery can result. In the case of diffuse airflow through the walls parallel, as shown in the simplified house model of Figure 3.7, analytical solutions have been developed to estimate the value of the heat recovery factor, *f*. For instance, Krarti (1994) estimated that the heat recovery factor, *f*, can be calculated as follows:

$$f = 1 - \left(\frac{PeBi_i e^{\frac{Pe}{2}}}{\left[2Bi_i Bi_o + Pe\left(Bi_i - Bi_o\right)\right]\sinh\left(\frac{Pe}{2}\right) + PE\left(Bi_i + Bi_o\right)\cosh\left(\frac{Pe}{2}\right)} \right) \tag{3.16}$$

where *Pe* is the Peclet number and is defined as $Pe = \dot{m}c_p/UA$, Bi_i is the biot number at the inside surface, $Bi_i = h_i/U$, and Bi_o is the biot number at the outside surface, $Bi_o = h_o/U$.

Note when the biot numbers, Bi_o and Bi_i, are large (i.e., the surface temperatures are the same as the air temperatures), then Equation (3.16) is reduced to the solution reported by Anderlind (1985):

$$f = \frac{2*\left(e^{Pe} - Pe - 1\right)}{Pe*\left(e^{Pe} - 1\right)} \tag{3.17}$$

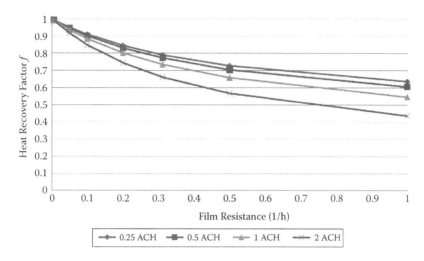

FIGURE 3.8 Variation of the heat recovery factor, *f*, with the air infiltration rate and convection heat transfer coefficients.

Figure 3.8 illustrates the variation of the heat recovery factor as a function of the convection heat transfer and infiltration rate expressed in ACH.

As the result of the coupling between conduction and air infiltration of the building envelope, the heating and cooling thermal loads may be reduced. The reduction of the building thermal load can be expressed as the ratio of the building thermal load when coupling occurs over the thermal load without coupling.

3.2.6 Calculation of Building Load Coefficient

From Equation (3.2), it is clear that in order to reduce the heat transfer from the above-grade building envelope components, its *R*-value should be increased or its *U*-value decreased. To achieve this objective, a thermal insulation can be added to the building envelope. In the next section, calculation methods of the energy savings due to addition of insulation are presented to determine the cost-effectiveness of such measure.

To characterize the total heat transmission of the entire building, a building load coefficient (BLC) is defined to account for heat transmission through all above-grade building envelope components (roofs, walls, doors, and windows) and air infiltration:

$$BLC = \sum_{i=1}^{N} UA_i + \dot{m}c_{\text{inf}} \tag{3.18}$$

where UA_i is the UA-value of each element of the above-grade building envelopes, including walls, roofs, windows, and doors, and $\dot{m}_{\text{inf}}$ is the mass flow rate of the infiltrating air.

Case Study

A house is located in Denver. It has a rectangular floor plan, 60 ft by 28 ft, with 8-ft ceilings. The roof has a single ridge with a 6/12 pitch. The walls, ceiling, and roof have the construction detailed in Table 3.9, with layers listed inside to outside.

TABLE 3.9   Construction Details for the Exterior Walls, Ceiling, Roof, and Attic End Walls

Walls	Ceiling	Roof	Attic End Walls
½" gypsum	½" gypsum	2 × 6 rafters	2 × 4 stud wall
2 × 4 stud wall with fiberglass insulation	2 × 6 rafters	¾" plywood sheathing	½" plywood sheathing
½" plywood sheathing	9" blown cellulose (k = 0.286 Btu-in./hr.°F.ft²)	Felt paper	Aluminum siding
Aluminum siding		Asphalt shingles	

The windows are operable double pane, clear, low ε (ε = 0.1 on one surface) with a ½ in. air space and wood frames in a rough opening area of 240 ft². Both front and back doors are 1¾ in. solid-core flush-wood doors. Each door has an area of 20 ft². The floor is slab on grade with R-5 perimeter insulation (covering 3 ft). Air infiltration is estimated to be 0.35 ACH.

Dimensional Calculations

Length = 60 ft
Width = 28 ft
Height = 8 ft
Volume = 13,440 ft³
DegPitch = ARCTAN(6/12)
WidthRoof = Width/COS(DegPitch)
Perim = 2*(Length + Width) = 176 ft
$AWall$ = Perim*Height-A_Window-A_Door = 1,128 ft²
$ACeil$ = Length*Width = 1,680 ft²
$ARoof$ = Length*WidthRoof = 1,878 ft²
$AEnd$ = Width*Width*TAN(DegPitch)/2 = 196 ft²

Calculation Approach for the BLC

Walls

Layer	R-Value between Framing	R-Value at Framing
Inside convection, still air, nonreflective vertical wall	0.68	0.68
Gypsum, ½"	0.45	0.45
Stud, 2 × 4 nominal		4.38
Insulation, 3.5"	11	
Plywood, ½"	0.62	0.62
Siding	0.61	0.61
Outside convection, 15 mph	0.17	0.17
Path R-value	13.53	6.91
Path $U = 1/(R\text{-value})$	0.07391	0.1447
Path area ratio	0.85	0.15

Effective U: 0.08453 Btu/h ft²F
Effective R-value: 11.83

Ceiling

Layer	R-Value between Framing	R-Value at Framing
Inside convection, still air, nonreflective, horizontal, up	0.61	0.61
Gypsum, ½ in.	0.45	0.45
Stud, 2 × 6 nominal		6.88
Insulation, 9 in.	31.47	
Insulation, 3.5 in.		12.24
Inside convection, still air, nonreflective, horizontal, up	0.61	0.61
Path R-value	31.14	20.78
Path $U = 1/(R\text{-value})$	0.0321	0.0481
Path area ratio	0.906	0.094

Effective U: 0.03362 Btu/h ft²F
Effective R-value: 29.75

Attic End Walls

Layer	R-Value between Framing	R-Value at Framing
Inside convection, still air, nonreflective vertical wall	0.68	0.68
Stud, 2 × 4 nominal		4.38
Plywood, ½ in.	0.62	0.62
Siding	0.61	0.61
Outside convection, 15 mph	0.17	0.17
Path R-value	2.08	6.46
Path U = 1/(R-value)	0.4808	0.1548
Path area ratio	0.85	0.15

Effective U: 0.4318 Btu/h ft²F
Effective R-value: 2.31

Roof

Layer	R-Value between Framing	R-Value at Framing
Inside convection, still air, nonreflective, horizontal, up	0.61	0.61
Stud, 2 × 6 nominal		6.88
Plywood, ¾ in.	0.94	0.94
Felt (vapor permeable)	0.06	0.06
Asphalt shingles	0.44	0.44
Outside convection, 15 mph	0.17	0.17
Path R-value	2.22	9.01
Path U = 1/(R-value)	0.4505	0.1099
Path area ratio	0.906	0.094

Effective U: 0.4184 Btu/h ft²F
Effective R-value: 2.39

Ceiling + Roof

$$R_{top} = \frac{1}{(UA)_{top}} = \frac{R_{v,ceil}}{A_{ceil}} + \frac{1}{U_{roof}A_{roof} + U_{end}A_{end}}$$

Thus, $(UA)_{top} = 53.03$ Btu/h.°F

Windows and Door

$$U_{window} = 0.51 \text{ Btu/h ft}^2.°F \text{ (ASHRAE, 2009)}$$

$$U_{door} = 0.40 \text{ Btu/h ft}^2.°F \text{ (ASHRAE, 2009)}$$

Floor

$$U = 0.77 \text{ Btu/h ft°F (refer to the next section)}$$

Infiltration

$$\dot{V}_{inf} = \text{ACH*V/60} = 78.4 \text{ cfm}$$

$$\rho = 0.061 \text{ lb}_m/\text{ft}^3 \text{ (at altitude)}$$

Design Heat Loss

From Equation (3.2):

$$BLC = (UA)_{top} + (UA)_{walls} + (UA)_{windows} + (UA)_{door} + U'P + 60.\rho c_p \dot{V}_{inf}$$

$$BLC = 53.03 + 95.35 + 122.4 + 16.0 + 135.52 + 68.87 = 491.17 \text{ Btu/h.°F}$$

3.3 Foundation Heat Transfer

The practice of insulating building foundations has become more common over the last few decades. However, the vast majority of existing residential buildings are not insulated. Globally, earth-contact heat transfer appears to be responsible for 1 to 3 quadrillion kJ of annual energy use in the United States. This energy use is similar to the impact due to infiltration on annual cooling and heating loads in residential buildings (Claridge, 1988). In addition to the energy savings potential, insulating building foundations can improve the thermal comfort, especially for occupants of buildings with basements or earth-sheltered foundations.

Typically, the foundation heat transfer is a major part of heating/cooling loads for low-rise buildings, including single-family dwellings. A detailed discussion of the insulation configurations for various building types as well as various calculation techniques to estimate foundation heat transfer can be found in Krarti (2000). In this section, a brief description of the common foundation insulation configurations is presented with an overview of the most important factors affecting ground-coupled heat transfer. Moreover, a simplified method is provided to calculate annual and seasonal foundation heat loss/gain from residential foundations.

It should be noted that in the United States, there are three common foundation types for residential buildings: slab-on-grade floors, basements, and crawlspaces. The basement foundations can be either deep or shallow. Typically, shallow basements and crawlspaces are unconditioned. Figure 3.9 shows the three common building foundation types. In some applications, the building foundation can include any combination of the three foundation types, such as a basement with a slab-on-grade floor. Among the factors that affect the selection of the foundation type are the geographical location and the speculative real estate market.

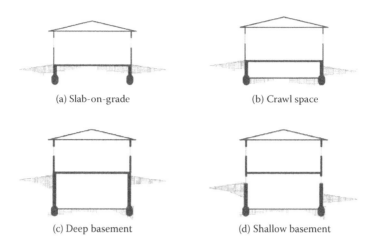

(a) Slab-on-grade   (b) Crawl space

(c) Deep basement   (d) Shallow basement

FIGURE 3.9   Foundation types for the buildings.

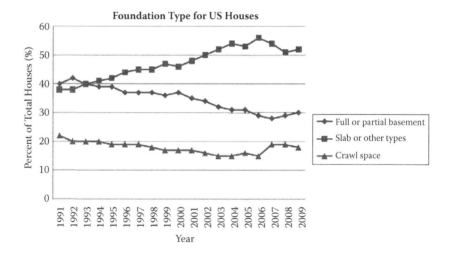

FIGURE 3.10   Annual distribution of foundation types for U.S. houses

A report from the U.S. Census Bureau indicates that the share of houses built with crawlspaces remained constant at about 20% over the last 20 years, as shown in Figure 3.10. However, the percentage of houses with slab foundations has increased from 38% in 1991 to 52% in 2009. Meanwhile, the share of houses built with basements has declined from a peak of 42% in 1992 to 30% in 2009. In 1993, houses were built with almost an equal number of basement and slab foundations. Moreover, data from the U.S. Census Bureau clearly indicate that the foundation type selection depends on the geographical location. In the Northeast and Midwest regions, the basement foundation is the most common, with a share of about 80%. Meanwhile, the slab foundation is more dominant in the South and the West.

3.3.1 Soil Temperature Variation

Heat transfer between buildings and ground is difficult to estimate and is rather complex to model. Indeed, ground-coupled heat transfer can occur through a variety of mechanisms, including:

- Conduction through soil grains or through water and ice in the soil voids
- Advection by ground water
- Convection of water vapor through voids in the soil matrix
- Radiation in the soil voids
- Mass transfer of moist air and water driven by pressure gradients, gravity, and thermal convection
- Phase change due to evaporation or freezing of water

Thermal properties of the ground are generally recognized to be the most important parameters that affect ground-coupled heat transfer. Unfortunately, data for soil thermal properties are often very difficult to obtain. Indeed, soil thermal properties are influenced by myriad factors, such as soil type, soil density, soil moisture content, and even soil temperature and soil depth. For instance, a single measurement of a particular soil parameter such as soil moisture content does not establish how that parameter might vary over an annual cycle. In an attempt to reduce the ambiguity in selecting reasonable soil thermal properties for building applications, Sterling et al. (1993) provided a database of typical values for *apparent* soil thermal conductivity. The apparent thermal conductivity lumps the effects of various modes of heat transport within the soil to only heat conduction transfer. Table 3.10 lists the low and high values for the apparent thermal conductivity for selected soil types (ASHRAE, 2009).

Away from any building, the soil daily temperature at any depth can be estimated based on the ground surface temperature as expressed by Equation (3.19):

$$T(z,t) = T_m + T_a e^{-z/d} \sin\left[\omega t - \theta - \frac{z}{d} - \frac{\pi}{2}\right] \tag{3.19}$$

where $\omega = \dfrac{2\pi}{8760}$ is annual angular frequency, $d = \sqrt{\dfrac{2a_s}{\omega}}$ is the dampening depth, t is time expressed in hours starting from January 1, T_m is the annual average soil surface

TABLE 3.10 Apparent Thermal Conductivity for Selected Soil Types

Soil Type	Btu·in./h·ft²·°F Recommended Values		W/(m·K) Recommended Values	
	Low	High	Low	High
Sands	5.4	15.6	0.78	2.25
Silts	11.4	15.6	1.64	2.25
Clays	7.8	10.8	1.12	1.56
Loams	6.6	15.6	0.95	2.25

TABLE 3.11   Soil Surface Temperatures for Selected U.S. Sites (Refer to Equation (3.20))

Site	State	IP Unit			SI Unit		
		Avg. Temp.	Amplitude	Phase Angle	Avg. Temp.	Amplitude	Phase Angle
		°F	°F	Radian	°C	°C	Radian
Auburn	Alabama	65	17	0.49	18.3	−8.3	0.49
Tempe	Arizona	70	20	0.47	21.1	−6.7	0.47
Davis	California	66	19	0.63	18.9	−7.2	0.63
Fort Collins	Colorado	50	24	0.54	10.0	−4.4	0.54
Moscow	Idaho	47	18	0.73	8.3	−7.8	0.73
Argonne	Illinois	51	23	0.7	10.6	−5.0	0.7
Burlington	Iowa	54	30	0.57	12.2	−1.1	0.57
Manhattan	Kansas	55	26	0.61	12.8	−3.3	0.61
Lexington	Kentucky	55	23	0.6	12.8	−5.0	0.6
Upper Marlboro	Maryland	56	25	0.56	13.3	−3.9	0.56
East Lansing	Michigan	50	24	0.6	10.0	−4.4	0.6
St. Paul	Minnesota	48	25	0.65	8.9	−3.9	0.65
State Univ.	Mississippi	67	21	0.58	19.4	−6.1	0.58
Kansas City	Missouri	54	22	0.56	12.2	−5.6	0.56
Bozeman	Montana	44	21	0.68	6.7	−6.1	0.68
Lincoln	Nebraska	54	28	0.52	12.2	−2.2	0.52
New Brunswick	New Jersey	53	21	0.69	11.7	−6.1	0.69
Ithaca	New York	49	19	0.69	9.4	−7.2	0.69
Columbus	Ohio	53	22	0.65	11.7	−5.6	0.65
Lake Hefner	Oklahoma	64	23	0.63	17.8	−5.0	0.63
Ottawa	Ontario	47	21	0.64	8.3	−6.1	0.64
Corvallis	Oregon	56	18	0.53	13.3	−7.8	0.53
Calhoun	South Carolina	64	22	0.49	17.8	−5.6	0.49
Madison	South Dakota	47	26	0.59	8.3	−3.3	0.59
Jackson	Tennessee	60	20	0.44	15.6	−6.7	0.44
Temple	Texas	70	21	0.58	21.1	−6.1	0.58
Salt Lake City	Utah	51	21	0.48	10.6	−6.1	0.48
Seattle	Washington	53	15	0.64	11.7	−9.4	0.64

temperature, T_a is the annual amplitude of soil surface temperature variation, and Θ is the phase lag between ambient air and soil surface temperatures.

The soil surface temperature variation over time can be expressed typically as a sine function of time:

$$T(0,t) = T_m + T_a \sin\left[\omega t - \theta - \frac{\pi}{2}\right] \qquad (3.20)$$

Table 3.11 provides the values of T_m, T_a, and Θ for select U.S. sites using the data provided by Kusuda and Achenbach (1965). There are some simplified models that can

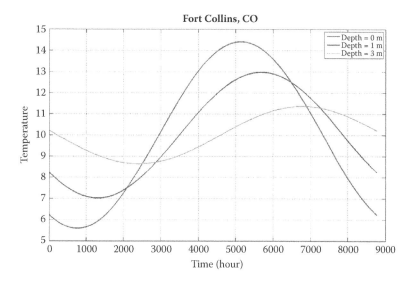

FIGURE 3.11 Monthly variation of ground temperatures at selected depths.

predict the annual soil surface temperature variation, including the three parameters T_m, T_a, and Θ using ambient air temperature (Krarti et al., 1994). It is very common to assume that the annual average soil temperature, T_m, is close to the deep ground water temperature.

Figure 3.11 illustrates the annual soil temperatures for Fort Collins, Colorado, at various depths from the ground surface. Figure 3.12 illustrates the soil temperature variation with depth for 4 days representative to four seasons. As shown in Figure 3.12, the soil temperature becomes constant at about 10 m or 30 ft.

3.3.2 Building Foundation Insulation Configurations

There are several options to place insulations along the building foundations. In some climates, there is no need to insulate building foundations in order to take advantage of the significant ground thermal capacitance. Figure 3.13 illustrates four common insulation configurations for slab-on-grade floor foundations, including no insulation, uniform insulation, edge insulation under slab, and outside vertical insulation along the foundation walls.

To determine the effectiveness of insulating building foundations, detailed analysis is typically needed to estimate the temperature distribution within the ground medium as well as the heat flux along the building foundation. The analysis involves generally solving two- and three-dimensional heat transfer problems (Krarti, 2000). A simplified one-dimensional analysis is not suitable to estimate ground-coupled heat transfer. Figures 3.14 and 3.15 illustrate the impact of adding insulation uniformly beneath the slab surface of a residence on the temperature field in the ground during the wintertime in Denver, Colorado.

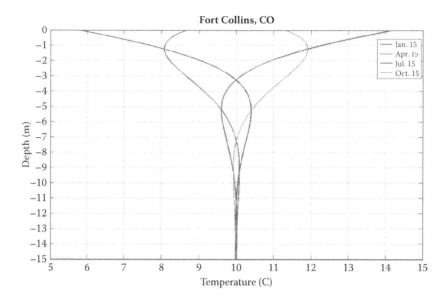

FIGURE 3.12 Soil temperature variation with depth for four seasons.

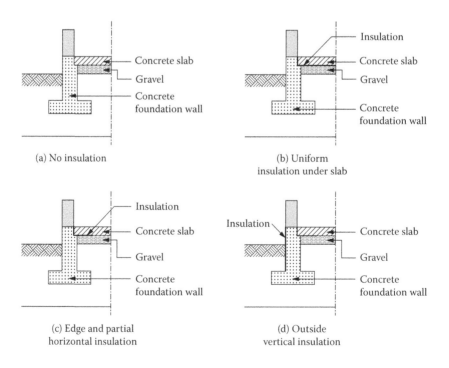

FIGURE 3.13 Slab-on-grade floor insulation configurations.

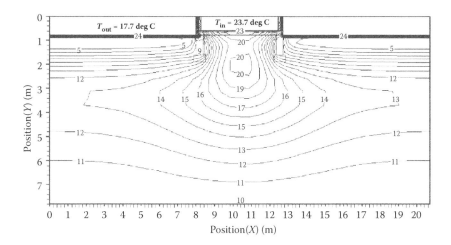

FIGURE 3.14 Soil temperature isotherms for uninsulated slab during wintertime (Denver, CO).

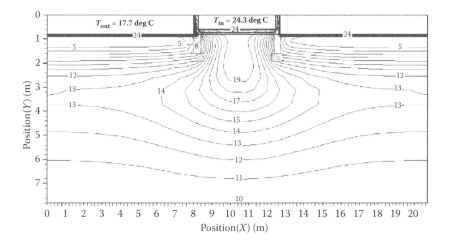

FIGURE 3.15 Soil temperature isotherms for insulated slab during wintertime (Denver, CO).

Figure 3.16 compares one- and two-dimensional analyses to estimate building foundation heat transfer. In particular, Figure 3.16 shows daily average building foundation heat loss during three different seasons in Denver. As expected, the results indicate that building foundation heat losses estimated using a two-dimensional heat transfer model are higher than those predicted with a one-dimensional model regardless of the floor insulation configurations. For the insulated floor, the difference in daily average heat loss is about 2.5 W/m² during wintertime.

Table 3.12 summarizes the foundation heat loss per unit floor area obtained for four slab floor insulation configurations applied to a conditioned building located in Denver, during summertime and wintertime. Foundation heat loss is significantly reduced when

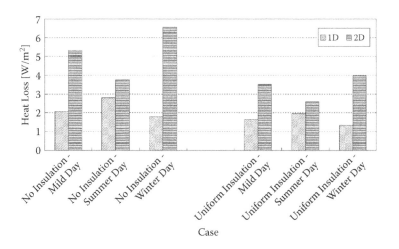

FIGURE 3.16   Daily average building foundation heat loss during three typical days in Denver, Colorado.

TABLE 3.12   Seasonal Foundation Heat Loss for Four Insulation Configurations Applied to a Conditioned Building in Denver, Colorado

Insulation Configuration	Summer		Winter	
	[W/m²]	Reduction [%]	[W/m²]	Reduction [%]
Uninsulated	2.61	0	3.07	0
Edge partial	1.98	24	2.34	24
Uniform	1.92	26	2.17	29
Outside vertical	2.11	19	2.53	17

edge and uniform insulation configurations are used. An outside vertical insulation configuration does not seem to significantly reduce foundation heat loss for a conditioned space. A simplified method to estimate heat loss or gain from the foundations of residential buildings is outlined in Section 3.3.3.

3.3.3 Simplified Calculation Method for Building Foundation Heat Loss/Gain

A simplified design tool for calculating heat loss for slabs and basements has been developed by Krarti and Chuangchid (1999). This design tool is easy to use and requires straightforward input parameters with continuously variable values, including foundation size, insulation *R*-values, soil thermal properties, and indoor and outdoor temperatures. The simplified method provides a set of equations, which are suitable for estimating the design, and the annual total heat loss for both slab and basement foundations as a function of a wide range of variables.

The heat loss and gain from the foundation of a residential building can be determined at any time based on Equation (3.21). Specifically, the simplified method can calculate the annual or seasonal foundation heat transfer using two equations to estimate, respectively, the mean Q_m and the amplitude Q_a of the annual foundation heat loss.

$$Q(t) = Q_m + Q_a \sin\left[\omega t - \varphi - \frac{\pi}{2} \right]$$ (3.21)

For the annual mean foundation heat loss:

$$Q_m = U_{eff,m} \cdot A \cdot (T_m - T_r)$$ (3.22)

where A is the foundation area in contact with the ground, T_m is the average annual soil surface temperature, T_r is the building indoor temperature, and $U_{eff,m}$ is the mean effective U-value of the foundation estimated as follows:

$$U_{eff,m} = m \cdot U_o \cdot D$$

For the annual amplitude foundation heat loss:

$$Q_a = U_{eff,a} \cdot A \cdot T_a$$ (3.23)

where

$$U_{eff,a} = a \cdot U_o \cdot D^{0.16} \cdot G^{-0.6}$$

The coefficients a and m depend on the insulation placement configurations and are provided in Table 3.13.

The normalized parameters U_o, D, and G, used by both Equations (3.21) and (3.22), are defined below:

$$U_o = \frac{k_s}{(A/P)_{eff,b}} \;; \qquad G = k_s . R_{eq} \cdot \sqrt{\frac{\omega}{\alpha_s}}$$

(3.24)

$$D = \ln\left[(1+H)\left(1+\frac{1}{H}\right)^H \right] \;; \qquad H = \frac{(A/P)_{eff,b}}{k_s . R_{eq}}$$

TABLE 3.13 Coefficients m and a to Be Used in Equations (3.21) and (3.22) for Foundation Heat Gain Calculations

Insulation Placement	m	a
Uniform—horizontal	0.40	0.25
Partial—horizontal	0.34	0.20
Partial—vertical	0.28	0.13

For partial insulation configurations (for slab foundations, the partial insulation can be placed either horizontally, extending beyond the foundation, or vertically along the foundation walls):

$$R_{eq} = R_f \times \cfrac{1}{\left[1 - \left(\cfrac{c}{A/P} \times \cfrac{R_i}{\left(R_i + R_f\right)}\right)\right]}$$

For uniform insulation configurations: $R_{eq} = R_f + R_i$, where

$$\left(A/P\right)_{eff,b,mean} = \left[1 + b_{eff} \times \left(-0.4 + e^{-H_b}\right)\right] \times \left(A/P\right)_b \qquad (3.25a)$$

$$\left(A/P\right)_{eff,b,amp} = \left[1 + b_{eff} \times e^{-H_b}\right] \times \left(A/P\right)_b \qquad (3.25b)$$

$$H_b = \frac{\left(A/P\right)_b}{k_s . R_{eq}} \text{ and } b_{eff} = \frac{B}{\left(A/P\right)_b}$$

In the equations above, the following parameters are used:

A	Basement/slab area (total of floor and wall) [m² or ft²]
B	Basement depth [m or ft]
b_{eff}	Term defined in Equations (3.25a) and (3.25b)
C_p	Soil specific heat [J/kg°C or Btu/lbm.°F]
c	Insulation length of basement/slab [m or ft]
D	Term defined in Equation (3.24)
G	Term defined in Equation (3.24)
H	Term defined in Equation (3.24)
k_s	Soil thermal conductivity [Wm⁻¹.°C⁻¹ or Btu/h.ft.°F]
P	Perimeter of basement/slab [m or ft]
Q	Total heat loss/gain [W or Btu/h]
Q_m	The annual mean of the total heat loss/gain [W or Btu/h]
Q_a	The annual amplitude of the total heat loss/gain [W or Btu/h]
R_{eq}	Equivalent thermal resistance R-value of entire foundation [m²K/W or ft².°F.h/Btu]
R_f	Thermal resistance R-value of floor [m²K/W or ft².°F.h/Btu]
R_i	Thermal resistance R-value of insulation [m²K/W or ft².°F.h/Btu]
T_a	Ambient or outdoor air temperature [°C or °F]
T_r	Room or indoor air temperature [°C or °F]
$U_{eff,m}$	Effective U-value for the annual mean [Wm⁻².°C⁻¹ or Btu/h.ft².°F] defined in Equation (3.22)
$U_{eff,a}$	Effective U-value for the annual amplitude [Wm⁻².°C⁻¹ or Btu/h.ft².°F] defined in Equation (3.23)
U_o	U-value [Wm⁻².°C⁻¹ or Btu/h.ft².°F] defined in Equation (3.24)
r	Soil density [kg/m³ or lbm/ft³]
w	Annual angular frequency [rad/s or rad/h]
a_s	Thermal diffusivity [m²/s or ft²/h]

It should be noted that the simplified model provides accurate predictions when A/P is larger than 0.5 m.

The annual average heat flux (heat loss or gain) from the building foundation is simply Q_m. The highest foundation heat flux under design conditions can be obtained as follows: $Q_{des} = Q_m + Q_a$.

To illustrate the use of the simplified models, Example 3.3 presents the calculation details of the heat loss from a residence basement insulated with uniform insulation.

EXAMPLE 3.3

Determine the annual mean and annual amplitude of total basement heat loss for a house. The basic geometry and construction details of the basement are provided (see data provided in step 1). The house is located in Denver, Colorado.

Solution: The calculation based on the simplified method outlined by Equations (2.21) through (2.25) is presented as a step-by-step procedure:

Step 1. Provide the required input data (from ASHRAE, 2009):

Dimensions

Basement width: 10.0 m (32.81 ft)

Basement length: 15.0 m (49.22 ft)

Basement wall height: 1.5 m (4.92 ft)

Basement total area: 225.0 m² (2,422.0 ft²)

Ratio of basement area to basement perimeter: $(A/P)_b = 3.629$ m (11.91 ft)

4 in. thick reinforce concrete basement, thermal resistance R-value: 0.5 m²K/W (2.84 h.ft²F/Btu)

Soil Thermal Properties

Soil thermal conductivity: $k_s = 1.21$ W/m.K (0.70 Btu/h.ft.F)

Soil thermal diffusivity: $a_s = 4.47 \times 10^{-7}$ m²/s (48.12 × 10⁻⁷ ft²/s)

Insulation

Uniform insulation R-value = 1.152 m²K/W (6.54 h.ft²F/Btu)

Temperatures

Indoor temperature: $T_r = 22°C$ (71.6°F)

Annual average ambient temperature: $T_a = 10°C$ (50°F)

Annual amplitude ambient temperature: $T_{amp} = 12.7$ K (23 R)

Annual angular frequency: $\omega = 1.992 \times 10^{-7}$ rad/s

Step 2. Calculate Q_m and Q_a values. Using Equations (3.21) through (3.25), the various normalized parameters are first calculated. Then the annual mean and amplitude of the basement heat loss are determined.

$$H_b = \frac{(A/P)_b}{k_s.R_{eq}} = \frac{3.629}{1.21 \times (0.5 + 1.152)} = 1.8155$$

$$b_{eff} = \frac{B}{(A/P)_b} = \frac{1.5}{3.629} = 0.4133$$

$$\left(A/P\right)_{eff,b,mean} = \left[1 + 0.4133 \times \left(-0.4 + e^{-1.8155}\right)\right] \times 3.629 = 3.2731$$

$$\left(A/P\right)_{eff,b,amp} = \left[1 + 0.4133 \times e^{-1.8155}\right] \times 3.629 = 3.8731$$

$$U_{o,m} = \frac{k_s}{\left(A/P\right)_{eff,b,mean}} = \frac{1.21}{3.2731} = 0.3697$$

$$U_{o,a} = \frac{k_s}{\left(A/P\right)_{eff,b,mean}} = \frac{1.21}{3.8731} = 0.3124$$

$$H_{mean} = \frac{\left(A/P\right)_{eff,b,mean}}{k_s.R_{eq}} = \frac{3.2731}{1.21 \times (0.5 + 1.152)} = 1.6374$$

$$H_{amp} = \frac{\left(A/P\right)_{eff,b,amp}}{k_s.R_{eq}} = \frac{3.8731}{1.21 \times (0.5 + 1.152)} = 1.9376$$

$$D_{mean} = \ln\left[(1+H)\left(1+\frac{1}{H}\right)^H\right] = 1.7503$$

$$D_{amp} = \ln\left[(1+H)\left(1+\frac{1}{H}\right)^H\right] = 1.8839$$

$$G = k_s.R_{eq} \cdot \sqrt{\frac{\omega}{\alpha_s}} = 1.21 \times (0.5 + 1.152) \times \sqrt{\frac{1.992 \times 10^{-7}}{4.47 \times 10^{-7}}} = 1.3344$$

Therefore,

$$Q_m = U_{eff,m} A(T_a - T_r) = 0.4 \times 0.3697 \times 1.7503 \times 225 \times (22.0 - 10.0)$$

$$= 698.85 \text{ W } (2384.48 \text{ Btu/h})$$

and

$$Q_a = U_{eff,a} A T_a = 0.25 \times 0.3124 \times 1.8839^{0.16} \times 1.3344^{-0.6} \times 225 \times 12.7$$

$$= 207.72 \text{ W } (708.74 \text{ Btu/h})$$

Table 3.14 shows the comparison of the results between the simplified and the ITPE solution (Krarti, 2000).

TABLE 3.14 Comparison of Predictions between the Simplified Method and the ITPE Solution (heat loss per unit area in W/m²)

Method	Mean (Q_m)	Amplitude (Q_a)
Simplified	699	208
ITPE solution	658	212

Note: For more details about the interzone temperature profile estimation (ITPE) solution technique for foundation heat transfer problems, refer to Krarti (2000).

3.4 Properties of Dry and Humid Air

In order to determine thermal comfort within buildings as well as to analyze the performance of heating and cooling systems, it is important to evaluate the properties of air, especially when it is humid. Humid or moist air is obtained when water vapor is mixed into dry air. The properties of moist air, often called psychrometric properties, can be determined using a psychrometric chart, as illustrated in Figure 3.17. The definitions of the commonly used psychrometric properties are provided below.

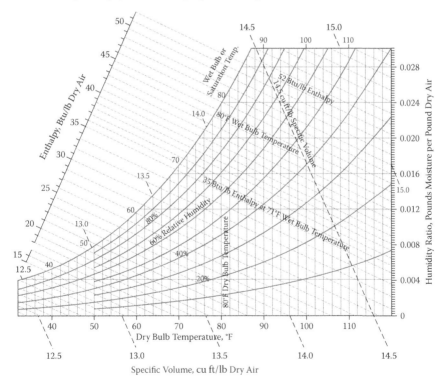

FIGURE 3.17 Psychrometric chart for moist air.

1. The *humidity ratio* is the mass of water per mass of dry air, (for example, kilograms of water per kilograms of dry air). This is sometimes expressed as grains of water per pound of dry air. There are 7,000 grains per pound.
2. The *specific humidity*, W, is the ratio of the water vapor to the total mass of the moist air sample. In terms of saturation pressure, P_s, of water vapor and the total pressure, P_t, of moist air, the specific humidity of moist air can be expressed as follows:

$$W = 0.622 \frac{P_s}{P_t - P_s} \quad (3.26)$$

Since $P_s \ll P_t$, the specific humidity, W, is almost a linear function of P_s.
3. The *relative humidity*, f, is the ratio of the mole fraction of water vapor in a given moist air sample to the mole fraction in an air sample at saturation at the same temperature and pressure. The relative humidity can be estimated using the following expression:

$$\varphi = \frac{P_v}{P_s} \quad (3.27)$$

where P_v is the partial pressure of water vapor. It should be noted that the total pressure of moist air, P_t, is the sum of partial pressure of water vapor, P_v, and the partial pressure of dry air, P_a:

$$P_t = P_v + P_a \quad (3.28)$$

4. The specific volume, v, is the volume occupied by a unit mass of dry air. Using the ideal gas law, the specific volume of dry air can be estimated using the ideal law equation:

$$v = \frac{R_a T}{P_t - P_s} \quad (3.29)$$

5. The *dew point* temperature is the temperature at saturation for a given humidity ratio and a given pressure.
6. The *wet bulb* temperature is the temperature caused by the adiabatic evaporation of water at a given pressure.
7. The *enthalpy* of the moist air is the sum of the enthalpies of the air and the water vapor. The specific enthalpy of moist air, h, can be estimated from the dry bulb temperature, T, and the specific humidity, W, and the enthalpy of saturated steam at temperature T, h_g, using the following expression:

$$h = c_p T + W.h_g \quad (3.30)$$

For moist air, it is common to use the following expressions to estimate specific enthalpy:

$$\begin{cases} (SI) & h = T + W(2501.3 + 1.86T) & [kJ/kg] \\ (IP) & h = 0.24T + W(1061.2 + 0.444T) & [Btu/lb] \end{cases} \tag{3.31}$$

Example 3.4 illustrates how to determine some properties of moist air.

EXAMPLE 3.4

Determine the values of the humidity ratio, specific volume, and enthalpy of moist air with a dry bulb temperature of 80°F and relative humidity of 90%. Assume that the atmospheric pressure is 14.696 psia.

Solution: From the steam tables, saturated vapor pressure at 80°F can be estimated: $P_s = 0.5069$ psia.

Thus, the vapor pressure of the moist air is computed from Equation (3.27): $P_v = P_{sf} = (0.5069)(0.90) = 0.4562$ psia.

The humidity ratio is estimated using Equation (3.26):

$$W = 0.622 \frac{P_v}{P_t - P_v} = 0.622 \frac{0.4562}{14.696 - 0.4562} = 0.02 \quad lbw/lba$$

The specific volume of dry air is obtained using Equation (3.29):

$$v = \frac{R_a T}{P_a} = \frac{(53.352)(80 + 460)}{(14.696 - 0.4562)(144)} = 14.050 \quad ft^3/lba$$

Note the factor 144 is used to convert ft^2 to $in.^2$.

Finally, the enthalpy of the moist air is determined using Equation (3.31):

$$h = 0.24T + W(1061.2 + .444T) = (0.24)(80) + (0.02)[1061.2 + (.444)(80)] = 41.13 \quad Btu/lb$$

Using the psychrometric chart, the same (or very close) values for W, v, and h can be obtained.

The psychrometric chart of Figure 3.17 has several applications, including the analysis of the weather data for a given site. Figure 3.18 illustrates the utilization of the psychrometric chart to assess the characteristics of site weather data through the use of Climate Consultant 5.0, freeware useful for assessing the best passive heating and cooling strategies for a building located at the site. Another application of the psychrometric chart is the evaluation of indoor thermal comfort within various building spaces. A brief outline of the main factors affecting indoor thermal comfort as well as metrics commonly used to quantify the level of thermal comfort is provided in Section 3.5.

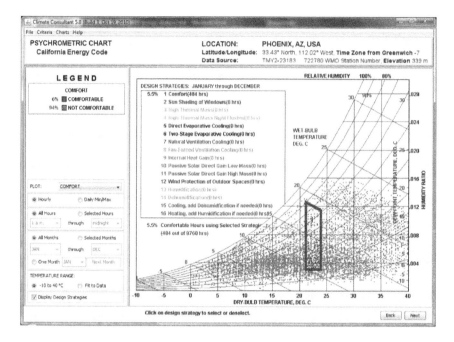

FIGURE 3.18   Application of psychrometric chart to evaluate the weather of Phoenix, Arizona, using Climate Consultant 5.0.

3.5 Thermal Comfort

Comfort is an elusive and subjective concept. ASHRAE Standard 55, "Thermal Environmental Conditions for Human Occupancy," defines *thermal comfort* as the "condition of mind that expresses satisfaction with the thermal environment." This definition emphasizes that the judgment of comfort is a cognitive process involving many inputs influenced by physical, physiological, psychological, and other processes. The conscious mind appears to reach conclusions about thermal comfort and discomfort from direct temperature and moisture sensations from the skin, deep body temperatures, and efforts necessary to regulate body temperatures.

Several thermal comfort models do exist, including effective temperature, operative temperature, mean radiant temperature, and predicted mean value. Most of these models are based on heat balance applied to the human body to indicate that heat generated by metabolic activities ($q_{met,heat}$) must be dissipated by convection (q_{conv}), radiation (q_{rad}), evaporative heat transfer (q_{evap}), and respiration ($q_{resp,sen}$ and $q_{resp,lat}$):

$$q_{met,heat} = q_{conv} + q_{rad} + q_{evap} + q_{resp,sen} + q_{resp,lat} \tag{3.32}$$

There are several thermal comfort indicators, including the following.

3.5.1 Mean Radiant Temperature

When only convection and radiation heat exchanges are considered, a mean radiant temperature can be defined and estimated using the globe temperature, air temperature, and air velocity as

$$T_{mrt}^4 = T_g^4 + 0.103 * 10^9 \overline{V}^{1/2}(T_g - T_a) \quad (IP)$$

(3.33)

$$T_{mrt}^4 = T_g^4 + 0.247 * 10^9 \overline{V}^{1/2}(T_g - T_a) \quad (SI)$$

3.5.2 Operative Temperature

The weighted average of air and radiant temperatures using convective and radiative heat transfer coefficients is called the *operative temperature*.

$$T_{op} = \frac{h_c T_a + h_r T_r}{h_c + h_r}$$

(3.34)

3.5.3 Effective Temperature

The *effective temperature*, ET^*, is defined as the operative temperature at 50% RH that gives the same total heat transfer as at the actual environmental conditions. The exact relationship is a function of many variables, including clothing levels, activity levels, and convection and radiation heat transfer coefficients.

3.5.4 Franger's Predicted Mean Value

The predicted mean value (PMV) is the average response of a large number of people. Given the subjective nature of comfort, there will actually be a distribution of satisfaction among a large group of people. Fanger developed a set of correlations between PMV and the thermal load:

$$PMV = 3.155\left(0.303e^{-0.114M} + 0.028\right)L \quad US$$

(3.35)

$$PMV = \left(0.303e^{-0.036M} + 0.028\right)L \quad SI$$

Figure 3.19 shows an empirical relationship of the percentage of people dissatisfied (PPD) with a thermal environment as a function of the PMV. As shown in Figure 3.19, a neutral PMV (i.e., $PMV = 0$) indicates that 5% of people are dissatisfied with the indoor thermal comfort.

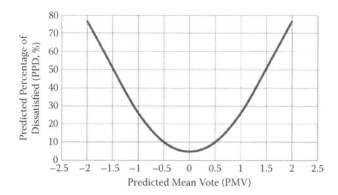

FIGURE 3.19   Correlation between PDD and PMV for Fanger's model.

3.5.5 Pierce Two-Node Model

The Pierce two-node model was developed by A.P. Gagge et al. (1970) at the John B. Pierce Foundation Laboratory. The model is a simplified mathematical lumped capacity analysis of a human thermoregulatory system. The model thermally lumps the human body as two isothermal concentric thermal compartments representing the skin and the core. The temperature within a compartment is assumed to be uniform, so that the only temperature gradient is between the compartments. This model uses Fanger's model for calculating respiratory heat losses. The core loses energy when the muscles do work. Heat is also lost from the core through respiration. In addition, heat is conducted passively from the core to the skin. Based on the experimental research, the predicted mean value modified by effective temperature (PMVET) is proposed to be similar to the Fanger comfort model (PMV), using a 9-point thermal sensation scale (4 to –4). The thermal sensation (TSENS) of sedentary people is represented using an 11-point thermal sensation scale (5 to –5).

3.5.6 KSU Two-Node Model

The KSU two-node model was developed at Kansas State University. It is basically similar to the Pierce two-node model, except that the calculation of thermal sensation (TSV) in the KSU model is different from that used to estimate the thermal sensation (TSENS) of the Pierce model. The empirical sensation equation was developed from a wide range of experimental conditions, ranging from very hot to very cold, activity levels from 1 to 6 met, and clothing levels from 0.05 to 0.7 clo. Thermal conductance between the core and the skin temperature is applied to calculate the thermal sensation in a cold environment. However, the change of skin wettedness is considered in the warm environment. In addition, the thermal sensation vote (TSV) has a scale similar to that of the Fanger comfort model (PMV) and Pierce two-node model (TSENS).

3.5.7 ASHRAE Comfort Standard

ASHRAE has developed an industry consensus standard to describe comfort requirements in buildings. The standard is known as ASHRAE 55-2004, "Thermal

Environmental Conditions for Human Occupancy." The purpose of this standard is to specify the combinations of indoor thermal environmental factors and personal factors that will produce thermal environmental conditions acceptable to a majority of the occupants within the space. Refer to Figure 3.20 to estimate acceptable conditions for sedentary activity (let < 1.2).

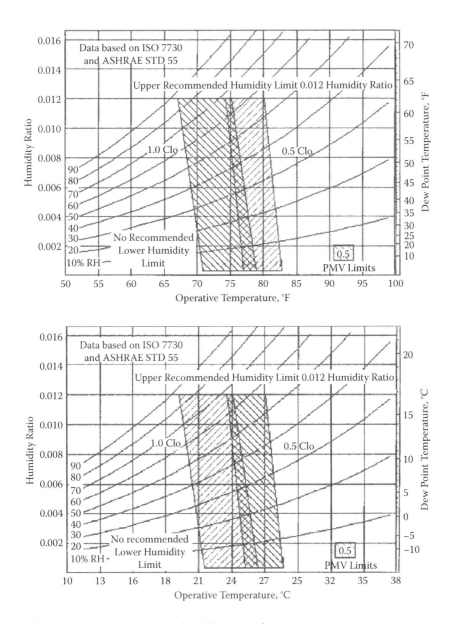

FIGURE 3.20 Psychromteric chart for ASHRAE comfort zone.

Acceptable operative temperatures depend on the activity of a person (expressed in *met*) and his/her clothing level (expressed in *clo*):

$$T_{op,active}(°F) = T_{o,sedentary} - 5.4(1 + clo)(met - 1.2) \qquad (IP)$$

(3.36)

$$T_{op,active}(°C) = T_{o,sedentary} - 3.0(1 + clo)(met - 1.2) \qquad (SI)$$

3.5.8 In-Class Exercises

1. a. Compute the mean radiant temperature in a room in which measurements showed that the dry bulb temperature is 70°F, 40 fpm air velocity, and the globe temperature is 78°F.
   b. Estimate the operative temperature if the convective and radiative heat transfer coefficients are about equal.
   c. Assess the thermal comfort level in the room.

2. Determine the summer and winter temperature settings in a gymnasium (where the activity level is on average 2.5 and the clothing level is 0.5) so people are comfortable.

3.6 Simplified Energy Analysis Methods

For residential buildings, simplified energy analysis methods can be used to evaluate the whole building energy performance. In this section two simplified analysis methods are described, including the variable-base degree-day method and the thermal network technique.

3.6.1 Variable-Base Degree-Day Method

The degree-day method provides an estimation of the heating and cooling loads of a building due to transmission losses through the envelope and any solar and internal heat gains. The degree-day method is based on steady-state analysis of the heat balance across the boundaries of the building. A building is typically subject to several heat flows, including conduction, infiltration, solar gains, and internal gains, as illustrated in Figure 3.21. The net heat loss or heat gain at any instant is determined by applying a heat balance (i.e., first law of thermodynamics) to the building. For instance, for heating load calculation, the instantaneous heat balance provides:

$$\dot{q}_h = BLC.(T_i - T_o) - \dot{q}_g \qquad (3.37)$$

where *BLC* is the building load coefficient as defined in Equation (3.38), but modified to include the effects of both transmission and infiltration losses. Thus, the BLC for any building can be calculated as follows:

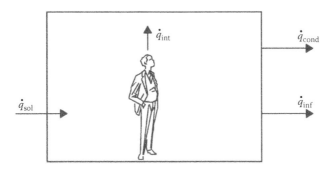

FIGURE 3.21 A simplified heat balance model for a building.

$$BLC = \sum_{j=1}^{N_E} U_{T,j}.A_j + \dot{m}_{inf}.c_{p,a} \tag{3.38}$$

q_g is the net heat gains due to solar radiation, q_{sol}; internal gains (people, lights, and equipment), q_{int}; and in some cases the ground losses, q_{grd}, if they are significant:

$$\dot{q}_g = \dot{q}_{sol} + \dot{q}_{int} - \dot{q}_{grd}$$

This equation can be rearranged to introduce the balance temperature, T_b, for the building:

$$q_H = BLC.\left[(T_i - \frac{q_g}{BLC}) - T_o\right] = BLC.(T_b - T_o) \tag{3.39}$$

with the balance temperature defined as

$$T_b = T_i - \frac{q_g}{BLC} \tag{3.40}$$

Therefore, the balance temperature adjusts the interior temperature set point by the amount of temperature increase due to a reduction in the building heating load resulting from the internal gains. Before the oil crisis, the transmission and the infiltration losses were significant (and thus the BLC value was high relative to the internal gains). It is estimated that the net internal gain contributes to about 3°C (or 5°F) in most buildings. Therefore, the balance temperature was assumed to be 18°C (or 65°F) for all the buildings. However, with the increase in thermal efficiency of the building envelope and the use of more equipment within the buildings, the internal heat gains are more significant, and thus can contribute to significantly reducing the heating load of the buildings.

By integrating the instantaneous heating load over the heating season, the total building heating load can be determined. Note only the positive values of Q_H are considered in the integration. In practice, the integration is approximated by the sum of the heating loads averaged over short time intervals (1 hour or 1 day).

If hourly averages are used, the seasonal total building heating load is estimated as

$$Q_H = \sum_{j=1}^{N_{h,H}} \dot{q}_{H,j}^+ = BLC. \sum_{i=1}^{N_{h,H}} (T_b - T_{o,j})^+ \tag{3.41}$$

The sum is performed over the number, $N_{h,H}$, of hours in the heating season. From Equation (3.42), a parameter that characterizes the heating load of the building can be defined as the heating degree-hours (*DHH*), which is a function of only the outdoor temperatures and the balance or base temperature, T_b, which varies with the building heating set-point temperature and the building internal gains:

$$DH_H(T_b) = \sum_{j=1}^{N_{h,H}} (T_b - T_{o,j})^+ \tag{3.42}$$

If daily averages are used, the seasonal total building heating load is estimated as:

$$Q_H = 24. \sum_{i=1}^{N_{d,H}} \dot{q}_{H,i}^+ = 24.BLC. \sum_{i=1}^{N_{d,H}} (T_b - T_{o,i})^+ \tag{3.43}$$

The sum is performed over the number, $N_{d,H}$, of days in the heating season. Similarly to the heating degree-hours defined by Equation (3.42), the heating degree-days (*DDH*) can be introduced for a base temperature, T_b:

$$DD_H(T_b) = \sum_{i=1}^{N_{d,H}} (T_b - T_{o,i})^+ \tag{3.44}$$

The total energy use, *EH*, to meet the heating load of the building can be estimated by assuming a constant efficiency of the heating equipment over the heating season (for instance, several heating equipment manufacturers provide the annual fuel use efficiency (AFUE) rating for their boilers or furnaces).

Using heating degree-hours, $DH_H(T_b)$, the total energy use, E_H, is estimated as follows:

$$E_H = \frac{Q_H}{\eta_H} = \frac{BLC.DH_H(T_b)}{\eta_H} \tag{3.45}$$

Using heating degree-days, $DD_H(T_b)$, E_H is calculated using Equation (3.46):

$$E_H = \frac{Q_H}{\eta_H} = \frac{24.BLC.DD_H(T_b)}{\eta_H} \tag{3.46}$$

The variable-base degree-hour and degree-day methods stated by Equations (3.45) and (3.46) can also be applied to determine the cooling load by estimating the cooling season degree-hours, DH_C, and degree-days, DD_C, using, respectively, Equations (3.47) and (3.48):

$$DH_C(T_b) = \sum_{j=1}^{N_{h,C}} (T_{o,j} - T_b)^+ \tag{3.47}$$

$$DD_C(T_b) = \sum_{i=1}^{N_{d,C}} (T_{o,i} - T_b)^+ \tag{3.48}$$

where $N_{h,c}$ and $N_{d,c}$ are the number of hours and days, respectively, in the cooling season.

Figure 3.22 illustrates the concept of both heating and cooling degree-days based on the variation of the average daily outdoor air temperature over 1 year.

It should be noted that the variable-base degree-hour and degree-day methods can provide remarkably accurate estimation of the annual energy use due to heating, especially for buildings dominated by losses through the building envelope, including infiltration. Unfortunately, the degree-hour or degree-day method is not as accurate for calculating the cooling loads (Claridge et al., 1987) due to several factors, including effects of building thermal mass that delays the action of internal gains, mild outdoor temperatures in summer resulting in large errors in the estimation of the cooling degree-days, and the large variation in infiltration or ventilation rates as occupants open windows or economizer cycles are used.

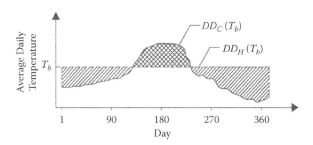

FIGURE 3.22 Heating and cooling degree-days associated with a balance temperature, T_b.

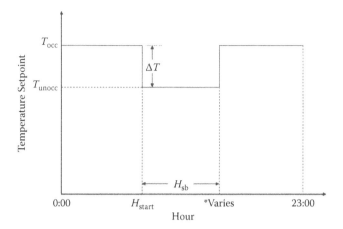

FIGURE 3.23   Heating temperature setting with a setback temperature, ΔT.

In several homes, the heating set point is set back during unoccupied hours, typically during daytime, as outlined in Figure 3.23. Instead of using Equation (3.42) with no setback, the heating degree-hours when temperature setback is implemented can be estimated as follows:

$$DH(T_{b,setback}) = \sum_{j=1}^{N_{h,OCC}} (T_b - T_{o,j})^+ + \sum_{j=1}^{N_{h,UNOCC}} (T_{b,setback} - T_{o,j})^+ \tag{3.49}$$

where T_b = base temperature during the occupied period (°F or °C), $T_{b,setback}$ = base temperature adjusted for setback period during the unoccupied period (°F or °C), $N_{h,UNOCC}$ = number of hours during the heating season when the home is unoccupied, and $N_{h,OCC}$ = number of hours during the heating season when the home is occupied.

As illustrated in Figure 3.23, the temperature setback, $T_{b,setback}$, can be defined as a function of the balance temperature, T_b, and the setback temperature difference:

$$T_{b,setback} = T_b - \Delta T \tag{3.50}$$

In residential buildings, setbacks are typically implemented during weekdays to accommodate typical daytime work schedules, as illustrated in Figure 3.24. In this case, the heating degree-hours can be estimated as follows:

$$HD(T_{b,setback}) = \sum_{j=1}^{N_{h,WE}} (T_b - T_{o,j})^+ + \sum_{j=1}^{N_{h,WD,OCC}} (T_b - T_{o,j})^+ + \sum_{j=1}^{N_{h,WD,UNOCC}} (T_{b,setback} - T_{o,j})^+ \tag{3.51}$$

where $N_{h,WD,UNOCC}$ = number of weekday hours during unoccupied period, $N_{h,WD,OCC}$ = number of weekday hours during occupied period, and $N_{h,WE}$ = number of weekend hours.

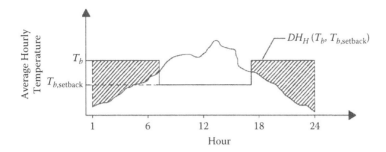

FIGURE 3.24 Heating degree-hours for a balance temperature, T_b, and a setback temperature $T_{b,setback}$.

The temperature setback reduces the heating degree-hours and thus the energy use required to heat the home. To estimate the reduction of the heating degree-hours due to a temperature setback, the fraction f_{DH} is defined as the ratio of the degree-hours with setback estimated using Equation (3.51) to the degree-days without setback calculated using Equation (3.42):

$$f_{DH} = \frac{DH(T_{b,setback})}{DH(T_b)} \tag{3.52}$$

It should be noted if there is no setback, then $f_{DH} = 1$.

Figures 3.25 and 3.26 show the variation fraction f_{DH} as a function of the temperature setback differential, $\Delta T = (T_b - T_{b,setback})$, for several temperature settings and starting times for the temperature setback when the house is located in Eagle, Colorado, and Savannah, Georgia, respectively.

Methods to estimate heating and cooling degree-days as well as the fraction, f, as defined in Equation (3.52) are provided in Chapter 5.

3.6.2 Transient Thermal Network Analysis

The degree-day method presented above assumes limited *steady-state* heat transfer processes. In most buildings, convection heat transfer is generally steady state; however, conduction and radiation heat transfer are often transient and vary with time. The timescales associated with these transients in residential buildings are on the order of hours, which is the same general timescale for changes in the driving potentials of outdoor temperature and solar radiation. Therefore, in calculating the heat transfer into a building through walls and roofs, the calculations *must* account for transient effects. Straight steady-state heat transfer calculations are a crude simplification, especially for a building with significant thermal mass. In residential buildings, simplified transient analysis techniques can be utilized to estimate the mass effect and predict the time variation of indoor temperature. In this section, the RC thermal network technique is first introduced, and then applied to estimate the thermal mass and the time constant of a residential building.

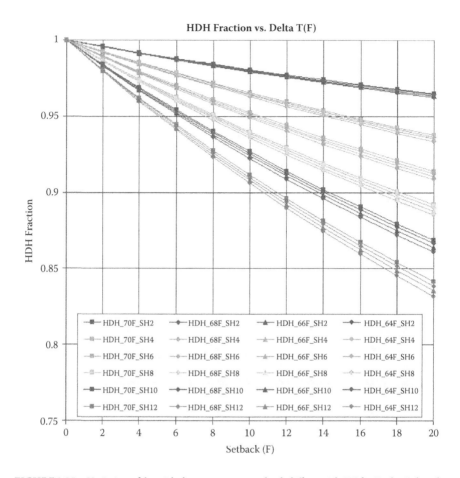

FIGURE 3.25   Variation of f_{DH} with the temperature setback differential, ΔT, for Eagle, Colorado.

3.6.2.1 Overview of the RC Thermal Network Analysis Technique

An entire building can be represented by a thermal RC network. In the simplest case, building heat transfer can be represented by two temperature nodes, a single capacitance and a single resistance, as shown in Figure 3.26. The two temperatures are the indoor and outdoor temperatures, the capacitance represents the entire thermal mass of the building, and the resistance represents the heat transfer path between the indoor and outdoor temperatures. Additional heat flow paths are added to account for internal heat gains and heat gain by the HVAC system.

Mathematically, the thermal network for the house shown in Figure 3.27 can be expressed using a simple expression:

$$C\frac{dT_{in}}{dt} = BLC(T_{out} - T_{in}) + Q_{gain} \tag{3.53}$$

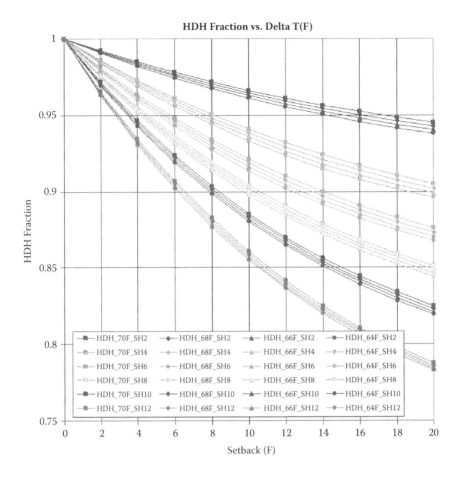

FIGURE 3.26 Variation of f_{DH} with the temperature setback differential, ΔT, for Savannah, Georgia.

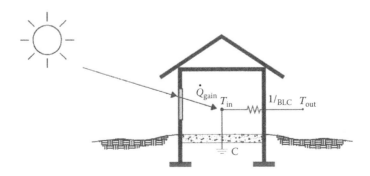

FIGURE 3.27 Thermal network model for a home.

where $C = mc_p$ is the capacitance of the home, including the mass of all the envelope components (i.e., floor and walls) as well as furniture and interior partitions. Notice that if the HVAC system is controlled to maintain the indoor temperature at a constant value, this equation reverts back to a simple and familiar steady-state equation:

$$C \frac{dT_b}{dt} = BLC(T_{out} - T_b) \tag{3.54}$$

where T_b is the balance temperature of the building defined earlier for the variable-base degree-day method as follows:

$$T_b = T_{in} - Q_{gain} / BLC \tag{3.55}$$

The time constant, τ, of a building is defined as the ratio of the building capacitance, C, over the building load coefficient (BLC):

$$\tau = \frac{C}{BLC} \tag{3.56}$$

Thus, the differential equation, Equation (3.54), can be expressed simply as:

$$\frac{dT_b}{dt} = -\frac{T_{in} - T_{out}}{\tau} \tag{3.57}$$

If the outdoor air temperature, T_{out}, is constant, the solution to the differential equation is given below with $T_{b,0} = T_b(t = 0)$:

$$\frac{T_b(t) - T_{out}}{T_{b,0} - T_{out}} = e^{-t/\tau} \tag{3.58}$$

Therefore, the indoor air temperature variation with time, $T_{in}(t)$, can be estimated as follows:

$$T_{in}(t) = T_{out} + \frac{Q_{gain}}{BLC} + \left[T_{in}(t=0) - T_{out} - \frac{Q_{gain}}{BLC} \right] e^{-t/\tau} \tag{3.59}$$

In particular, when there are no internal gains (i.e., $Q_{gain} = 0$), the indoor air temperature variation with time, $T_{in}(t)$, can be simply estimated as follows:

$$T_{in}(t) = T_{out} + \left[T_{in}(t=0) - T_{out} \right] e^{-t/\tau} \tag{3.60}$$

EXAMPLE 3.5

Consider a home with BLC = 380 Btu/h.°F and a capacitance of C = 5,270 Btu/°F. During a cold night when the outdoor air temperature is constant at T_{out} = 0°F with no internal gains (i.e., Q_{gain} = 0), determine the time for the indoor temperature to decrease from 68°F to 50°F.

Solution: First, the time constant of the home is estimated using Equation (3.5):

$$\tau = \frac{5,270}{380} = 13.86 \text{ hr}$$

Then, using Equation (3.8), the time t can be obtained as shown below:

$$t = -\tau * Ln\left[\frac{T_{in}(t) - T_{out}}{T_{in}(t=0) - T_{out}}\right] = 13.86 * Ln\left[\frac{60-0}{68-0}\right] = 1.73 \text{ hr}$$

3.6.2.2 Procedure for Estimating Home Capacitance

The RC thermal network model presented in this section can be used to estimate the thermal mass of a building by measuring the indoor air temperature during a period where the internal gains are small and the outdoor air temperature remains constant. A cold winter night during a period when the heating system is shut off (or during a setback period) can be an adequate cooldown testing period, especially if the outdoor temperature does not vary significantly. Figure 3.28 shows the time variations of both outdoor and indoor temperature variations for a home during a cooldown period.

The time constant can be estimated as the slope of the time variation of the variable Y defined as

$$Y = Ln\left[\frac{T_{in}(t) - T_{out}}{T_{in}(t=0) - T_{out}}\right] \tag{3.61}$$

Based on Equation (3.61), Y varies linearly with t with a constant of variation $a = -1/\tau$, where τ is the time constant of the home. Using linear regression, slope a can be obtained as illustrated by Figure 3.29 for the testing data presented in Figure 3.28 for the period between 2:15 and 3:45 a.m., when the outdoor temperature remains constant at about 56°F.

Based on the value of the slope a ($a = -0.0018$ min^{-1}), as shown in Figure 3.29, the time constant can be estimated to be $\tau = -1/a = -1/(0.0018$ min$^{-1}) = 555.6$ min $= 9.25$ h. This time constant indicates that the home has a rather significant level of thermal mass. The capacitance of the home can be estimated using Equation (3.56) based on the BLC value.

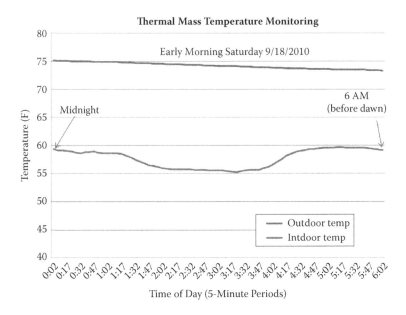

FIGURE 3.28   Indoor and outdoor temperature variation during a cooldown period.

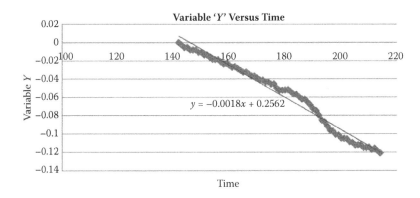

FIGURE 3.29   Estimation of the home time constant through regression of *Y* vs. time.

3.7 Whole Building Energy Models

Generally, whole building energy analysis tools can be classified into either forward or inverse methods. In the forward approach, as depicted in Figure 3.30, the energy predictions are based on a physical description of the building systems, such as geometry, location, construction details, and HVAC system type and operation. Most of the existing detailed energy simulation tools, such as DOE-2, TRNSYS, and EnergyPlus, follow the forward modeling approach. In the inverse approach illustrated in Figure 3.31, the

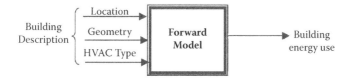

FIGURE 3.30 Basic approach of a typical forward energy analysis model.

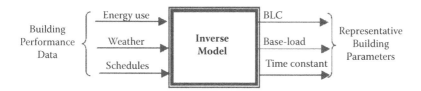

FIGURE 3.31 Basic approach of a typical inverse energy analysis model.

energy analysis model attempts to deduce representative building parameters (such as the building load coefficient (BLC), the building base load, or the building time constant) using existing energy use, weather, and relevant performance data. In general, the inverse models are less complex to formulate than the forward models. However, the flexibility of inverse models is typically limited by the formulation of the representative building parameters and the accuracy of the building performance data. Most of the existing inverse models rely on regression analysis (such as the variable-base degree-day models (Fels, 1986) or the change-point models (Kissock et al., 1998)) tools or the connectionist approach (Kreider et al., 1997) to identify the building parameters. In Chapter 4, a screening methodology is introduced and is based on inverse modeling of residential buildings using the degree-day method described in Section 3.6.

It should be noted that tools based on the forward or inverse approaches are suitable for other applications. Among the common applications are verification of energy savings actually incurred from energy conservation measures (for more details about this application, refer to Chapter 16), diagnosis of equipment malfunctions, and efficiency testing of building energy systems.

Energy analysis tools can also be classified based on their ability to capture the dynamic behavior of building energy systems. Thus, energy analysis tools can use either steady-state or dynamic modeling approaches. In general, the steady-state models are sufficient to analyze seasonal or annual building energy performance. However, dynamic models may be required to assess the transient effects of building energy systems, such as those encountered for thermal energy storage systems and optimal start controls.

In this chapter, selected energy analysis tools commonly used in the United States and Europe are described. These tools are grouped into two categories:

- *Inverse methods*: Use both steady-state and dynamic modeling approaches and include variable-base degree-day methods.
- *Forward methods*: Include either steady-state or dynamic modeling approaches and are often the basis of detailed energy simulation computer programs.

3.7.1 Inverse Modeling Methods

As discussed in the introduction, methods using the inverse modeling approach rely on existing building performance data to identify a set of building parameters. The inverse modeling methods can be valuable tools in improving the building energy efficiency. In particular, the inverse models can be used to:

- Help detect malfunctions by identifying time periods or specific systems with abnormally high energy consumption
- Provide estimates of expected savings from a defined set of energy conservation measures
- Measure and verify savings achieved by energy retrofits

Typically, regression analyses are used to estimate the representative parameters for the building or its systems (such as building load coefficient or heating system efficiency) using measured data. In general, steady-state inverse models are based on monthly or daily data and include one or more independent variables. Dynamic inverse models are usually developed using hourly or subhourly data to capture any significant transient effect, such as the case where the building has a high thermal mass to delay cooling or heating loads.

3.7.1.1 Steady-State Inverse Models

These models generally attempt to identify the relationship between the building energy consumption and selected weather-dependent parameters, such as monthly or daily average outdoor temperatures, degree-hours, or degree-days. As mentioned earlier, the relationship is identified using statistical methods (based on linear regression analysis). The main advantages of the steady-state inverse models are:

- Simplicity: Steady-state inverse models can be developed based on a small data set such as energy data obtained from utility bills.
- Flexibility: Steady-state inverse models have a wide range of applications. They are particularly valuable in predicting the heating and cooling energy end uses for both residential and small commercial buildings.

However, steady-state inverse models have some limitations since they cannot be used to analyze transient effects such as thermal mass effects and seasonal changes in the efficiency of the HVAC system.

Steady-state inverse models are especially suitable for measurement and verification (M&V) of energy savings accrued from energy retrofits. Chapter 10 provides a more detailed discussion about the inverse models that are commonly used for M&V applications. In this section, only simplified methods based on steady-inverse modeling are briefly presented. These simplified models have been used to determine the energy impact of selected energy efficiency measures and are based on the degree-day method.

Using steady-state analysis, the building energy use per billing period is correlated to either the average outdoor temperatures or the heating or cooling degree-days (obtained for the billing period). Thus, the energy consumption is estimated for each billing period using the following expression, depending on whether outdoor temperature or degree-day is used.

Equation (3.62) provides a temperature-based model, while Equation (3.63) outlines a degree-day-based model for evaluating residential building heating loads:

$$E_H = 24. f_H . \frac{BLC}{\eta_H}.(T_{b,H} - \overline{T}_o) + E_{base,H} \tag{3.62}$$

$$E_H = 24. f_H \frac{BLC}{\eta_H}.DD_H(T_{b,H}) + E_{base,H} \tag{3.63}$$

Equation (3.64) provides a temperature-based model, while Equation (3.65) outlines a degree-day-based model for evaluating residential building cooling loads:

$$E_C = 24. f_C . \frac{BLC}{COP_C}.(\overline{T}_o - T_{b,C}) + E_{base,C} \tag{3.64}$$

$$E_C = 24. f_C . \frac{BLC}{COP_C}.DD_C(T_{b,C}) + E_{base,C} \tag{3.65}$$

where $E_{H/C}$ represents the annual building energy use during the heating or cooling season, BLC is the building loss coefficient, η_H is the average seasonal energy efficiency of the heating system, COP_C is the average seasonal coefficient of performance of the cooling system, $T_{b,H/C}$ is the building balance temperature for heating or cooling energy use, DD_H is the heating or cooling degree-days (based on the balance temperature), and $E_{base,H/C}$ is the base load for building energy use. It represents the nonheating or noncooling energy use.

Through regression analysis, the balance temperature and the building load coefficient (assuming known heating or cooling system efficiency) can be determined. Once the building parameters are estimated, the models of Equations (3.62) to (3.65) can be used to establish an energy use model for the building and to determine any energy savings attributed to measures that affect one of the three parameters: balance temperature, building load coefficient, or heating/cooling system efficiency. Figure 3.32 illustrates the use of the heating model outlined by Equation (3.62) to correlate the natural gas usage as a function of the heating degree-day for a residential building (Kalinic, 2009).

3.7.1.2 Dynamic Models

Steady-state inverse models are only suitable for predicting long-term building energy use. Therefore, energy use data are collected for a relatively long time period (at least one season or 1 year) to carry out the regression analysis. On the other hand, dynamic inverse models can be used to predict short-term building energy use variations using data collected for a short period of time, such as 1 week. Generally, a dynamic inverse model is based on a building thermal model that uses a specific set of parameters. These building model parameters are identified using typically some form of a regression analysis.

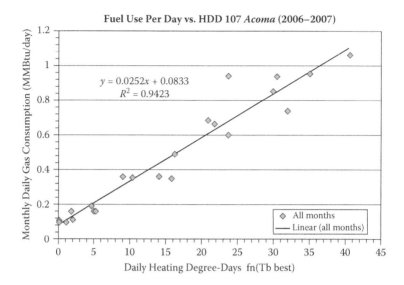

FIGURE 3.32   Analysis of gas consumption as a function of monthly heating degree-days.

An example of a dynamic model relating building cooling energy use to the outdoor air temperatures at various time steps (typically hours) is presented by Equation (3.66):

$$E_C^n + b_1 E_C^{n-1} + ... + b_N E_C^{n-N} = a_0 T_o^n + a_1 T_o^{n-1} + ... + a_M T_0^{n-M} \tag{3.66}$$

Other examples of dynamic inverse models include equivalent thermal network analysis, Fourier series models, and artificial neural networks. These models are capable of capturing dynamic effects such as building thermal mass dynamics. The main advantages of the dynamic inverse models include the ability to model complex systems that depend on several independent parameters. Their disadvantages include their complexity and the need for more detailed measurements to fine-tune the model. Unlike steady-state inverse models, dynamic inverse models usually require a high degree of user interaction and knowledge of the modeled building or system.

3.7.2 Forward Modeling Methods

Forward modeling methods are generally based on the physical description of the building energy systems. Typically, forward models can be used to determine the energy end uses as well as predict any energy savings incurred from energy conservation measures. Selected existing U.S. energy analysis tools that use the forward modeling approach are described in the following sections. For a more detailed discussion, the reader is referred to ASHRAE *Handbook of Fundamentals* (ASHRAE, 2009).

3.7.2.1 Steady-State Methods

Steady-state energy analysis methods that use the forward modeling approach are generally easy to use since most of the calculations can be performed by hand or using spreadsheet programs. Most steady-state forward tools are based on one form of degree-day methods. As described in Section 3.5, the degree-day method uses a seasonal degree-day computed at a specific set-point temperature (or balance temperature) to predict the energy use for building heating. Typically, these degree-day methods are not suitable for predicting building cooling loads. In the United States, the traditional degree-day method using a base temperature of 65°F has been replaced by the variable-base degree-day method and is applied mostly to residential buildings. In Europe, heating degree-days using 18°C as the base temperature are still used for both residential and commercial buildings.

The variable-base degree-day methods predict seasonal building energy used for heating with one variation of the following formulation:

$$FU = \frac{24.BLC.f. \, DD_H(T_b)}{\eta_H} \qquad (3.67)$$

where FU represents the fuel use (gas, fuel oil, or electricity, depending on the heating system); BLC is the building loss coefficient, including transmission and infiltration losses through the building envelope; f is a correction factor, to include various effects, such as night setback effects, and free heat gains (Chapter 5 discusses a simplified method to estimate this factor); T_b is the building heating balance temperature; and $DD_H(T_b)$ is the heating degree-days calculated at the balance temperature, T_b. Chapter 5 provides simplified methods to estimate degree-days at any balance temperature. Monthly and annual degree-days for selected balance temperatures are also provided in the appendices.

Variable-base degree-day methods provide generally good predictions of the fuel use for residential buildings dominated by transmission loads. However, they are not recommended for buildings dominated by internal loads or with involved HVAC system operation strategies.

3.7.2.2 Dynamic Methods

Dynamic analytical models use numerical or analytical methods to determine energy transfer between various building systems. These models consist generally of simulation computer programs with hourly or subhourly time steps) to estimate adequately the effects of thermal inertia—due, for instance, to energy storage in the building envelope or its heating system. The important characteristic of the simulation programs is their capability to account for several parameters that are crucial to accurate energy use, especially for buildings with significant thermal mass, thermostat setbacks or setups, explicit energy storage, or predictive control strategies. A typical calculation flowchart of detailed simulation programs is presented in Figure 3.33.

Detailed computer programs require a high level of expertise and are generally suitable to simulate large buildings with complex HVAC systems and involved control strategies that are difficult to model by simplified energy analysis tools.

In general, an energy simulation program requires a detailed physical description of the building (including building geometry, building envelope construction details,

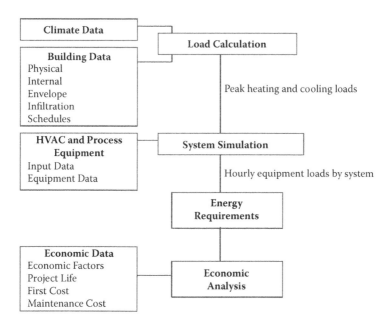

FIGURE 3.33   Flowchart of complete building model.

HVAC equipment type and operation, and occupancy schedules). Thermal load calculations are based on a wide range of algorithms, depending on the complexity and the flexibility of the simulation program. To adequately estimate energy savings from energy efficiency measures, energy simulation tools have to be calibrated using existing measured energy data (utility bills, for instance). A basic calibration procedure is discussed in detail in Chapter 10.

While energy simulation programs are generally capable of modeling most of the building energy systems, they are often not sufficiently flexible and have inherent limitations. To select the appropriate energy simulation program, it is important that the user be aware of the capabilities of each simulation available to him or her. Some of the well-known simulations programs are briefly presented below:

- DOE-2 (version DOE-2.1). DOE-2 was developed at the Lawrence Berkeley National Laboratory (LBNL) by the U.S. Department of Energy and is widely used because of its comprehensiveness. It can predict hourly, daily, monthly, and annual building energy use. DOE-2 is often used to simulate complex buildings. Typically, significant efforts are required to create DOE-2 input files using a programming language called Building Description Language (BDL). Several tools are currently available to facilitate the process of developing DOE-2 input files. Among energy engineers and professionals, DOE-2 has become a standard building energy simulation tool in the United States, and several other countries using interfaces such as eQUEST and VisualDOE. Figure 3.34 shows a three-dimensional rendering of a residential building modeled using eQUEST.

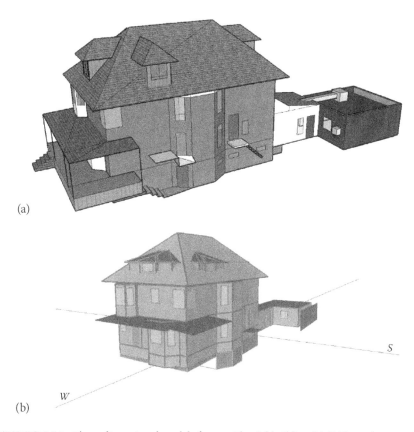

(a)

(b)

FIGURE 3.34 Three-dimensional models for a residential building (a) CAD rendering and (b) eQUEST rendering. (From Kalinic, N., *Measurement and Verification of Savings from Implemented ECMs*, MS report, University of Colorado, Boulder, 2009.)

- EnergyPlus builds on the features and capabilities of both DOE-2 and BLAST. EnergyPlus uses new integrated solution techniques to correct one of the deficiencies of both BLAST and DOE-2—the inaccurate prediction of space temperature variations. Accurate prediction of space temperatures is crucial to properly analyze energy-efficient systems. For instance, HVAC system performance and occupant comfort are directly affected by space temperature fluctuations. Moreover, EnergyPlus has several features that should aid engineers and architects to evaluate a number of innovative energy efficiency measures that cannot be simulated adequately with either DOE-2 or BLAST. These features include:
  - Free cooling operation strategies using outdoor air
  - Realistic HVAC systems controls
  - Effects of moisture adsorption in building elements
  - Indoor air quality with a better modeling of contaminant and air flows within the building

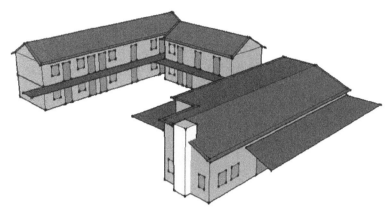

FIGURE 3.35   Three-dimensional model for a large building using EnergyPlus.

Several interfaces for EnergyPlus have been developed over the last few years. A complete list of the interfaces is periodically updated on the EnergyPlus website (EnergyPlus, 2009). Figure 3.35 illustrates a three-dimensional rendering of a large residential building modeled using EnergyPlus.

- TRNSYS provides a flexible energy analysis tool to simulate a number of energy systems using user-defined modules. While TRNSYS can be used to model heating and cooling thermal loads for homes, it is widely used for modeling active solar systems, including photovoltaic (PV) and solar hot water heating systems. Figure 3.36 illustrates a thermal model for a home using TRNSYS. A good knowledge of computer programming (Fortran) is required to properly use the TRNSYS simulation tool.

3.8 Summary

In this chapter, suitable thermal analysis methods and techniques are presented to assess the energy performance of residential buildings. In particular, fundamental heat transfer concepts as well as thermal comfort metrics are reviewed and applied to estimate the characteristics of various energy systems of a building. Moreover, simplified as well as detailed analysis methods are described, with a brief discussion of the general analysis procedures used by these tools. The energy auditor should select the proper tool to carry out the energy analysis of the building and to estimate the potential energy and cost savings for retrofit measures.

Problems

3.1   A 20 ft by 10 ft flat roof of a residential building, located at sea level, is comprised of the following layers from outside to inside: 3/8 in. buildup roofing, 2 in. expanded polystyrene (molded beads), 1/8 in. steel roof pan (assume $R = 0$), 3.5 in. air gap, and 5/8 in. drywall. The indoor air is maintained at 70°F and 30% RH.

   a.   Calculate the overall R-value of the roof and the heat loss through the roof under winter design conditions when the outdoor temperature is 0°F.

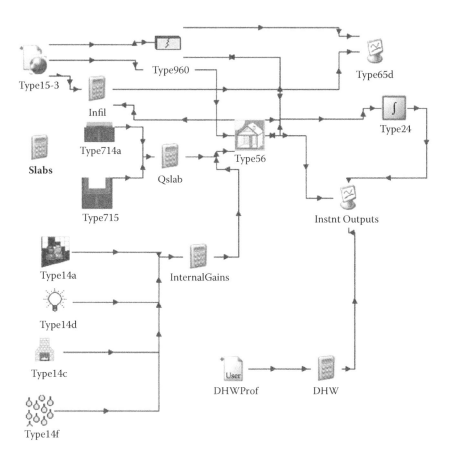

FIGURE 3.36 A thermal model for a house using TRNSYS simulation tool.

b. Calculate the temperature of the steel roof pan (top of air gap) when the outdoor temperature is 0°F. May condensation occur at the surface of the steep roof pan?

c. Estimate the overall *R*-value of the roof and the heat gain from the roof under summer design conditions of 110°F. The indoor temperature is now maintained at 75°F.

3.2 Consider a 10 m by 5 m flat roof of a home, located at sea level, is made up of 0.5 cm of plaster, 10 cm of concrete, and 1.0 cm of drywall. The indoor air is maintained at 20°C and 50% RH.

a. Calculate the overall *R*-value of the roof and the heat loss through the roof under winter design conditions when the outdoor temperature is –5°C.

b. Calculate the temperature of the interior surface of the roof when the outdoor temperature is –5°C. May condensation occur at the surface?

c. Estimate the overall *R*-value of the roof and the heat gain from the roof under summer design conditions of 35°C. The indoor temperature is now maintained at 24°C.

    d.   Determine the reduction of the *R*-value under both winter and summer conditions when 2 cm of polystyrene insulation is added about the concrete layer.

3.3 A homeowner in Boise, Idaho, has an attached garage with the following construction characteristics:

- Dimensions: 15 ft by 20 ft by 10 ft with a flat roof.
- One wall (15 ft by 10 ft) adjacent to the house with an overall *R*-value of 10 h.ft².°F/Btu.
- The other walls, roof, and doors (there are no windows) are in contact with outdoor air and have an average *R*-value of 5 h.ft².°F/Btu.
- The floor is made up of a 4 in. concrete slab with no insulation.
- The infiltration rate is 0.40 ACH.

The homeowner has decided to install a space heater to maintain the garage at a temperature of 68°F, which is the same as the house indoor air temperature. Determine:

    a.   The heat transfer through the wall separating the garage from the house.

    b.   The total heat transfer from the garage to the outdoors assuming an outdoor air temperature of 9°F. Include the effect of the slab as well as the air infiltration.

    c.   Estimate the temperature of the garage if the heater was not installed. Assume that the outdoor air temperature is 9°F. Show all the calculation steps.

3.4 A large homebuilder is developing a new line of two-story residential homes in Colorado. Some of these homes will be built in a typical, suburban environment with moderate shielding, while other homes will be built in the plains of eastern Colorado with no wind shielding. An infiltration test showed that the effective leakage area for a typical home is 300 in.².

    a.   Estimate how much larger the infiltration load will be out in the plains compared to the suburbs. Use the outdoor temperature that corresponds to the 99.6% winter design conditions for Denver, Colorado. Assume the wind speed is 10 mph and that the indoor temperature is 70°F.

    b.   By a better design, the infiltration rate area for the prototypical house was reduced to 0.4 ACH when the indoor-outdoor temperature difference is 30°F and the wind speed is 10 mph. Calculate the reduction in infiltration load for the houses in the suburbs and the plains under the same winter conditions defined in question (a).

3.5 A blower door test on a 30 ft by 50 ft by 9 ft ranch house located in a moderately shielded area of Denver, Colorado, revealed that the leakage area is about 175 in.². The indoor temperature is kept at 68°F during the winter and 74°F during the summer.

    a.   Estimate in air changes per hour the infiltration rate of the house for the heating season (average outdoor temperature is 35.2°F and wind speed is 6.5 mph).

    b.   Estimate in air changes per hour the infiltration rate of the house for the cooling season (average outdoor temperature is 77.9°F and wind speed is 4.0 mph).

    c.   It was decided to weather strip the house so that the infiltration rate is reduced to just 0.25 ACH during the winter. Determine the reduction in average heating load and cooling load in Btu/h and in percent.

3.6 A blower door test on a 15 m by 10 m by 3 m two-story house located in a lightly shielded area of Chicago revealed that the leakage area is about 600 cm². The indoor temperature is kept at 20°C during the winter and 24°C during the summer.

   a. Estimate in air changes per hour the infiltration rate of the house for the heating season.

   b. Estimate in air changes per hour the infiltration rate of the house for the cooling season.

   c. It was decided to weather strip the house so that the infiltration rate is reduced to just 0.35 ACH during the winter. Determine the reduction in average heating load and cooling load in percent.

   Use the weather data for Chicago provided in the appendices to estimate the average outdoor temperatures and wind speeds.

3.7 Both pressurization and depressurization tests were conducted on a residence. The residence has 950 ft² (95 m²) of floor area with an average ceiling height of 8 ft. The results of these tests are shown below:

Pressure (Pa)	Pressurization Test (cfm)	Depressurization Test (cfm)
10	177	237
20	324	384
30	411	471
40	531	561
50	590	650
60	676	710
70	735	765
80	795	825
90	885	887
100	942	942

   a. Determine the leakage area of the house.

   b. Determine the effect of the shielding level on the annual average infiltration rate. Assume the house is located in Denver, Colorado.

3.8 For a house with a 50 ft by 25 ft slab-on-grade floor made up of a 4 in. concrete floor:

   a. Estimate the annual energy ground-coupled loss/gain from the building located in Denver, Colorado.

   b. Estimate the reduction in annual foundation heat loss if the slab is insulated uniformly with R-10 rigid insulation.

3.9 Determine the mean effective U-value for a basement foundation of a house made up of only 4 in. concrete. Assume the soil is sandy. The dimensions of the basement are 10 m long, 5 m wide, and 2 m high (below grade).

3.10 Determine the values of humidity ratio, specific volume, and enthalpy of moist air with a dry bulb temperature of 80°F and relative humidity of 90%. Assume that the atmospheric pressure is 14.696 psia. Check the calculations using the psychrometric chart.

3.11 Determine the values of humidity ratio, specific volume, and enthalpy of moist air with a dry bulb temperature of 35°C and relative humidity of 80%. Assume that the atmospheric pressure is 1 atm. Check the calculations using the psychrometric chart.

3.12 Model a one-floor 42 ft by 24 ft ranch house using any detailed simulation tool such eQUEST, EnergyPlus, TRNSYS, and ESP-r. Use the following assumptions to model the house:

- The whole house is one thermal zone with a floor-to-ceiling height of 9 ft.
- The attic is unheated and is modeled as a separate zone with a 3 ft height.
- The roof is flat.
- The temperature is kept at 70°F (heating) and 75°F (cooling) throughout the year.
- Assume an infiltration rate of 0.75 ACH.
- Assume that the DWH usage is 0.2 gallon/person.
- The wall has R-11 fiberglass insulation and the ceiling has an insulation of R-19. Use material library to select the proper construction materials.
- The house is heated and cooled through a central residential system with a gas furnace efficiency of 0.80.
- The house is located in Denver, Colorado.

a. Determine the design heating and cooling loads for the house. Select the size of the required gas furnace (Btu/h) and the air conditioning system (in tons).

b. Determine the total annual gas and electricity usage for the house. Compare the gas usage for heating (only) predicted by simulation to that calculated from the DD method assuming $T_b = 65°F$. What should be T_b for the house so the DD method matches the simulation prediction for the annual heating gas usage?

c. Determine the energy savings (using the simulation tool) and the cost-effectiveness (use both simple payback period and LCC analyses) of each of the following energy conservation measures (assume the house located in Denver, Colorado):

   i. The infiltration rate is cut to 0.25 ACH with a total cost of $200.
   ii. The insulation is increased to R-19 in the walls with a cost of 0.25/ft².
   iii. The insulation is increased to R-30 in the ceiling with a cost of $0.35/ft².
   iv. Install a high-efficiency furnace with a seasonal efficiency of 0.90 and an incremental cost of $750.

   For the economic analysis, assume 25 years and a 5% discount rate. Assume the electricity cost is $0.08/kWh and $0.95/therm.

   Briefly discuss if the results make sense to you (explain why).

4

Energy Efficiency Screening Approaches

Abstract

In this chapter screening techniques to identify energy-inefficient residential buildings are introduced. These screening tools are especially useful to carry out retrofit programs on a community scale since they help identify and prioritize homes that are good candidates for energy efficiency upgrades. Screening and diagnostic procedures can further retrofit efforts by providing the preliminary energy efficiency level of homes before a walkthrough audit is conducted. The presented screening tools are based on monthly utility data analysis and the development of reference metrics for energy efficiency levels associated with residential buildings.

4.1 Introduction

Based on the statistics compiled for the 2005 Residential Energy Consumption Survey (RECS), the total number of U.S residences is estimated at 111 million, including 72.1 million single-family detached homes (EIA, 2009). These single-family detached homes consumed 7.81 quadrillion Btus of site energy in 2005. As indicated in Chapter 1, the vast majority of U.S. existing residential homes can benefit from cost-effective energy efficiency measures, including building envelope retrofits. Several federal and state programs are especially interested in improving the energy efficiency of low-income housing units. Indeed, it is estimated that there are 45 million low-income housing units, including 24 million single-family homes. These low-income residential buildings are projected to consume 1.54 quadrillion end-use Btus in 2020. By implementing basic retrofit measures, including shell upgrades, energy savings of 610 trillion Btus are expected in 2020. Low-income weatherization assistance programs are often used as an effective approach to achieve significant energy savings for low-income communities (McKinsey, 2009).

To ensure the effective implementation of community-scale conservation programs, reliable screening tools are needed to identify energy-inefficient homes and retrofit opportunities. Comprehensive energy audits are traditionally conducted. However, these audits are relatively costly and time-consuming. Instead, readily available data such as monthly energy consumption, conditioned building area, and local weather data can be used as screening and diagnostic tools to identify energy-inefficient homes before a walkthrough energy audit is conducted. Further, the screening tools can help the auditor infer design and operational characteristics of the building. The fundamental concept of the screening tools presented in this chapter is based on the analysis of utility data to determine the energy efficiency characteristics of homes based on inverse modeling techniques discussed in Chapter 3.

Several diagnostic and screening techniques have been proposed to estimate building parameters using inverse modeling techniques based on the change-point models (Fels, 1986; Reddy et al., 1999; Kissock et al., 2003). Moreover, Raffio et al. (2007) have shown that the proactive use of inverse modeling methods can identify retrofit candidates when compared among several buildings. The screening methodology presented in this chapter uses the variable-base degree-day (VBDD) method presented in Chapter 3 instead of change-point models. In particular, the screening tool allows comparison of the evaluated building parameters against reference values and identification of specific design and operational characteristics.

First, the general screening approach is presented. Then, some applications of the approach are outlined to assess the energy efficiency of individual homes as well as to identify the most energy-inefficient homes suitable for energy auditing as part of a community-scale weatherization program.

4.2 Screening Methodology

4.2.1 Description of the Screening Approach

Since residential buildings are dominated by building envelope thermal loads, their heating and cooling energy use can be estimated using the variable-base degree-day method outlined in Chapter 3, especially when buildings have little thermal mass. Moreover, energy sources for heating are typically different from those of cooling. Indeed, most U.S. residential buildings typically use electricity for cooling and natural gas or fuel oil for heating. As noted in Chapter 3, natural gas or fuel use can be correlated to the heating variable-base degree-day:

$$E_H = 24. f_H \frac{BLC}{\eta_H}.DD_H(T_{b,H}) + E_{base,H} \qquad (4.1)$$

Similarly, the electricity use can be correlated to the cooling variable-base degree-day:

$$E_C = 24. f_C.\frac{BLC}{COP_C}.DD_C(T_{b,C}) + E_{base,C} \qquad (4.2)$$

where $E_{H/C}$ represents the annual building energy use during the heating or cooling season; BLC is the building loss coefficient and is defined based on the air infiltration mass flow rate, $\dot{m}_{inf}$, and the building shell *UA*-values for all the exterior surfaces, including the walls, roof, windows, and the floor; η_H is the average seasonal energy efficiency of the heating system; COP_C is the average seasonal coefficient of performance of the cooling system; $T_{b,H/C}$ is the building balance temperature for heating or cooling energy use; and $DD_{H/C}$ is the heating or cooling degree-days (based on the balance temperature). Both $DD_H(T_b)$ and $DD_C(T_b)$ can be estimated using Equations (4.3) and (4.4), respectively:

$$DD_H(T_b) = \sum_{i=1}^{N_{d,H}} (T_b - T_{o,i})^+$$ (4.3)

$$DD_C(T_b) = \sum_{i=1}^{N_{d,C}} (T_{o,i} - T_b)^+$$ (4.4)

where $N_{d,H}$ and $N_{d,C}$ are the number of days during, respectively, the heating and the cooling months or billing periods, and $E_{base,H/C}$ is the base load for building energy use. It represents the nonheating or noncooling energy use.

Using regression analysis (i.e., inverse modeling), the fuel use of Equation (4.1) can be correlated to heating degree-days and the electricity use of Equation (4.2) to the cooling-degree days, as expressed by Equations (4.5) and (4.6), respectively:

$$E_H = \alpha_H + \beta_H.DD_H(T_{b,H})$$ (4.5)

$$E_C = \alpha_C + \beta_C.DD_C(T_{b,C})$$ (4.6)

where α_H is the base load fuel use, β_H is the heating slope or HS, α_C is the base load electricity use, and βC is the cooling slope or CS.

Using the variable-base degree-day models of Equations (4.1) and (4.2), the coefficients α_H, α_C, β_H, and β_C can be estimated as follows:

$$\alpha_H = E_{base,H}$$ (4.7)

$$\beta_H = 24.f_H \frac{BLC}{\eta_H}$$ (4.8)

$$\alpha_C = E_{base,C}$$ (4.9)

$$\beta_C = 24.f_C.\frac{BLC}{COP_C}$$ (4.10)

Therefore, the coefficients α_H and α_C represent, respectively, the nonheating and noncooling base load of the building. The coefficients β_H and β_C account for the characteristics of the building shell (BLC), the heating or cooling energy efficiency (η_H and COP_C), as well as operational schedule ($24.f_H$ and $24.f_C$). These coefficients, β_H and β_C, the slopes of the linear regressions in Equations (4.3) and (4.4), are rich with information about the building energy systems and can be used as metrics to assess the building energy efficiency level.

Using the monthly utility data for a 12-month period, an iterative procedure can be used to find the balance temperature $T_{b,H}$ or $T_{b,C}$ that provides the best fit with the VBDD method predictions. The iterative procedure can be summarized in the following steps:

1. Select an initial value for the balance temperature of 40°F (4°C).
2. Calculate the degree-days for each of the 12 months.
3. Perform a linear regression of the monthly energy use with the monthly degree-days at the selected balance temperature.
4. Record the regression R².
5. Repeat the process with the next balance-point temperature.

The upper balance temperature value is set at 70°F (21°C). After completing the iterative procedure, then:

1. Select the optimal balance-point temperature with the highest R².
2. Report the regression parameters, including the values of the base load (i.e., α_H or α_C) and the slope (i.e., β_H or β_C).
3. Continue the analysis for the next 12-month data set.

The search for the best-fit balance temperature is illustrated graphically in Figure 4.1, where the optimal value is obtained with the highest R². In the case of Figure 4.1, the best-fit balance temperature is 57°F.

Figure 4.2 shows the iterative analysis results from three homes located in Boulder, Colorado. Each regression analysis is performed for 12 months of simulated natural gas

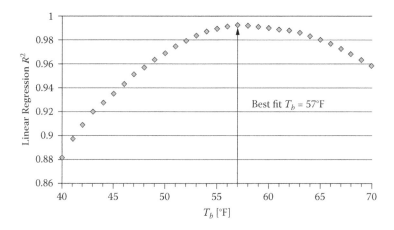

FIGURE 4.1   Best-fit search for the balance temperature based on R² correlation values.

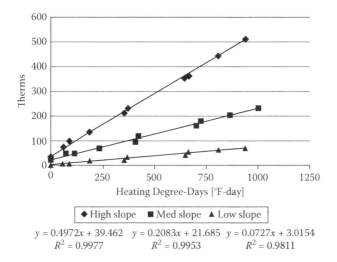

FIGURE 4.2 Linear regressions of natural gas consumption and heating degree-days for three homes.

consumption vs. monthly heating degree-days to the individual best-fit balance temperature. The linear regression results listed in Figure 4.2 indicate that the base load and the heating slope values vary from home to home. In particular, homes with smaller heating slopes (i.e., β_H values) use less heating energy and may indicate that they are more energy efficient.

4.2.2 Verification of the Screening Approach

Casey et al. (2010) have performed a verification analysis of the screening methodology described in the preceding section. In particular, they generated synthetic monthly utility bills using detailed simulation modeling to avoid the uncertainty of the actual utility data. Several home prototypes have been defined and modeled with a wide range of construction characteristics, including thermal, operational, and mechanical information, as well as occupant behavior using different thermostat settings and miscellaneous load schedules. Homes were classified into three possible size archetypes:

- A 1,058 ft^2 (95 m^2) one-story ranch with crawlspace
- A 2,116 ft^2 (190 m^2) one-story ranch with conditioned basement
- A 3,174 ft^2 (285 m^2) two-story home with conditioned basement

For each archetype, three possible construction conditions were defined (IECC, 2009):

- Compliant with IECC 2009 for residential buildings in climate zone 5B (Boulder, Colorado)
- Moderately below code
- Substantially below code

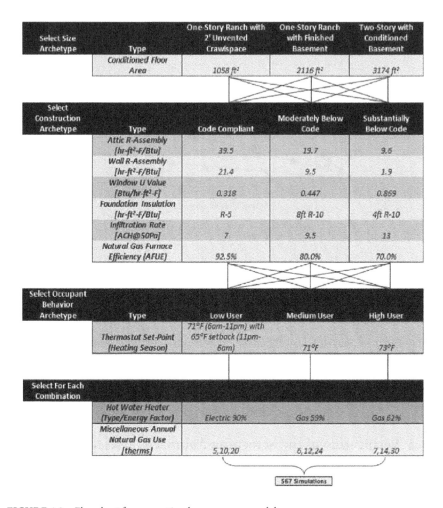

Select Size Archetype	Type	One-Story Ranch with 2' Unvented Crawlspace	One-Story Ranch with Finished Basement	Two-Story with Conditioned Basement
	Conditioned Floor Area	1058 ft²	2116 ft²	3174 ft²

Select Construction Archetype	Type	Code Compliant	Moderately Below Code	Substantially Below Code
	Attic R-Assembly [hr-ft²-F/Btu]	39.5	19.7	9.6
	Wall R-Assembly [hr-ft²-F/Btu]	21.4	9.5	1.9
	Window U Value [Btu/hr-ft²-F]	0.318	0.447	0.869
	Foundation Insulation [hr-ft²-F/Btu]	R-5	8ft R-10	4ft R-10
	Infiltration Rate [ACH@50Pa]	7	9.5	13
	Natural Gas Furnace Efficiency (AFUE)	92.5%	80.0%	70.0%

Select Occupant Behavior Archetype	Type	Low User	Medium User	High User
	Thermostat Set-Point (Heating Season)	71°F (6am-11pm) with 65°F setback (11pm-6am)	71°F	73°F

Select For Each Combination				
	Hot Water Heater (Type/Energy Factor)	Electric 90%	Gas 59%	Gas 62%
	Miscellaneous Annual Natural Gas Use [therms]	5,10,20	6,12,24	7,14,30

567 Simulations

FIGURE 4.3   Flowchart for generating home energy models.

For each home, three occupant behavior schedules were defined:

- High user
- Medium user
- Low user

Figure 4.3 presents the combinatorial procedure to generate about 567 home models. Using each home model, hourly energy use data were obtained for 1 year. A postprocessing routine was used to combine the hourly energy use to form monthly energy use representing typical utility bill data.

Based on their characteristics, the home models are identified as either compliant or incompliant with the prescriptive measures of the IECC 2009 standard. Three screening approaches are considered to determine the energy efficiency (or inefficiency) of the 567 home models, including:

Fuel use (FU)-based screening: In this approach, the annual fuel use is estimated for all the home models and compared to the compliant archetype home model. Therefore, the screening metric used in this approach is simply:

$$FU = E_H \qquad (4.11)$$

Degree-day normalized fuel use: The annual fuel use for each home model is normalized by heating the degree-day based on the heating balance temperature (which depends on the home). Thus, the metric used is simply:

$$FU/HDD = \frac{E_H}{DD_H\left(T_{b,H}\right)} \qquad (4.12)$$

Heating slope: In this approach, the regression analysis of the monthly data is carried for each home to find the heating slope, HS [i.e., HS = $\beta_H/(24.f_H)$]. Note that according to Equation (4.8), HS is related to BLC:

$$HS = \frac{\text{BLC}}{\eta_H} \qquad (4.13)$$

For each archetype home, reference values for each screening approach are estimated using code-compliant models. Therefore, for each archetype home, three reference values are found: FU_{ref}, $[FU/HDD]_{ref}$, and HS_{ref}. Any home model associated to the archetype with a metric value above the reference value associated to each approach is labeled noncompliant.

Casey et al. (2010) have conducted a comparison analysis of the three screening approaches described above using simulated utility data generated for 567 home models (based on three archetype homes). Figure 4.4 illustrates the comparative analysis results for the three approaches (FU, FU/HD, and HS) using home models for the one-story ranch with crawlspace archetype (189 home models). The black bars represent noncompliant, inefficient homes, and the gray bars represent compliant, efficient homes. The x-axis shows the 189 home models, and the y-axis shows the ranking metric. The reference or threshold value for each approach is indicated by the arrow below the x-axis. Table 4.1 lists the reference values for each screening approach.

As shown in Figure 4.3, both the FU and FU/HDD rankings classify several code-compliant homes as noncompliant. The same homes, characterized by high base load and miscellaneous gas use, however, were correctly identified by the HS ranking. This result highlights the ability of the HS-based screening approach to disaggregate weather-independent energy use from space conditioning energy use. The HS_{ref} values can be used as a metric for ranking and even for identifying the inefficiency level of homes. The HS_{ref} values for three home archetypes are estimated and listed in Table 4.2. For instance, if a one-story ranch with a crawlspace archetype simulation resulted in an HS of 450 Btu/h-°F (230 W/°C), according to Table 4.2, this value exceeds the HS_{ref}

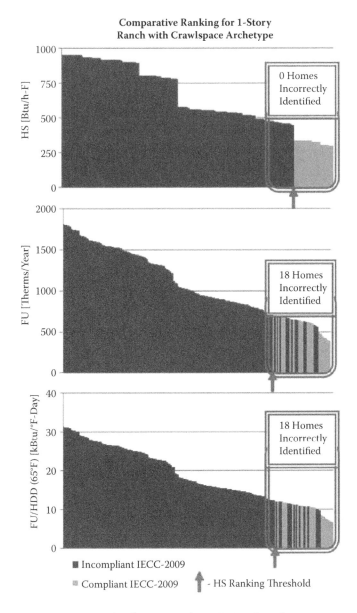

FIGURE 4.4   Comparative results of HS, FU, and FU/HDD ranking for a one-story ranch with crawlspace.

threshold of 365 Btu/h-°F (186 W/°C) and the home would be classified as noncompliant and a possible candidate for energy efficiency improvements.

Moreover, the HS-based screening approach helps to identify some specific features of the building characteristics using the correlation coefficients α_H and β_H as functions of the building characteristics as indicated by Equations (4.7) and (4.8). As shown in

TABLE 4.1 Summary of Ranking Threshold Values for a One-Story Ranch with Crawlspace Archetype

Screening Approach	Reference Value
HS	365 [Btu/h-°F]
FU	704 [therms/year]
FU/HDD	12 [kBtu/°F-day]

TABLE 4.2 Summary of HS_{ref} Values for All Archetypes

Home Archetype	HS_{ref} [Btu/h.°F]	HS_{ref} [W/°C]
1 story with crawlspace	365	186
1 story with basement	480	245
2 stories with basement	1,002	512

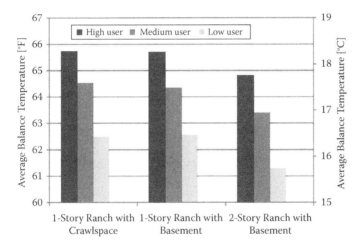

FIGURE 4.5 Variation of the balance temperatures with temperature settings.

Figure 4.5, the balance temperature of the home may provide some insights about the behavior of the occupants and their temperature settings. Higher heating thermostat set points result in higher balance temperature values for all home archetypes considered in the study by Casey et al. (2010). Therefore, and as noted by Raffio et al. (2007), homes with high balance temperatures are candidates for programmable thermostats. Similarly, high values of the coefficient α_H indicate high monthly base loads due to the use of hot tubs, energy-inefficient hot water heaters, and other large energy-consuming appliances. Figure 4.6 indicates that the coefficient α_H can be used to rank and identify homes with high base load energy uses.

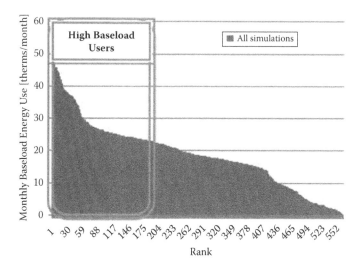

FIGURE 4.6   Use of α_H as a parameter to rank base loads.

4.3 Screening Methodology Applications

In this section, two applications of the screening methodology described in Section 4.2 are presented. In the first application, a community-scale screening is considered to rank, based on their energy inefficiency, several homes that would benefit from energy auditing. The second application involves single homes to identify their energy efficiency level.

4.3.1 Application 1: Community-Scale Screening

In this application, it is proposed to identify and rank among 174 homes the best candidates for energy audits. For these homes, utility bills are available as well as a county tax assessor's database, which provides each home archetype classification, age, and size. Table 4.3 lists the number of homes per archetype classification.

Using the HS-based screening approach outlined in Section 4.2, reference homes for each archetype are defined according to the prescriptive requirements of IECC 2009. From these reference homes, reference HS values, HS_{ref}, are estimated using the calculated BLC and the efficiency of the heating system and by setting $f_H = 1$ in Equation (4.13) with no temperature setback. Based on these HS_{ref} values, the homes within each archetype are ranked by the difference of their HS and HS_{ref}, as illustrated by Figure 4.7 for selected homes with a one-story/finished basement archetype classification. As shown in Figure 4.7, the HS values for all homes are above HS_{ref}, and thus they are not energy efficient. The difference of HS values, ΔHS, provides a metric to rank the various homes by their energy-inefficient level. Figures 4.8 and 4.9 illustrate the results of the HS-based screening for homes with respectively, one-story crawlspace and two-story/unfinished basement archetype classifications.

TABLE 4.3 Summary of House Construction Archetype Classifications

Archetype (above-grade stories/foundation type)	Number of Homes
1 story/crawlspace	34
1 story/finished basement	28
1 story/unfinished basement	5
2 story/crawlspace	24
2 story/finished basement	60
2 story/unfinished basement	23
Total	174

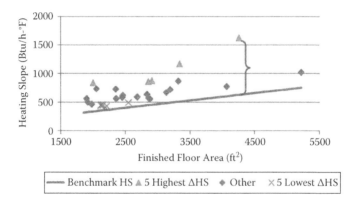

FIGURE 4.7 HS-based screening methodology applied to selected homes with a one-story/finished basement archetype.

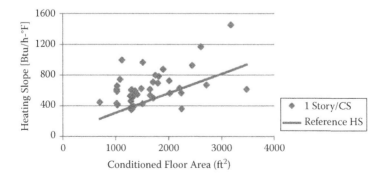

FIGURE 4.8 HS-based screening methodology applied to homes with one story with a crawlspace archetype.

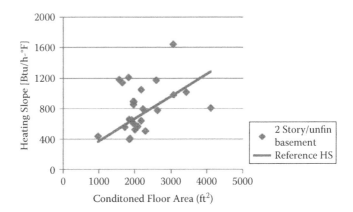

FIGURE 4.9   HS-based screening methodology applied to homes with two stories with an unfinished basement archetype.

4.3.2 Application 2: Screening of Individual Homes

In this application, the energy efficiency level of a specific home is assessed to justify if any retrofits are needed to improve its energy performance and reduce its energy cost. The one-story house, considered for this application, is a 970 ft² single-family residence located in Boulder, Colorado, built in the 1950s with a crawlspace. Figure 4.9 shows the results of the linear regression analysis to correlate the monthly natural gas use to the heating degree-days. The table below summarizes the basic findings of the regression analysis.

β_H (Btu/DDh)	α_H (therms)	T_b (°F)
7,469	12.190	62.5

The house has a temperature setback for 15 hours during the day. Using the simplified method presented in Chapter 5, f_H can be estimated as shown below:

$$f_H \cong \frac{N_{occ}}{24} + \frac{\left(24 - N_{occ}\right)}{24} \times \frac{T_{b,setback} - T_{ave,winter}}{T_b - T_{ave,winter}} = \frac{9}{24} + \frac{15}{24} \times \left(\frac{54.5°F - 44.9°F}{62.5°F - 44.9°F}\right) = 0.716$$

The heating slope, HS, is then determined based on the slope of the regression, β_H determined as outlined in Figure 4.10:

$$HS = \frac{\beta_H}{24.f_H} = 435 \frac{\text{Btu}}{\text{hr}°\text{F}}$$

Based on the value of the HS_{ref} value of 365 Btu/h.°F for a one-story home with a crawlspace as shown in Table 4.2, the energy efficiency of the house is below the desired threshold level and can benefit from some improvements. Figure 4.11 illustrates the comparison of the house HS to the reference value, HS_{ref}, based on IECC 2009.

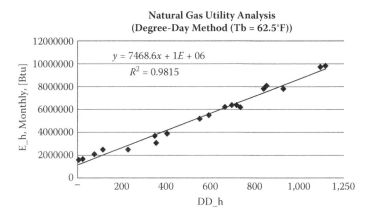

FIGURE 4.10 Regression analysis of the monthly natural gas use with heating degree-days.

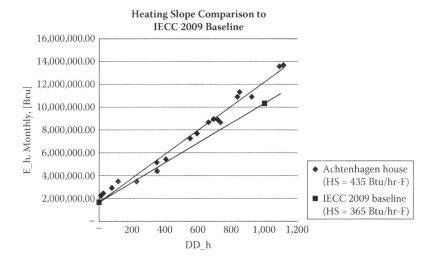

FIGURE 4.11 Audited house heating slope compared to HS_{ref} based on IECC 2009.

To verify the basic features of the house, a detailed energy audit has been conducted, including a pressurization test, temperature monitoring, and a detailed modeling analysis. The walkthrough audit has indicated that the homeowners have implemented some retrofit measures consisting of 10 in. of loose insulation blown in the attic, ½ in. of polyisocyanurate foam board insulation added to the exterior wall, some interior insulation installed along the crawlspace walls, and a double-pane window upgrade with vinyl frames. Table 4.4 summarizes the construction details and their *R*-values.

The results of the detailed audit have indicated that indeed the energy efficiency of a house can be improved and that several energy efficiency measures are cost-effective, as outlined in Table 4.5, including installing wall insulation, decreasing the lighting power

TABLE 4.4 Summary of Home Construction Details and Associated *R*-Values

Building Component	Construction	Equivalent *R*-Value (hr-ft²-F)
Foundation/crawlspace walls	Poured concrete; interior insulation	7.65
Walls	½ in. gypsum; 2 × 4 studs; board sheathing; cement board siding; ½ in. polyisocyanurate board insulation; vinyl siding	6.83
Walls: Attic	Minimal 2 × 4 framing; board sheathing; vinyl siding	2.068
Floors: Crawlspace	Primarily dirt, some sections concrete	12.5
Floors: Main floor	¾ in. finished hardwood; 1 in. board subfloor; 2 × 8 joists	N/A
Ceiling	½ in. gypsum; 2 × 6 joists; 10 in. blown-in insulation	35.59
Roof	2 × 6 rafters; 1 in. board sheathing; tar felt paper; asphalt shingles	1.97
Windows	Double pane, ½ in. air gap, vinyl frame, clear glazing, sliding sash	1.96
Doors	Solid wood with some single pane glazing	2.08

TABLE 4.5 List of Cost-Effective Energy Efficiency Measures for the Audited House

Energy Efficiency Measure	Total Energy Cost ($/year)	Initial Cost ($)	Savings Energy Cost ($/year)	Simple Payback Period (years)	Life Cycle Cost ($)
Base case	**$1,692**	**$0**	**$0**	—	**$21,089**
Increase wall insulation	$1,428	$584	$264	2.2	$18,380
Reduce lighting power density	$1,449	$656	$243	2.7	$18,713
Use Energy Star appliances for EPD reduction	$1,276	$3,900	$416	9.4	$19,808
Reduce hot water usage	$1,632	$195	$60	3.3	$20,539
Install insulating window shades	$1,664	$128	$28	4.6	$20,863
Reduce infiltration by weatherizing	$1,653	$150	$39	3.8	$21,057

density, installing Energy Star appliances, reducing DHW use, and reducing infiltration. The life cycle cost (LCC) analysis was conducted based on a discount rate of 5% and a life cycle of 20 years.

4.4 Summary

A systematic methodology for screening the energy efficiency level of the residential buildings has been described in this chapter. The screening approach is based on monthly utility data analysis against degree-days with variable-base temperature. The methodology has been applied for evaluating the energy performance of homes using heating energy use analysis. In particular, a heating slope is defined as the ratio of building load coefficient and the heating system efficiency, and is found to be a good metric for the energy efficiency level of homes. Reference HS values can be used to rank and identify homes that need energy retrofits and can benefit from either a walkthrough audit or a detailed energy audit. Two applications of the screening approach have been presented, including ranking the inefficiency level of several homes in a residential community and assessing the energy performance of individual homes.

Problems

4.1 Determine if an energy audit is needed for a one-story house with a crawlspace with a BLC of 353 Btu/h.°F and a heating efficiency of 80%. Use the HS concept introduced in this chapter.

4.2 Assess the energy efficiency level for a one-story house with a finished basement with a BLC of 280 Btu/h.°F and a heating efficiency of 80%.

4.3 Determine the heating base load and the heating slope of a house located in Boulder, Colorado, based on the monthly utility bills provided in the table below. Use variable-base degree-day regression analysis to estimate the HS. The house is two stories with a finished basement. Using the HS_{ref} defined in this chapter, assess the energy efficiency of the house. For the analysis assume that the house is heated 24 h/day and that the seasonal efficiency of the gas furnace is about 78%. It should be noted that the house is generally unoccupied during the period of May 25 to July 24 (for both 2003 and 2004).

Date of Bill	Billed Days	Therms Used	Outdoor Temp. (°F)
Dec. 20, 2004	34	230	33
Nov. 21, 2004	30	71	51
Oct. 17, 2004	29	50	63
Sep. 19, 2004	31	21	72
Aug. 20, 2004	28	18	77
Jul. 24, 2004	35	6	79
Jun. 20, 2004	30	10	65
May 22, 2004	30	61	56
Apr. 18, 2004	30	118	45
Mar. 22, 2004	30	163	38
Feb. 16, 2004	27	186	26
Jan. 23, 2004	35	230	32
Jan. 02, 2004	30	199	32
Nov. 17, 2003	29	132	41
Sep. 19, 2003	29	16	73
Sep. 5, 2003	31	4	78
Jun. 20, 2003	30	9	67
May 24, 2003	31	53	56
Apr. 25, 2003	30	87	49
Mar. 22, 2003	27	107	41
Feb. 24, 2003	32	159	36

4.4 Assess the energy efficiency level of a two-story house with a finished basement based on the following utility data and average outdoor temperature. You may use Erbs method (refer to Chapter 5) to estimate the heating degree-day for any balance temperature. The house is heated 24 h/day.

Bill Date	Usage (Therms)	Average Temp. (°F)
12/2/04	86	39.7
1/4/05	102	36.5
2/3/05	117	35.5
3/8/05	84	37.9
4/5/05	55	42
5/4/05	50	48.4
6/3/05	27	57.6
7/5/05	19	65.4
8/3/05	7	75.1
9/1/05	11	69.7
10/4/05	17	66.2
11/1/05	32	52.6
12/2/05	59	45
1/5/06	126	33.3
2/3/06	70	40.7
3/6/06	93	33.7
4/4/06	74	39.4
5/3/06	29	53.9
6/2/06	22	61
7/5/06	16	71.6
8/2/06	15	74.4
8/31/06	15	71.6
9/29/06	17	58.4
10/31/06	39	51
12/1/06	73	43.4
1/5/07	125	35.3
2/2/07	137	27.2
3/5/07	99	34.6
4/3/07	39	47.6
5/3/07	46	47.9
6/4/07	22	58
7/3/07	14	67.7
8/2/07	17	74.8
8/31/07	29	73.7
10/2/07	17	64.5
10/31/07	27	55.2
12/3/07	84	44.9

4.5 Using the HS-based screening approach, determine the base load, balance temperature, and heating slope, as well as the BLC of a one-story house located in Boulder, Colorado, based on the following utility data and average outdoor temperature. The house is heated with an electric furnace and has an evaporative cooler.

Month	Year	No. Days	Consumption$_{total}$ (kWh)	Consumption/Day (kWh)
Oct	2000	29	1,547	53.34
Nov	2000	28	2,770	98.93
Dec	2000	29	1,843	63.55
Jan	2001	30	1,894	63.13
Feb	2001	28	1,340	47.86
Mar	2001	28	1,321	47.18
Apr	2001	32	1,284	40.13
May	2001	31	945	30.48
Jun	2001	31	1,155	37.26
Jul	2001	30	1,281	42.70
Aug	2001	30	1,360	45.33
Sep	2001	30	1,257	41.90
Oct	2001	30	1,083	36.10
Nov	2001	34	1,798	52.88
Dec	2001	30	1,663	55.43
Jan	2002	29	1,816	62.62
Feb	2002	30	1,435	47.83
Mar	2002	27	1,225	45.37
Apr	2002	33	1,077	32.64
May	2002	30	1,329	44.30
Jun	2002	28	1,153	41.18
Jul	2002	29	1,074	37.03
Aug	2002	30	1,114	37.13
Sep	2002	34	1,282	37.71
Oct	2002	28	1,056	37.71
Nov	2002	28	1,221	43.61
Dec	2002	33	2,420	73.33
Jan	2003	34	1,905	56.03
Feb	2003	30	2,548	84.93
Mar	2003	28	2,489	88.89
Apr	2003	30	1,668	55.60
May	2003	30	1,731	57.70
Jun	2003	31	1,797	57.97
Jul	2003	30	1,445	48.17
Aug	2003	30	1,419	47.30
Sep	2003	30	1,143	38.10
Oct	2003	29	1,061	36.59

5

Building Envelope Retrofit

Abstract

This chapter provides some energy conservation measures related to the residential building thermal envelope. The measures discussed in this chapter tend to improve the comfort level within the buildings as well as the energy efficiency of the building envelope components. Specifically, this chapter provides the auditor with selected measures to improve the thermal performance of building envelope components, such as adding thermal insulation to roofs and walls, weather stripping to improve the leakage characteristics of the building shell, and installing energy efficiency features for windows.

5.1 Introduction

Generally, the envelope of a structure is designed by architects to respond to many considerations, including structural and esthetics. Before the oil crisis of 1973, the energy efficiency of the envelope components was rarely considered as an important factor in the design of a building. However, since 1973 several standards and regulations have been developed and implemented to improve the energy efficiency of various components of building envelopes. For energy retrofit analysis, it is helpful to determine if the building was constructed or modified to meet certain energy efficiency standards. If this is the case, retrofitting of the building envelope may not be cost-effective. However, improvements to the building envelope can be cost-effective if the building was built without any concern for energy efficiency, such as the case with homes constructed with no insulation provided in the walls or roofs.

Moreover, the building envelope retrofit should be performed after careful assessment of the home thermal loads. For instance, for several residential buildings, envelope transmission losses and infiltration loads are dominant and the internal loads within these buildings are typically low. Meanwhile in high-rise commercial buildings, the internal heat gains due to equipment, lighting, and people are typically dominant and the transmission loads affect only the perimeter spaces.

125

As noted in Chapter 3, accurate assessment of the energy savings incurred by building envelope retrofits generally requires detailed hourly simulation programs since the heat transfer in buildings is complex and involves several mechanisms. In this chapter, simplified calculation methods are presented to estimate the energy savings for selected retrofit measures of the residential building envelope commonly proposed to improve not only the energy efficiency of the building, but also the thermal comfort of its occupants.

5.2 Simplified Calculation Tools for Building Envelope Audit

To determine the cost-effectiveness of any energy conservation measure for a residential building envelope component, the energy use savings has to be estimated. In this section, a general calculation procedure based on the variable-base degree-days method is provided with some recommendations to determine the values of the parameters required to estimate the energy use savings.

5.2.1 Estimation of the Energy Use Savings

When an energy conservation measure is performed to improve the efficiency of the building envelope (for instance, by adding thermal insulation to a roof or by reducing the air leakage area for the building envelope), the building load coefficient (BLC) is reduced. Assuming no change in the indoor temperature set point and in the internal gains within the building, the heating balance temperature actually decreases due to the envelope retrofit, as can be concluded from the definition of the heating balance temperature defined in Chapter 3 by Equation (3.40). Therefore, the envelope retrofit reduces the heating load. The energy use savings due to the retrofit can be generally calculated as follows:

$$\Delta E_{H,R} = E_{H,E} - E_{H,R} = \frac{24.f_{DH}.\left(BLC_E.DD_H(T_{b,E}) - BLC_R.DD_H(T_{b,R})\right)}{\eta_H} \qquad (5.1)$$

where $E_{H,E}$ and $E_{H,R}$ are the annual heating energy use for the existing and retrofitted building, B_{LCE} and B_{LCR} are the BLC values for the existing and the retrofitted building, $T_{b,E}$ and $T_{b,R}$ are the balance temperatures for the existing and the retrofitted building, $D_{DH(T_{b,E})}$ and $DD_H(T_{b,R})$ are the annual heating degree-days based on, respectively, $T_{b,E}$ and $T_{b,R}$, f_{DH} is the correction factor to adjust for any temperature setback (a simplified method to estimate this factor is provided in Section 5.2.3), and η_H is the heating system seasonal efficiency.

The efficiency of the heating system is assumed to remain the same before and after the retrofit. It is generally the case unless the heating system is replaced or retrofitted. In many applications, the variation caused by the retrofit of the balance temperature is rather small. In these instances, the degree-days (DD_H) can be considered constant before and after the retrofit so that the energy use savings can be estimated more easily with the following equation:

$$\Delta E_{H,R} = \frac{24.f_{DH}.\left(BLC_E - BLC_R\right).DD_H(T_{b,E})}{\eta_H} \quad (5.2)$$

A similar method can be used to estimate the savings in cooling energy use due to a building envelope retrofit as noted by Equation (5.3):

$$\Delta E_{C,R} = \frac{24.f_{DC}.\left(BLC_E - BLC_R\right).DD_C(T_{b,E})}{COP_C} \quad (5.3)$$

where $DD_C(T_{b,E})$ and $DD_H(T_{b,R})$ are the annual cooling degree-days based on, respectively, $T_{b,E}$ and $T_{b,R}$, f_{DC} is the correction factor to adjust for any temperature setup, and COP_C is the seasonal average coefficient of performance for the cooling system.

Note that when only one element of the building envelope is retrofitted (for instance, the roof), the difference $(BLC_E - BLC_R)$ is equivalent to the difference in the roof *UA*-values before and after the retrofit (i.e., $UA_{roof,E} - UA_{roof,R}$).

To use either Equation (5.2) or Equation (5.3), it is clear that the auditor needs to estimate the heating or cooling degree-days, DD_H or DD_C, and the existing overall building load, BLC. Some recommendations on how to calculate these parameters are summarized in the following sections.

5.2.2 Estimation of the BLC for the Building

The building load coefficient can be estimated using two approaches as briefly described below. Depending on the data available, the auditor should select the appropriate approach.

1. *Direct calculation*: The auditor should have all the data (either through the architectural drawings or from observation during a site walkthrough) needed to estimate the R-value or U-value of all the components of the building envelope and their associated surface areas. Several references are available to provide the R-value of various construction layers commonly used in buildings (ASHRAE, 2009). Appendix B summarizes thermal properties for construction materials common in residential buildings. In addition, the auditor should estimate the infiltration/ventilation rates either by rules of thumb or by direct measurement, such as pressurization or depressurization techniques, as discussed in Chapter 3. With these data, the building load coefficient can be calculated using Equation (3.38) and reproduced as Equation (5.4):

$$BLC = \sum_{j=1}^{N_E} U_{T,j}.A_j + \dot{m}_{inf}.c_{p,a} \quad (5.4)$$

where $U_{T,j}$ is the overall U-value for an exterior surface, j, for the building, such as a wall, roof, window, or door; A_j is the total area of the exterior surface, j; N_E is the total number of the exterior surfaces; $\dot{m}_{inf}$ is the mass flow rate for infiltrating

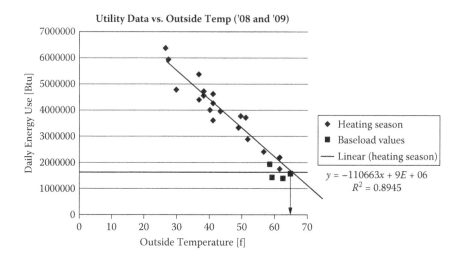

FIGURE 5.1   Determination of a residential building BLC based on a linear regression of the gas consumption as function of average outdoor temperature.

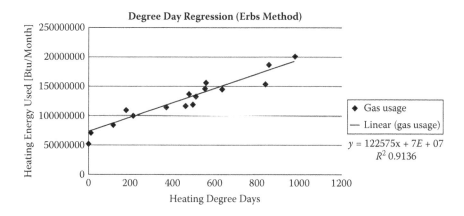

FIGURE 5.2   Determination of the BLC for a home based on a linear regression of the gas consumption as function of monthly heating degree-day.

air to the building; $c_{p,a}$ is the specific heat of the infiltrating air; f_H is the correction factor to adjust for any temperature setback (a simplified method to estimate this factor is provided in Section 5.3); and η_H is the heating system seasonal efficiency.

2. *Indirect estimation*: In this method, the auditor can rely on the utility energy use and its correlation with the average outdoor temperature or the degree-day to provide an accurate estimation of the BLC using linear regression analysis. Examples of the use of the method to determine the BLC are illustrated by Figure 5.1, using temperature-based regression analysis, and Figure 5.2, using degree-day-based regression

TABLE 5.1 Heating Degree-Days for Base Temperatures 65°F (18°C),
55°F (13°C), and 45°F (7°C) for Selected Locations in the United States

Location	DD_H (65°F)	DD_H (55°F)	DD_H (45°F)
Albuquerque, New Mexico	4,292	2,330	963
Bismarck, North Dakota	9,044	6,425	4,374
Chicago, Illinois	6,127	3,912	2,219
Dallas/Ft. Worth, Texas	2,290	949	250
Denver, Colorado	6,016	3,601	1,852
Los Angeles, California	1,245	158	0
Miami, Florida	206	8	0
Nashville, Tennessee	3,696	1,964	1,338
New York, New York	4,909	2,806	1,311
Seattle, Washington	4,727	2,091	602

analysis. In both figures, the BLC is determined by the slope of the regression line correlating the building energy use to either the outdoor air temperature or the degree-day. It should be noted that the outdoor air temperature should be averaged over the same periods for which the utility data are available. Moreover, the balance temperature for the degree-day method should be determined by optimizing the correlation regression of the energy use as a function of the monthly heating degree-day. As described in Section 5.2.3, the Erbs or the Shoenau and Kehrig method can be used to calculate monthly degree-day for any balance temperature.

5.2.3 Estimation of the Degree-Days

Data for heating degree-days can be found in several sources for various values of balance temperature. Table 5.1 provides heating degree-days for selected U.S. cities for various balance temperatures: 65°F (18°C). Additional degree-days data published by the ASHRAE handbook (ASHRAE, 2009) for several U.S. and international locations can be found in Appendices C (SI units) and D (IP units).

However, rarely the balance temperature for an audited building is estimated to be equal to 65°F (18°C). Therefore, there is often a need to estimate heating and cooling degree-days at a wide range of balance temperature values. Several simplified methods are available to estimate the degree-days for any balance-point temperature from limited climatic data. Two of these simplified methods are described in this section.

5.2.3.1 Erbs Method

The first simplified method, proposed by Erbs et al. (1983), is based on the assumption that the outdoor ambient temperature for each month follows a probability distribution with a standard deviation σ_m, an average temperature $\bar{T}_{o,m}$, and a frequency distribution F function of the balance temperature T_b:

$$F(T_b) = 1/\left(1 + e^{-2.a.\theta_m}\right) \tag{5.5}$$

where θ_m is the normalized average outdoor temperature and is defined as

$$\theta_m = \frac{T_b - \overline{T}_{o,m}}{\sigma_m . N_m^{1/2}} \tag{5.6}$$

where N_m is the number of days in the month considered.

Using the variation of the frequency distribution F, the heating degree-days can be obtained as a function of T_b:

$$DD_H(T_b) = \sigma_m . N_m^{3/2} . \left[\frac{\theta}{2} + \frac{Ln(e^{-a\theta} + e^{a\theta})}{2.a} \right] \tag{5.7}$$

For locations spanning most climates in the United States and Canada, Erbs et al. (1983) found that the coefficients a and σ_m can be estimated using the following expressions:

$$a = 1.698\sqrt{N_m} \tag{5.8}$$

and

$$\sigma_m = 3.54 - 0.029 * \overline{T}_{o,m} + 0.0664 * \sigma_{yr} \tag{5.9}$$

where σ_{yr} is the standard deviation of the monthly temperatures relative to the annual average temperature, $\overline{T}_{o,yr}$:

$$\sigma_{yr} = \sqrt{\frac{\sum_{m=1}^{12} (\overline{T}_{o,m} - \overline{T}_{o,yr})^2}{12}} \tag{5.10}$$

5.2.3.2 Shoenau and Kehrig Method

The second simplified method, established by Schoenau and Kehrig (1990), uses the cumulative normal probability function:

$$F(Z) = \int_{-\infty}^{Z} f(z)dz \tag{5.11}$$

where $f(z)$ is the normal probability density function defined as

$$f(z) = \frac{1}{\sqrt{2\pi}} \exp(-\frac{z^2}{2}) \tag{5.12}$$

Both $f(z)$ and $F(Z)$ can be readily computed using built-in functions available in several scientific calculators or spreadsheet programs.

The monthly degree-days based on any balance temperature T_b are then estimated as follows:

$$DD_H(T_b) = \sigma_m.N_m.\left[Z_m F(Z_m) + f(Z_m)\right] \qquad (5.13)$$

where Z_m is the normalized outdoor temperature relative to the balance temperature and is defined as

$$Z_m = \frac{T_b - T_{o,m}}{\sigma_m} \qquad (5.14)$$

With this simplified method, the monthly cooling degree-days can also be determined using a similar expression of Equation (5.13):

$$DD_C(T_b) = \sigma_m.N_m.\left[Z_m F(Z_m) + f(Z_m)\right] \qquad (5.15)$$

where Z_m is defined instead as

$$Z_m = \frac{T_{o,m} - T_b}{\sigma_m} \qquad (5.16)$$

Annual degree-days can be obtained by simply adding the monthly degree-days over the 12 months of the year. Example 5.1 shows the calculation procedure for estimating the heating and cooling degree-days for 1 month using the Schoenau and Kehrig method.

EXAMPLE 5.1

Estimate the heating and cooling degree-days for the base temperature, $T_b = 60°F$, for Denver, Colorado, for the month of May. Use the data provided in Appendix B.

Solution: Based on the weather data of Appendix B, the average temperature and standard deviation in Denver for the month of May are, respectively, $T_{o,m} = 57.2°F$ and $\sigma_{yr} = 8.7°F$.

For heating degree-days, Equation (5.14) is used and provides $Z_m = 0.321$. From normal distribution tables, $f(Z_m) = 0.379$ and $F(Z_m) = 0.626$. Then, Equation (5.13) gives

$$DD_H(T_b) = (8.70).(31).\left[(0.321)(0.626) + 0.379\right] = 156.5°F\text{-days}$$

Similarly, for cooling degree-days, Equation (5.14) is used and provides $Z_m = -0.321$. From normal distribution tables, $f(Z_m) = 0.379$ and $F(Z_m) = 0.374$. Then, Equation (5.13) gives

$$DD_C(T_b) = (8.70).(31).\left[(-0.321)(0.374) + 0.379\right] = 69.7°F\text{-days}$$

5.2.3.3 Estimation of Heating Degree-Hours Associated with a Setback Temperature

In several homes, the heating set point is set back during unoccupied hours, typically during daytime. As discussed in Chapter 3, the temperature setback reduces the heating degree-hours and thus the energy use required to heat the home. To estimate the reduction of the heating degree-hours due to a temperature setback, the fraction f_{DH} is defined as the ratio of the degree-days with setback estimated to the degree-days without setback calculated:

$$f_{DH} = \frac{DD(T_{b,setback})}{DD(T_b)} \tag{5.17}$$

It should be noted when the house is heated with the same set point throughout the heating season with no temperature setback, then $f_{DH} = 1$.

In other cases, the fraction f_{DH} can be estimated for typical daytime temperature setback schedules used in residential buildings. Indeed, a correlation that has been established based on a series of regression analyses (Kearns and Krarti, 2011) provides the fraction f_{DH} as a function of the differential temperature setback and the duration of setback period (expressed in hours):

$$f_{DH} = a\exp(-m\Delta T_{setback}) + b\exp(-nh_{setback}) \tag{5.18}$$

where $\Delta T_{setback}$ = setback temperature difference (°F or °C), $h_{setback}$ = the number of setback hours (it is assumed that the setback starts at 8:00 a.m.), and a, b, n, and m = correlation coefficients defined in Table 5.2 for several U.S. climate zones and sites.

TABLE 5.2   Model Coefficients for Equation (5.18)

Zone	Location	R^2	a	b	m	n
1A	Honolulu, Hawaii	0.783	0.0623	0.9481	0.5229	0.0016
1A	Miami, Florida	0.951	0.1491	0.8975	0.1600	0.0075
2A	Savannah, Georgia	0.914	0.1970	0.8686	0.0613	0.0113
2B	Tucson, Arizona	0.937	0.1773	0.8751	0.0730	0.0087
3A	Charleston, South Carolina	0.906	0.1997	0.8711	0.0594	0.0121
3A	Little Rock, Arkansas	0.894	0.2410	0.8284	0.0412	0.0126
3B	San Diego, California	0.885	0.1336	0.9476	0.2127	0.0126
4A	Kansas City, Missouri	0.887	0.2991	0.7637	0.0268	0.0124
4B	Albuquerque, New Mexico	0.896	0.2624	0.8011	0.0337	0.0119
4C	Eugene, Oregon	0.895	0.2434	0.8317	0.0464	0.0135
5A	Chicago, Illinois	0.876	0.3277	0.7337	0.0217	0.0126
5A	Hartford, Connecticut	0.881	0.2840	0.7793	0.0273	0.0122
5B	Boulder, Colorado	0.888	0.2357	0.8198	0.0310	0.0100
6B	Eagle, Colorado	0.892	0.2674	0.7805	0.0234	0.0091
4–6	Any location	0.851	0.2623	0.7990	0.0307	0.0115

A simplified but less accurate method than the correlation presented by Equation (5.18) to estimate the fraction f_{DH} can be used:

$$f_{DH} = \frac{24 - h_{setback}}{24} + \frac{h_{setback}}{24} \frac{(T_{b,setback} - T_{H,o})}{(T_b - T_{H,o})}$$ (5.19)

where $T_{H,o}$ = average heating season outdoor ambient air temperature (°F or °C).

5.3 Analysis of Building Envelope Retrofit

Generally, improvements in the energy efficiency of the building envelope are expensive since labor-intensive modifications are typically involved (such as addition of thermal insulation and replacement of windows). As a consequence, the payback periods of most building envelope retrofits are rather long. In these instances, the building envelope retrofits can still be justified for reasons other than energy efficiency, such as increase in occupant thermal comfort or reduction of moisture condensation to avoid structural damages. However, there are cases where retrofits of the building envelope can be justified based solely on improvement in energy efficiency. Some of these retrofit measures are discussed in this section, with some examples to illustrate how the energy savings and the payback periods are calculated.

5.3.1 Insulation of Poorly Insulated Walls and Roofs

When an element of a building envelope is not insulated or poorly insulated, it may be cost-effective to add insulation in order to reduce transmission losses. While the calculation of the energy savings due to such a retrofit may require a detailed simulation tool to account for effects of the building thermal mass or the building HVAC systems, Equations (5.2) and (5.3) can be used to determine the energy savings during the heating and cooling seasons. The total energy savings due to adding insulation to the building envelope can be estimated by summing the energy savings obtained from a reduction in heating loads and those obtained from a decrease (or increase) in cooling loads, as outlined in Example 5.2.

In most cases where the building envelope is already adequately insulated, the addition of thermal insulation is not cost-effective based only on energy cost savings.

EXAMPLE 5.2

An apartment building has a 500 m² uninsulated roof with an *R*-value of 1.44 W/m².°C. Determine the payback period of adding insulation (*R* = 2.0°C. m²/W). The apartment building is electrically heated. The cost of electricity is $0.07/kWh. The house is in a location where the heating degree-days are 2,758°C-day (based on 18°C) and is heating 24 h/day, 7 days/week throughout the heating season. Assume that the installed cost of the insulation is $15/m².

Solution: To determine the energy savings due to the addition of insulation, we will assume that the annual heating degree-days before and after the retrofit remained unchanged and are close to 18°C. Using Equation (6.21) with a retrofitted roof U-value of 0.37 W/m²×°C and heating system efficiency set to unity (electrical system), the energy savings are calculated to be

$$\Delta E = 24.500m^2 * [(1.44 - 0.37)W/m^2{}^\circ C] * 2758^\circ C.day/yr = 35,413kWh/yr$$

Thus, the payback period for adding insulation on the roof can be estimated to be:

$$Payback = \frac{500m^2 * 15\$/m^2}{35,413kWh/yr * 0.07\$/kWh} = 3.0\,years$$

Therefore, the addition of insulation seems to be cost-effective. Further analysis is warranted to determine more precisely the cost-effectiveness of this measure.

It should be noted that traditionally, the thermal performance of building envelope assemblies is typically rated using steady-state thermal resistances or R-values without accounting for any thermal shorts or thermal bridges associated with connection elements between various envelope components. Several studies have shown that these center-of-cavity R-values fail to provide the actual thermal performance of an opaque building envelope (ASHRAE, 2009; Christian and Kosny, 1996). Indeed, experimental and numerical analyses have demonstrated that the R-value of a building envelope assembly can be reduced significantly (by more than 50% for some configurations) when the thermal impacts of all envelope components accounted for include the thermal shorts associated with farming, corners, and connections. A systematic approach proposed by ASHRAE to estimate the R-value for a wall assembly is presented in Chapter 3 (ASHRAE, 2009).

A rating system for opaque building envelope assemblies has been proposed and utilizes the concept of whole wall R-value to account for thermal shorts inherent to each wall/roof construction configuration (Christian and Kosny, 1996). This whole wall R-value is estimated under steady-state conditions and provides a fair and systematic comparison method between various alternatives for building envelope assemblies, including the conventional wood frame construction and more advanced assemblies, such as metal frame construction, insulated concrete forms, and structural insulated panels (SIPs). In particular, Oak Ridge National Laboratory has developed a whole wall R-value calculation tool that allows the user to estimate the whole wall R-value for several opaque building envelope wall and roof assemblies (ORNL, 2005). The whole wall R-value calculation is based on a series of experimental and numerical analyses conducted by ORNL.

Table 5.3 summarizes the center-of-cavity R-values, the clear wall R-values, and the whole wall R-values for both wood frame wall construction and structural insulated panels suitable for residential building wall constructions.

TABLE 5.3 Reported *R*-Value for Selected Wood Frame and SIP Constructions

Wall Assembly Details	Center-of-Cavity *R*-Value (h.°F.ft²/Btu)	Clear Wall *R*-Value (h.°F.ft²/Btu)	Whole Wall *R*-Value (h.°F.ft²/Btu)
Wood frame: 2 × 4, 16 O.C., R-11 fiberglass insulation	13.60	10.60	9.60
Wood frame: 2 × 6, 16 O.C., R-19 fiberglass insulation	21.60	16.90	13.70
SIP 4.5 in.	19.60	16.40	15.70
SIP 6.5 in.	28.5	24.70	21.60

Source: ORNL, 2005.

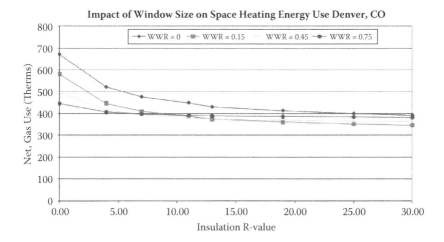

FIGURE 5.3 Annual space heating energy use as a function of wall *R*-value for various WWR values for a ranch house located in Denver, Colorado.

It is important to note that the effectiveness of adding a wall or roof insulation on energy use of residential buildings can depend significantly on the window size and glazing type of the house. Figures 5.3 and 5.4 illustrate the impact of window-to-wall ratio on, respectively, the space heating and the space cooling annual energy use for a ranch house located in Denver, Colorado. It is clear that space heating can be reduced significantly as the window-to-wall ratio (WWR) increases when the walls have low *R*-values. However, when the wall is well insulated (*R*-value higher than 19), the impact of WWR on space heating is rather small.

The impact of the insulation depends also on other factors, such as the thermal mass level as well as the placement of the location of the insulation layer within the construction assembly. Figures 5.5 and 5.6 show the impact of insulation *R*-value and the mass level on both heating and cooling energy use for a house located in Kingsman, Nevada, when the insulation is placed on the inside layer and outside layer relative to the massy wall. In particular, Figure 5.5 shows the contours of the percent savings as function of both *R*-value and mass level of the exterior wall. It is clear that placing the insulation on the outside wall would have more beneficial impact since it allows the wall mass

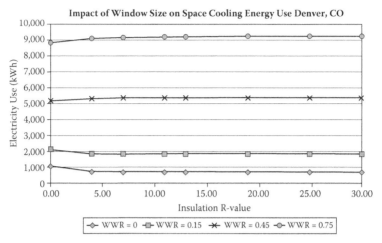

FIGURE 5.4   Annual space cooling energy use as a function of wall *R*-value for various WWR values for a ranch house located in Denver, Colorado.

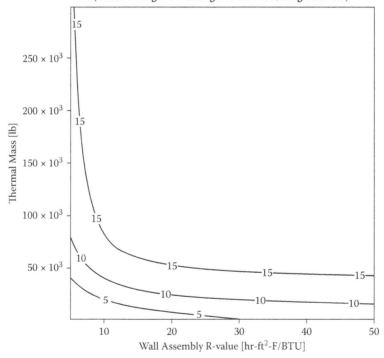

FIGURE 5.5   Total cooling and heating energy savings when the insulation is placed on the outside.

to thermally interact with the indoor air of the building. Indeed, when it is placed as an outside layer of the exterior wall assembly, insulation can reduce additional cooling and heating energy use compared to the case where the insulation is placed on the inside of the wall assembly. Moreover, placement of the insulation on the outside has the least impact of the indoor space in terms of floor area and occupant disturbance during implementation for retrofit applications. However, it is typically easier to place the insulation on the inside layer of the wall assembly.

5.3.2 Simplified Calculation Method for Building Foundation Heat Loss/Gain

As discussed in Chapter 3, estimation of building foundation heat loss and gain is rather complex and requires at least a two-dimensional solution of the ground-coupled heat transfer problem. A simplified calculation method, developed by Krarti and Chuangchid (1999), can be used to determine the heat loss or gain for slabs and

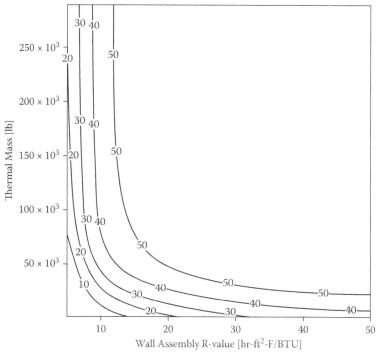

FIGURE 5.5 *(Continued)*

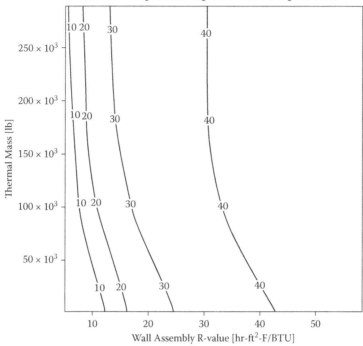

**The Total Heating Energy %Savings Relative to
Non-Thermal Mass Exterior Wall
as a Function of R-value and Thermal Mass
for Exterior Wall with Mass Material in Outside Layer
(Guest Cottage in Walking Box Ranch w/Kingman with)**

FIGURE 5.6   Total cooling and heating energy savings when insulation is placed on the inside.

basements. In particular, the annual average heat transfer from the building foundation can be estimated using the following expression:

$$Q_m = U_{eff,m} \cdot A \cdot (T_a - T_r) \tag{5.20}$$

The amplitude of the annual variation of the building foundation heat transfer can be estimated as indicated by Equation (5.21):

$$Q_a = U_{eff,a} \cdot A \cdot T_a \tag{5.21}$$

where $U_{eff,m}$ and $U_{eff,a}$ are, respectively, the effective annual average U-value and the effective annual amplitude U-value of the building foundation (Chapter 3 outlines the step-by-step approach to estimate this effective U-value); A is the total foundation area in contact with the ground medium; T_m and T_a are, respectively, the average and amplitude of the annual soil surface temperature variations (Chapter 3 provides these values for selected sites); and T_r is the indoor temperature of the building.

Using Equations (5.20) and (2.21), the heating thermal load associated with the building foundation can be estimated. Example 5.3 outlines the step-by-step calculation procedure to estimate the effective U-values for a slab-on-grade foundation.

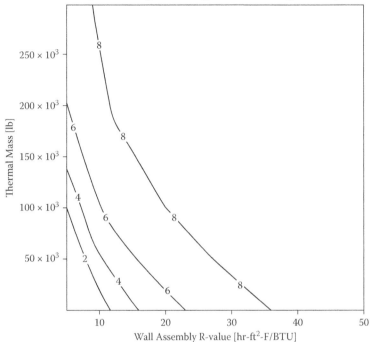

The Total Cooling Energy %Savings Relative to
Non-Thermal Mass Exterior Wall
as a Function of R-value and Thermal Mass
for Exterior Wall with Mass Material in Outside Layer
(Guest Cottage in Walking Box Ranch w/Kingman with)

FIGURE 5.6 *(Continued)*

EXAMPLE 5.3

Determine the annual mean and annual amplitude of total slab heat loss for the 10 m by 15 m slab foundation illustrated in Figure 5.7. The building is located in Denver, Colorado.

Solution: Using the method outlined in Chapter, the calculation procedure is presented using a step-by-step approach:

Step 1: Provide the required input data.
Dimensions
Slab width: 10.0 m (32.81 ft)
Slab length: 15.0 m (49.22 ft)
Ratio of slab area to slab perimeter: $A/P = 3.0$ m (9.84 ft)
4 in. thick reinforce concrete slab, thermal resistance R-value = 0.5 m^{2}K/W (2.84 h.ft^{2}F/Btu)

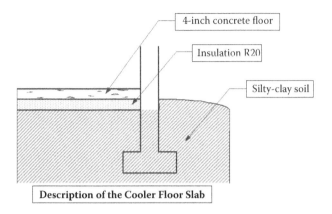

FIGURE 5.7   Slab foundation construction details.

Soil Thermal Properties
  Soil thermal conductivity: $k_s = 1.21$ W/m.K (0.70 Btu/h.ft.F)
  Soil density: $\rho = 2{,}700$ kg/m^3 (168.56 lb$_m$/ft^3)
  Soil thermal diffusivity: $a_s = 5.975 \times 10^{-7}$ m^2/s (64.3 $\times$ 10^{-7} ft^2/s)
Insulation
  Uniform insulation R-value = 3.52 m^{2}K/W (20.0 h.ft^{2}F/Btu)
Temperatures
  Indoor temperature: $T_r = 20°$C (68°F)
  Annual average ambient temperature: $T_a = 6.3°$C (43°F)
  Annual amplitude ambient temperature: $T_{amp} = 20$ K (36 R)
  Annual angular frequency: $w = 1.992 \times 10^{-7}$ rad/s

Step 2: Calculate Q_m and Q_a values.
The various normalized parameters are first calculated. Then the annual mean and amplitude of the foundation slab heat loss/gain are determined.

$$U_o = \frac{k_s P}{A} = \frac{1.21}{3.0} = 0.4033$$

$$H = \frac{A}{P.k_s.R_{eq}} = \frac{3.0}{1.21\times(0.5+3.52)} = 0.6168$$

$$D = \ln\left[(1+H)\left(1+\frac{1}{H}\right)^H\right] = 1.0748$$

$$G = k_s.R_{eq}\cdot\sqrt{\frac{\omega}{\alpha_s}} = 1.21\times(0.5+3.52)\times\sqrt{\frac{1.992\times10^{-7}}{5.975\times10^{-7}}} = 2.8086$$

Therefore,

$$\frac{Q_m}{A} = U_{eff,m}(T_r - T_a) = 0.4 \times 0.4033 \times 1.0748 \times (20 - 6.3) = 2.38 \text{ W/m}^2 \ (0.75 \text{ Btu/h.ft}^2)$$

and

$$\frac{Q_a}{A} = U_{eff,a}T_a = 0.25 \times 0.4033 \times 1.0748^{0.16} \times 2.8086^{-0.6} \times 20 = 1.10 \text{ W/m}^2 \ (0.35 \text{ Btu/h.ft}^2)$$

5.3.3 Window Improvements

Window improvements such as installation of high-performance windows, window films and coatings, or storm windows can save energy through reductions in the building heating and cooling thermal loads. Improvements in windows can impact both the thermal transmission and solar heat gains. In addition, energy-efficient windows create more comfortable environments with evenly distributed temperatures and quality lighting. Energy efficiency improvements can be made to all the components of a window assembly, including:

- Insulating the spacers between glass panes to reduce conduction heat transfer
- Installing multiple coating or film layers to reduce heat transfer by radiation
- Inserting argon or krypton gas in the space between the panes to decrease the convection heat transfer
- Providing exterior shading devices to reduce the solar radiation transmission to the occupied space

To determine accurately the annual energy performance of window retrofits, dynamic hourly modeling techniques are generally needed since fenestration can impact the building thermal loads through several mechanisms. However, simplified calculations based on Equations (5.2) and (5.3) to account for both heating and cooling savings can be used to provide a preliminary assessment of the cost-effectiveness of window retrofits, as outlined in Example 5.4.

EXAMPLE 5.4

A window upgrade is considered for an apartment building from double-pane metal frame windows ($U_E = 4.61$ W/m²·°C) to double-pane with low e film and wood frame windows ($U_R = 2.02$ W/m²·°C). The total window area to be retrofitted is 200 m². The building is located in Nantes, France, and is conditioned 24 h/day, 7 days/week throughout the heating season. An electric baseboard provides heating, while a window A/C provides cooling (EER = 8.0). Assume that the cost of electricity is \$0.10/kWh.

Solution: To determine the energy savings due to the addition of insulation, we will assume that the annual heating and cooling degree-days before and after the retrofit remained unchanged (this assumption is justified by the fact that the window contribution to the BLC is relatively small) and are, respectively, $DD_H =$ 2,244°C-day/year and $DD_C = 255$°C-day/year. The energy savings during heating (assume that the system efficiency is 1.0 for electric heating) are calculated to be

$$\Delta E = 24.200 m^2 * [(4.61 - 2.02)W/m^2 {}^\circ C] * 2244 {}^\circ C.day/yr = 27,897 kWh/yr$$

If an energy efficiency ratio (EER) value of 8.0 is assumed for the A/C system, the energy savings during cooling are estimated as follows:

$$\Delta E = 24.200 m^2 * [(4.61 - 2.02)W/m^2] * 255 {}^\circ C.day/yr * 1/8.0 = 396 kWh/yr$$

Therefore, the total energy savings due to upgrading the windows is 28,293 kWh, which corresponds to about $2,829 when electricity cost is $0.10/kWh. The cost of replacing the windows is rather high (it is estimated to be $150/m² for this project). The payback period of the window retrofit can be estimated to be

$$Payback = \frac{200 m^2 * \$150/m^2}{28,293 kWh/yr * \$0.10/kWh} = 10.4 \, years$$

Therefore, the window upgrade is not cost-effective based solely on thermal performance. The investment on new windows may, however, be justifiable based on other factors, such as increase in comfort within the space.

Windows by their size (often expressed in terms of WWR) and their glazing properties (including the U-factor and solar heat gain coefficient of SHGC) can affect significantly the energy use of a home. Figures 5.8 through 5.11 summarize the impact of window glazing selection on annual total source energy use as well as total heating and cooling energy use for the prototypical housing unit for three U.S. sites when the WWR = 10% and for a Korean site with three WWR values (10, 20, and 30%). The results are expressed using the percentage reduction relative to the annual total source energy use obtained for the baseline case with adiabatic and opaque windows (i.e., SHGC = 0 and U-factor = 0). In Figures 5.8 through 5.10, the x- and y-axes represent variations of the SHGC and U-factor, respectively, while the contour lines indicate the percentage reduction level. The performance of the six glazing types is also shown in Figures 5.8 through 5.10, as identified by their number listed in Table 5.4.

The zero percentage reduction contour line in Figures 5.8 to 5.10 determines the glazing properties that lead to neutral windows thermally performing as adiabatic and opaque windows. Negative percentage reduction translates the fact that the windows increase the overall energy consumption of the housing unit. Similar results for the impact of windows on energy performance of a residential building are obtained for two locations in Korea, as shown in Figures 5.11 and 5.13. As indicated by the results shown in Figures 5.11 through 5.13, several observations can be made:

TABLE 5.4 Properties of Six Window Glazing Types

Glazing	U-factor Btu/hr.ft².°F	U-factor W/m².K	SHGC	Description
#1	1.03	5.62	0.82	Single clear, vinyl frame
#2	0.54	3.07	0.70	Double clear, Air, vinyl frame
#3	0.41	2.33	0.53	Double low-e clear, Air, vinyl frame
#4	0.31	1.76	0.26	Double low-e green, Air, vinyl frame
#5	0.29	1.64	0.60	Double low-e clear, Argon, vinyl frame
#6	0.21	1.19	0.48	Triple low-e, Air, vinyl frame

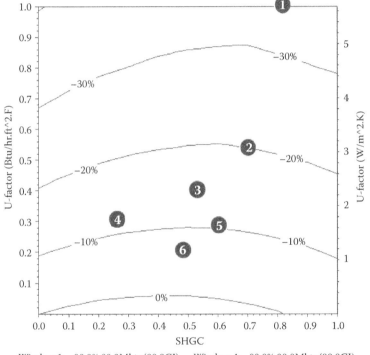

Window 1: −00.0% 00.0Mbtu(00.0GJ)   Window 4: −00.0% 00.0Mbtu(00.0GJ)
Window 2: −00.0% 00.0Mbtu(00.0GJ)   Window 5: −00.0% 00.0Mbtu(00.0GJ)
Window 3: −00.0% 00.0Mbtu(00.0GJ)   Window 6: −00.0% 00.0Mbtu(00.0GJ)

FIGURE 5.8 Percentage increase or decrease over baseline for annual source energy (combined heating and cooling) for Washington, DC.

- The heating-dominated climate of Inchon is more suitable than the milder climate of Ulsan to select windows with neutral impact. In fact, for the case when the housing unit has small windows (i.e., WWR = 10%), glazing type 6 leads to only a 3% increase in annual total housing unit source energy use when the housing unit

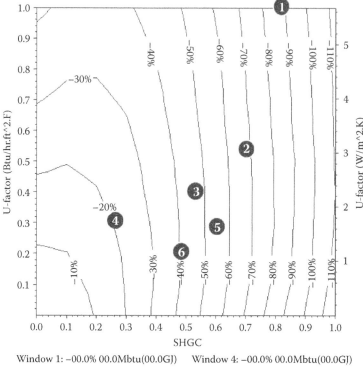

Window 1: −00.0% 00.0Mbtu(00.0GJ)   Window 4: −00.0% 00.0Mbtu(00.0GJ)
Window 2: −00.0% 00.0Mbtu(00.0GJ)   Window 5: −00.0% 00.0Mbtu(00.0GJ)
Window 3: −00.0% 00.0Mbtu(00.0GJ)   Window 6: −00.0% 00.0Mbtu(00.0GJ)

FIGURE 5.9   Percentage increase or decrease over baseline for annual source energy (combined heating and cooling) for Riverside, California.

is located in Inchon. However, the same glazing type results in a 7% increase for the same housing unit and window size located in Ulsan.

- As the window size increases, the housing unit uses more total source energy independent of the glazing type selected. In fact, there is no glazing type that can reduce the energy use when WWR = 30%, especially when the housing unit is located in Ulsan.
- As expected, glazing type 1 provides the most increase in total energy use, and glazing types 3 and 4 have similar thermal performance when the housing unit is located in Inchon, independent of the window size. Glazing type 2 provides similar thermal impact as glazing types 3 and 4 when WWR = 10%, but is noticeably less effective for larger windows. For climates and window sizes, glazing type 6, followed by glazing type 5, provides the best thermal performance.
- For the case of small windows (WWR = 10%), glazing types with the same *U*-factor but increasing SHGC values result in lower annual total source energy use when the housing unit is located in the cold climate of Inchon. However, when the unit is located in the milder climate of Ulsan or when the windows are

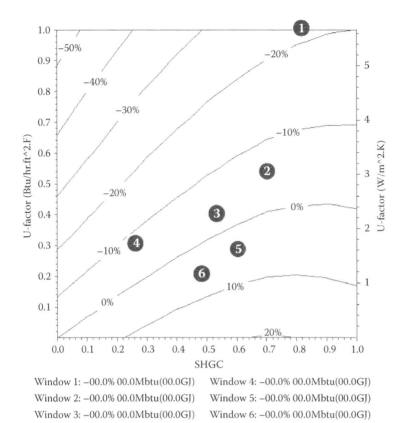

Window 1: –00.0% 00.0Mbtu(00.0GJ)   Window 4: –00.0% 00.0Mbtu(00.0GJ)
Window 2: –00.0% 00.0Mbtu(00.0GJ)   Window 5: –00.0% 00.0Mbtu(00.0GJ)
Window 3: –00.0% 00.0Mbtu(00.0GJ)   Window 6: –00.0% 00.0Mbtu(00.0GJ)

FIGURE 5.10   Percentage increase or decrease over baseline (SHGC 0 and *U*-factor 0) for annual source energy (combined heating and cooling) for Denver, Colorado.

large (WWR = 20 or 30%), higher SHGC values leads to higher energy use. This result indicates that an upper threshold level for the SHGC should be imposed for housing buildings located in mild climates or having large windows. Therefore, the building energy code of South Korea, which requires threshold values only for the window *U*-factor, should be improved to consider threshold levels for SHGC values.

5.3.4 Reduction of Air Infiltration

In several low-rise facilities, the thermal loads due to air infiltration can be significant. It is estimated that for a well-insulated residential building, the infiltration can contribute up to 40% to the total building heating load. Tuluca and Steven Winter Associates (1997) reported that measurements in eight U.S. office buildings found average air leakage rates of 0.1 to 0.5 air changes per hour (ACH). This air infiltration accounted for an estimated 10 to 25% of the peak heating load. Sherman and Matson (1993) have shown that the stock of housing in the United States is significantly overventilated from air infiltration,

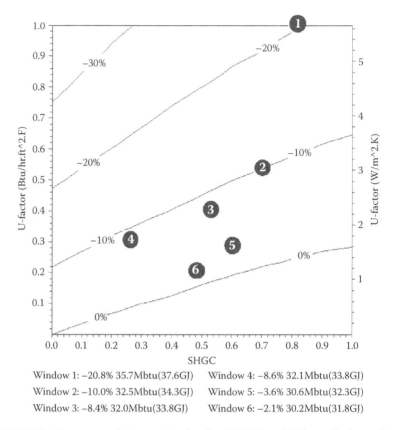

Window 1: −20.8% 35.7Mbtu(37.6GJ)   Window 4: −8.6% 32.1Mbtu(33.8GJ)
Window 2: −10.0% 32.5Mbtu(34.3GJ)   Window 5: −3.6% 30.6Mbtu(32.3GJ)
Window 3: −8.4% 32.0Mbtu(33.8GJ)   Window 6: −2.1% 30.2Mbtu(31.8GJ)

FIGURE 5.11   Percentage reduction against baseline (SHGC 0 and U-factor 0) of annual source energy for WWR = 10%, uniform window distribution, and Inchon.

and that there are 2 EJ of potential annual savings that could be captured. As described in Chapter 3, two measurement techniques can be used to evaluate the existing amount of infiltrating air. The blower technique is the most suitable to set up for detached homes to locate air leaks, while the tracer gas technique is more expensive and time-consuming and is appropriate to measure the outdoor air flow rate from both ventilation and infiltration entering large buildings, including apartments.

Several studies have evaluated the leakage distribution for residential buildings (Dickerhoff et al., 1982; Harrje and Born, 1982). In particular, it was found that leaks in walls (frames of windows, electrical outlets, plumbing penetrations) constitute the major sources of air leakage for residential buildings. To improve the air tightness of the building envelope several methods and techniques are available, including:

1. *Caulking*: Several types of caulking (urethane, latex, and polyvinyl) can be applied to seal various leaks, such as those around the window and door frames, and any wall penetrations, such as holes for water pipes.
2. *Weather stripping*: By applying foam rubber with adhesive backing, windows and doors can be air sealed.

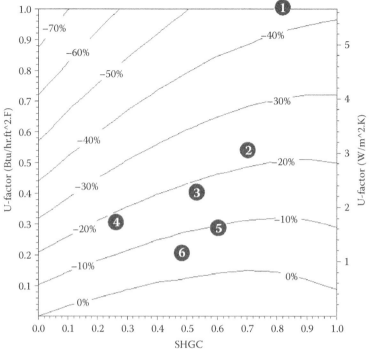

Window 1: −44.5% 41.3Mbtu(43.6GJ)   Window 4: −18.1% 33.8Mbtu(35.6GJ)
Window 2: −22.8% 35.1Mbtu(37.6GJ)   Window 5: −9.8% 31.4Mbtu(33.1GJ)
Window 3: −17.9% 33.7Mbtu(35.6GJ)   Window 6: −6.0% 30.3Mbtu(32.0GJ)

FIGURE 5.12 Percentage reduction against baseline (SHGC 0 and *U*-factor 0) of annual source energy for WWR = 20%, uniform window distribution, and Inchon.

3. *Landscaping*: This is a rather long-term project and consists of planting shrubs or trees around the building to reduce the wind effects and reduce air infiltration.
4. *Air retarders*: These systems consist of one or more air-impermeable components that can be applied around the building exterior shell to form a continuous wrap around the building walls. There are several air retarder (AR) types, such as liquid-applied bituminous, liquid-applied rubber, sheet bituminous, and sheet plastic. The AR membranes can be applied to impede the vapor movement through the building envelope, and thus act as vapor retarders. Unless they are part of an overall building envelope retrofit, these systems are typically expensive to install for existing buildings.

EXAMPLE 5.5
Consider a heated apartment building with a total conditioned volume of 1,000 m^3. A measurement of the air leakage characteristics of the building showed an infiltration rate of 1.5 ACH. Determine the energy savings due to caulking and weather stripping improvements of the exterior envelope of the building

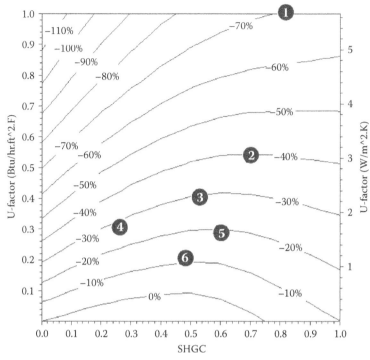

Window 1: −71.6% 47.8Mbtu(50.4GJ)   Window 4: −27.9% 35.6Mbtu(37.6GJ)
Window 2: −39.9% 39.0Mbtu(41.1GJ)   Window 5: −18.9% 33.1Mbtu(34.9GJ)
Window 3: −30.2% 36.3Mbtu(38.3GJ)   Window 6: −11.8% 31.1Mbtu(32.9GJ)

FIGURE 5.13   Percentage reduction against baseline (SHGC 0 and U-factor 0) of annual source energy for WWR = 30%, uniform window distribution, and Inchon.

to reduce air infiltration by half. Assume the apartment building is located in Seattle, Washington, and is heated by a gas-fired boiler with a seasonal efficiency of 80%.

Solution: To determine the energy savings due to the addition of insulation, we will assume that the annual heating degree-days before and after the retrofit remained unchanged and are close to 18°C. For Seattle, the degree-days (with base 65°F or 18°C) are about 2,656°C-days.

The existing air infiltration has an equivalent UA-value of $UA_{inf} = m_{cp,a} = 500$ W/°C. From Equation (5.2), with the new air infiltration equivalent UA-value of 250 W/°C and the heating system efficiency set to be 80% (gas-fired boiler), the energy savings are calculated to be

$$\Delta E = \frac{24}{0.80}.[(500-250)W / °C]* 2656°C.day / yr = 19,920kWh / yr$$

The cost of caulking and weather stripping is estimated to be about $1,500 (if only material costs are included). For a gas price of $0.10/kWh, the payback period for reducing the infiltration rate can be estimated to be

$$Payback = \frac{\$1,500}{19,920kWh / yr * 0.10\$ / kWh} = 0.75 \, years$$

Therefore, the caulking and weather stripping can be justified based only on energy cost savings. An additional benefit of reducing infiltration is improved thermal comfort.

5.4 Summary

Energy efficiency improvements of building envelope systems are generally expensive and are not cost-effective, especially for large commercial buildings. However, increasing the energy performance of a building shell can be justified for residential buildings based on energy cost savings, but also based on improvement in indoor thermal comfort and integrity of the building structure. For residential buildings, while the addition of insulation to walls and roofs may not always be cost-effective, weather stripping to reduce infiltration losses is almost always economically justifiable.

Problems

5.1 Find the monthly heating degree-days using the Erbs method for Denver, New York City, and Los Angeles. Consider the balance temperatures of 67, 65, 60, and 55°F. Use the following monthly average outdoor temperatures in your calculations:

Month	Denver	New York City	Los Angeles
Jan.	29.9	32.2	54.5
Feb.	32.8	33.4	55.6
Mar.	37.0	41.1	56.5
Apr.	47.5	52.1	58.8
May	57.0	62.3	61.9
Jun.	66.0	71.6	64.5
Jul.	73.0	76.6	68.5
Aug.	71.6	74.9	69.6
Sep.	62.8	68.4	68.7
Oct.	52.0	58.7	65.2
Nov.	39.4	47.4	60.5
Dec.	32.6	35.5	56.9

For the case of a balance temperature of 65°F, compare your results to the heating degree-days for the three cities directly computed from hourly weather data. (These degree-days are widely reported in the existing literature. Appendix B provides the degree-days for selected U.S. locations.) Comment on the results of your comparative analysis.

5.2 Repeat Problem 5.1 using the simplified calculation method of Schoenau and Kehrig. Use the required statistical data for the various U.S. locations provided in Appendix B. In addition to the heating degree-days, find the cooling degree-days based on 60, 65, and 70°F.

5.3 Determine the heating base load, the balance temperature, and the building load coefficient (BLC) of a house located in Boulder, Colorado, based on the monthly utility bills provided in the table below. For the analysis assume that the house is heated 24 h/day and that the seasonal efficiency of the gas furnace is about 78%. It should be noted that the house is generally unoccupied during the period of May 25 to July 24 (for both 2003 and 2004).

Date of Bill	Billed Days	Therms Used	Outdoor Temp. (°F)
Dec. 20, 2004	34	230	33
Nov. 21, 2004	30	71	51
Oct. 17, 2004	29	50	63
Sep. 19, 2004	31	21	72
Aug. 20, 2004	28	18	77
Jul. 24, 2004	35	6	79
Jun. 20, 2004	30	10	65
May 22, 2004	30	61	56
Apr. 18, 2004	30	118	45
Mar. 22, 2004	30	163	38
Feb. 16, 2004	27	186	26
Jan. 23, 2004	35	230	32
Jan. 02, 2004	30	199	32
Nov. 17, 2003	29	132	41
Sep. 19, 2003	29	16	73
Sep. 5, 2003	31	4	78
Jun. 20, 2003	30	9	67
May 24, 2003	31	53	56
Apr. 25, 2003	30	87	49
Mar. 22, 2003	27	107	41
Feb. 24, 2003	32	159	36

Provide any pertinent comments on the thermal performance of this house.

5.4 An energy audit of the roof indicates the following:

Roof area	10,000 ft^2
Existing roof *R*-value	5
Degree-days (winter)	6,000
Degree-hours (summer)	17,000
Fuel cost	$4/MMbtu
Boiler efficiency	0.70
Electric rate	$0.07/kWh
Air condition requirement	0.7 kW/ton

It is proposed to add R-19 insulation in the roof.

    a. Comment on the potential energy savings of adding insulation.
    b. If the insulation costs $0.35/ft^2, determine the payback of adding insulation.
    c. Is the expenditure on the additional insulation justified based on a minimum rate of return of 12%? Assume a 30-year life cycle. Solve the problem using both the present worth and annual cost methods.

5.5 Determine the annual energy savings due to replacing a single-pane window (R-1) by a double-pane window (R-2). The total area of the glazing is 200 ft^2. Assume the following:

Heating degree-days	6,000
Cooling degree-days	500
Fuel cost	$8/MMBtu
Boiler efficiency	0.80
Electric rate	$0.12/kWh
Refrigeration requirement	0.70 kW/ton
Insulation cost	$0.225/ft2

5.6 Do you recommend adding R-20 insulation to a 25,000 ft^2 uninsulated roof (R-5) when you consider a 25-year life cycle with a rate of return of 8%? Assume the following:

Heating degree-days	5,000
Cooling degree-hours	10,000
Fuel cost	$5/MMBtu
Boiler efficiency	0.70
Electric rate	$0.10/kWh
Refrigeration requirement	0.75kW/ton
Insulation cost	$0.225/ft2

5.7 A blower door test on a 30 ft × 50 ft × 9 ft house located in Denver, Colorado, revealed that under 4 Pa pressure differential between indoors and outdoors, the leakage area is about 200 in.^{2}.

    a. Determine in air changes per hour the annual average infiltration rate for the house.
    b. It was decided to weather strip the house so that the infiltration is reduced to just 0.25 ACH (for 4 Pa pressure differential). Determine the payback period of weather stripping the house given the following parameters:

Heating degree-days	6,000
Cooling degree-hours	5,000
Fuel cost	$5/MMBtu
Boiler efficiency	0.70
Electric rate	$0.10/kWh
Refrigeration requirement	0.75 kW/ton
Cost of weather stripping	$150

5.8 Both pressurization and depressurization tests were conducted on a residence. The residence has 950 ft² (95 m²) of floor area with an average ceiling height of 8 ft. The results of these tests are shown below:

Pressure (Pa)	Pressurization Test (cfm)	Depressurization Test (cfm)
10	177	237
20	324	384
30	411	471
40	531	561
50	590	650
60	676	710
70	735	765
80	795	825
90	885	887
100	942	942

a. Determine the leakage area of the house.

b. Determine the effect of shielding level on the infiltration rate. Assume the house is located in Denver, Colorado.

5.9 Consider a house with a total conditioned volume of 10,000 ft³. A measurement of the air leakage characteristics of the house showed an infiltration rate of 1.0 ACH. Determine the energy use and cost savings due to caulking and weather stripping improvements of the house exterior envelope to reduce air infiltration by half. Assume the apartment building is located in (a) Denver, Colorado, and (b) Atlanta, Georgia, and is heated by a gas-fired boiler with a seasonal efficiency of 85% and cooled by an air conditioner with a COP of 4.0. Assume the cost of electricity is $0.10/kWh and of natural gas $2/therm.

5.10 For a building with a 200 ft by 100 ft slab-on-grade floor made up of a 4 in. concrete floor:

a. Estimate the annual energy ground-coupled loss/gain from the building located in Denver, Colorado.

b. Estimate the annual heating energy use savings if the slab is insulated uniformly with R-10 rigid insulation. Assume a furnace efficiency of 85%.

c. Implement the calculation procedure for slab-on-grade floor heat loss in a spreadsheet and determine if there is an optimal cost-effective insulation level (i.e., R-value) for the foundation. Assume that the fuel cost is $10.0/MMBtu, the efficiency of the heating system is 0.80, the cost of adding insulation is 0.5/ft² of R-5 rigid insulation, and the discount rate is 5%.

6

Electrical
Systems Retrofit

Abstract

This chapter describes energy efficiency measures specific to appliances and electrical systems for residential buildings. First, a review of basic characteristics of an electric system operating under alternating current is provided. Then, the energy performance of electric equipment, including motors and appliances, is described. Finally, measures to improve the energy efficiency of an electrical distribution system common in residential buildings are outlined with specific examples. Throughout this chapter, simplified calculation methods are presented to evaluate the cost-effectiveness of the proposed energy efficiency measures.

6.1 Introduction

Lighting and heating, ventilating, and air conditioning (HVAC) for residential buildings account each for approximately 20% of total U.S. electricity use. Refrigerators represent another important energy end use in the U.S. residential sector, with about 16% of electricity. In several countries where residences are often not air conditioned, electrical systems, including lighting and appliances, can represent over 80% of the total energy used in a typical home.

In this chapter, a brief description of the electrical systems, such as motors, lighting fixtures, and appliances, is presented with a list of some measures to reduce the electrical energy use. First, a brief review of basic characteristics of an electric system is provided to highlight the major issues that should be considered when designing, analyzing, or retrofitting electrical equipment.

6.2 Review of Basics

The vast majority of buildings, including dwellings, utilize alternating current (AC) for operating electrical systems using either one-phase power sources, as is the case of detached single-family homes, or three-phase sources for large apartment buildings. In this section, a review of the fundamental building electrical power systems is provided.

6.2.1 Alternating Current Systems

For a linear electrical system subject to an AC, the time variation of the voltage and current can be represented as a sine function:

$$v(t) = V_m \cos \omega t \tag{6.1}$$

$$i(t) = I_m \cos(\omega t - \phi) \tag{6.2}$$

where V_m and I_m are the maximum instantaneous values of, respectively, voltage and current. These maximum values are related to the effective or root mean square (rms) values, as follows:

$$V_m = \sqrt{2} * V_{rms} = 1.41 * V_{rms}$$

$$I_m = \sqrt{2} * I_{rms} = 1.41 * I_{rms}$$

In the United States, the values of V_{rms} are typically 120 V for most electrical equipment and 240 V for select appliances (water heaters, ranges, and ovens) commonly used in residential buildings. In most other countries, V_{rms} ranges between 220 and 240 V in residences.

Above ω is the angular frequency of the alternating current and is related to the frequency f as follows:

$$\omega = 2\pi f$$

In the United States, the frequency f is 60 Hz, that is, 60 pulsations or oscillations in 1 s. In other countries, the frequency of the alternating current is $f = 50$ Hz.

Above ϕ is the phase lag between the current and the voltage. In the case, the electrical system is a resistance (such as an incandescent lamp), the phase lag is zero, and the current is on phase with the voltage. If the electrical system consists of a capacitance load (such as a capacitor or a synchronous motor), the phase lag is negative and the current is in advance relative to the voltage. Finally, when the electrical system is dominated by an inductive load (such as fluorescent fixture or an induction motor), the phase lag is positive and the current lags the voltage.

Figure 6.1 illustrates the time variation of the voltage for a typical electric system. The concept of root mean square (also called effective value) for the voltage, V_{rms}, is also indicated in Figure 6.1. It should be noted that in the United States the cycle for the voltage waveform repeats itself every 1/60 s (since the frequency is 60 Hz).

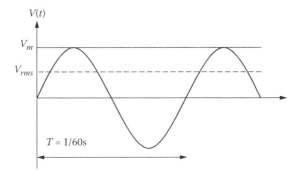

FIGURE 6.1 Time variation of the voltage and the concept of V_m and V_{rms}.

The instantaneous power, $p(t)$, consumed by the electrical system operated on one-phase AC power supply can be calculated using the ohm law:

$$p(t) = v(t).i(t) = V_m I_m \cos\omega t.\cos(\omega t - \phi) \tag{6.3}$$

The above equation can be rearranged using some basic trigonometry and the definition of the rms values for voltage and current:

$$p(t) = V_{rms}.I_{rms}\left(\cos\phi.(1 + \cos 2\omega t) + \sin\phi.\sin 2\omega t\right) \tag{6.4}$$

Two types of power can be introduced as a function of the phase lag angle ϕ, the real power P_R, and the reactive power P_X, as defined below:

$$P_R = V_{rms}.I_{rms}\cos\phi \tag{6.5}$$

$$P_X = V_{rms}.I_{rms}\sin\phi \tag{6.6}$$

Note that both types of power are constant and are not a function of time. To help understand the meaning of each power, it is useful to note that the average of the instantaneous power actually consumed by the electrical system over one period is equal to P_R:

$$\bar{p} = \frac{1}{T}\int_0^T p(t)dt = P_R \tag{6.7}$$

Therefore, P_R is the actual or real power consumed by the electrical system over its operation period (which consists typically of a large number of periods $T = 1/2\pi f$). P_R is typically called real power and is measured in kW. Meanwhile, P_X is the power required to produce a magnetic field to operate the electrical system (such as an induction motors). P_X is stored and then released by the electrical system; this power is typically

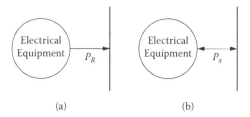

(a)   (b)

FIGURE 6.2   Illustration of the direction of electricity flow for (a) real power and (b) reactive power.

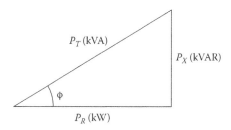

FIGURE 6.3   Power triangle for an electrical system.

called reactive power and is measured in kVAR. A schematic diagram is provided in Figure 6.2 to help illustrate the meaning of each type of power.

While the user of the electrical system consumes actually only the real power, the utility or the electricity provider has to make available to the user both the real power, P_R, and the reactive power, P_X. The algebraic sum of P_R and P_X constitutes the total power, P_T. Therefore, the utility has to know, in addition to the real power needed by the customer, the magnitude of the reactive power, and thus the total power.

As mentioned earlier, for a resistive electrical system, the phase lag is zero and thus the reactive power is also zero (see Equation (6.6)). For several motors, lighting fixtures, and electronic devices, the reactive power can be significant. In fact, the higher the phase lag angle ϕ, the larger the reactive power P_X. To illustrate the importance of the reactive power relative to the real power P_R and the total power P_T consumed by the electrical system, a power triangle is typically used to represent the power flow, as shown in Figure 6.3.

In Figure 6.3, it is clear that the ratio of the real power to the total power represents the cosine of the phase lag. This ratio is widely known as the power factor, *pf*, of the electrical system:

$$pf = \frac{P_R}{P_T} = \cos\phi \qquad (6.8)$$

Ideally, the power factor has to be as close to unity as possible (i.e., $pf = 1.0$). Typically, however, power factors above 90% are considered to be acceptable. If the power factor is

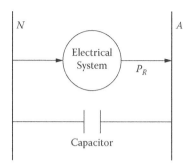

FIGURE 6.4 The addition of a capacitor can improve the power factor of an electrical system.

low, that is, if the electrical system has a high inductive load, capacitors can be added in parallel to reduce the reactive power, as illustrated in Figure 6.4.

Unlike the case for commercial buildings and industrial facilities, electric rates for residential buildings do not specifically have penalties for low power factor (Krarti, 2010). Therefore, any device that attempts to improve the power factor would not save any energy cost for residential homes. Indeed, the vast majority of utilities charge residential customers through fixed or flat rates (i.e., $/kWh). Some utilities use the tiered or block rates as optional rates for selected residential buildings. Typically, tiered rates for residential customers apply in the summer months, when the electrical demands are high. Standard flat rates apply during all other months. With tiered rates, the price paid by customers for electricity increases as customers' electrical energy use increases. For instance, a utility may charge the first 500 kilowatt-hours (kWh) of electricity used during a summer month at a lower rate, with all subsequent kilowatt-hours billed at a higher rate. With the implementation of smart grid, utilities may extend the time of use rates to their residential customers.

6.2.2 Power Quality

Power quality can be defined as the level of distortion of the voltage or current sine waveform due to an electrical system operation. The voltage and current for an electrical system with ideal power quality vary as a simple sine function of time, often referred to as the fundamental harmonic, and are expressed by Equations (6.1) and (6.2), respectively. When the power is distorted due, for instance, to electronic ballasts (which change the frequency of the electricity supplied to the lighting systems), several harmonics need to be considered in addition to the fundamental harmonic to represent the voltage or current-time variation, as shown in Equations (6.9) and (6.10):

$$v(t) = \sum_{k=1}^{N_V} V_k \cos(k\omega - \theta_k) \tag{6.9}$$

$$i(t) = \sum_{k=1}^{N_I} I_k \cos(k\omega - \phi_k) \tag{6.10}$$

Highly distorted waveforms contain numerous harmonics. While the even harmonics (i.e., second, fourth, etc.) tend to cancel each other's effects, the odd harmonics (i.e., third, fifth, etc.) have their peaks coincide and significantly increase the distortion effects. To quantify the level of distortion for both voltage and current, a dimensionless number referred to as the total harmonic distortion (THD) is determined through a Fourier series analysis of the voltage and current waveforms. The THDs for voltage and current are, respectively, defined as follows:

$$THD_V = \sqrt{\frac{\sum_{k=2}^{N_V} V_k^2}{V_1^2}} \tag{6.11}$$

$$THD_I = \sqrt{\frac{\sum_{k=2}^{N_V} I_k^2}{I_1^2}} \tag{6.12}$$

Table 6.1 provides current THD for selected but specific lighting and office equipment loads (NLPIP, 1995). Generally, it is found that devices with high-current THD contribute to voltage THD in proportion to their share to the total building electrical load. Therefore, the engineer should consider higher-wattage devices before lower-wattage devices to reduce the voltage THD for the entire building or facility. Example 6.1 shows a simple calculation procedure that can be used to assess the impact of an electrical device on the current THD. Thus, the engineer can determine which devices need to be corrected first to improve the power quality of the overall electric system. Typically, harmonic filters are added to electrical devices to reduce the current THD values.

EXAMPLE 6.1

Assess the impact on the current THD of a building of two devices: 13 W compact fluorescent lamp (CFL) with electronic ballasts and a laser printer while printing. Use the data provided in Table 6.1.

Solution: Both devices have an rms voltage of 120 V (i.e., V_{rms} = 120 V); their rms current can be determined using the real power used and the power factor given in Table 6.1 and Equation (6.5):

$$I_{rms} = \frac{P_R}{V_{rms} \cdot pf}$$

The above equation gives an rms current of 0.22 A for the CFL and 6.79 A for the printer. These values correspond actually to the rms of each device's fundamental current waveform and can be used in the THD equation, Equation (6.12), to estimate the total harmonic current of each device:

TABLE 6.1 Typical Power Quality Characteristics (Power Factor and Current THD) for Selected Electrical Loads

Electrical Load	Real Power Used (w)	Power Factor	Current THD (%)
Incandescent Lighting Systems			
100 W incandescent lamp	101	1.0	1
Compact Fluorescent Lighting Systems			
13 W lamp with magnetic ballast	16	0.54	13
13 W lamp with electronic ballast	13	0.50	153
Full-Size Fluorescent Lighting Systems (2 lamps per ballast)			
T12 40 W lamp with magnetic ballast	87	0.98	17
T12 40 W lamp with electronic ballast	72	0.99	5
T10 40 W lamp with magnetic ballast	93	0.98	22
T10 40 W lamp with electronic ballast	75	0.99	5
T8 32 W lamp with electronic ballast	63	0.98	6
T5 28 W lamp with electronic ballast	62	0.95	15
High-Intensity Discharge Lighting Systems			
400 W high-pressure sodium lamp with magnetic ballast	425	0.99	14
400 W metal halide lamp with magnetic ballast	450	0.94	19
Office Equipment			
Desktop computer without monitor	33	0.56	139
Color monitor for desktop computer	49	0.56	138
Laser printer (in standby mode)	29	0.40	224
Laser printer (printing)	799	0.98	15
External fax/modem	5	0.73	47

Source: Adapted from NLPIP, *Newsletter by the National Lighting Product Information Program*, 2(2), 5, 1995 (updated report on 2005).

$$I_{tot} = I_{rms} . THD_I$$

The resultant values of 0.33 A for the CFL and 1.02 A for the printer show that although the printer has a relatively low current THD (15%), the actual distortion current produced by the printer is more than three times that of the CFL because the printer uses more power.

IEEE (1992) recommends a maximum allowable voltage THD of 5% at the building service entrance (i.e., point where the utility distribution system is connected to the building electrical system). Based on a study by Verderber et al. (1993), the voltage THD reaches the 5% limit when about 50% of the building electrical load has a current THD of 55%, or when 25% of the building electrical load has a current THD of 115%.

It should be noted that when the electrical device has a power factor of unity (i.e., $pf = 1$), there is little or no current THD (i.e., THDI = 0%) since the device has only a resistive load and effectively converts input current and voltage into useful electric power. As shown in Table 5.11, the power factor and the current THD are interrelated and both define the characteristics of the power quality. In particular, Table 6.1 indicates that lighting systems with electronic ballasts have a typically high power factor and low current THD. This good power quality is achieved using capacitors to reduce the phase lag between the current and voltage and filters to reduce harmonics.

The possible problems that have been reported due to poor power quality include:

1. Overload of neutral conductors in three phases with four wires. In a system with no THD, the neutral wire carries no current if the system is well balanced. However, when the current THD becomes significant, the currents due to the odd harmonics do not cancel each other, and rather add up on the neutral wire, which can overheat and cause a fire hazard.
2. Reduction in the life of transformers and capacitors. This effect is mostly caused by distortion in the voltage.
3. Interference with communication systems. Electrical devices that operate with high frequencies such as electronic ballasts (that operate at frequencies ranging from 20 to 40 kHz) can interfere and disturb the normal operation of communication systems, such as radios, phones, and energy management systems (EMS).

6.3 Electrical Motors

6.3.1 Introduction

Motors convert electrical energy to mechanical energy and are typically used to drive machines. The driven machines can serve a myriad of purposes in the building, including moving air (supply and exhaust fans), moving liquids (pumps), and compressing gases (refrigerators). To select the type of motor to be used for a particular application, several factors have to be considered, including:

1. The form of the electrical energy that can be delivered to the motor: direct current (DC) or alternating current (AC), single or three phase.
2. The requirements of the driven machine, such as motor speed and load cycles.
3. The environment in which the motor is to operate: normal (where a motor with an open type ventilated enclosure can be used), hostile (where a totally enclosed motor must be used to prevent outdoor air to infiltrate inside the motor), or hazardous (where a motor with an explosion-proof enclosure must be used to prevent fires and explosions).

The basic operation and the general characteristics of AC motors are discussed in the following section. In addition, simple measures are described to improve the energy efficiency of existing motors.

6.3.2 Overview of Electrical Motors

There are basically two types of electric motors used in buildings and industrial facilities: (1) induction motors and (2) synchronous motors. Induction motors are the more common type, accounting for about 90% of the existing motor horsepower. Both types use a motionless stator and a spinning rotor to convert electrical energy into mechanical power. The operation of both types of motor is relatively simple and is briefly described below.

Alternating current is applied to the stator, which produces a rotating magnetic field in the stator. A magnetic field is also created in the rotor. This magnetic field causes the rotor to spin in trying to align with the rotating stator magnetic field. The rotation of the magnetic field of the stator has an angular speed that is a function of the number of poles, N_P, and the frequency, f, of the AC, as expressed in Equation (6.13):

$$\omega_{mag} = \frac{4\pi.f}{N_P} \qquad (6.13)$$

One main difference between the two motor types (synchronous vs. induction) is the mechanism by which the rotor magnetic field is created. In an induction motor, the rotating stator magnetic field induces a current, and thus a magnetic field, in the rotor windings, which are typically of the squirrel cage type. In a magnetic motor, the rotor cannot rotate at the same speed as the magnetic field (if the rotor spins with the same speed as the magnetic field, no current can be induced in the rotor, since effectively the stator magnetic field remains at the same position relative to the rotor). The difference between the rotor speed and the stator magnetic field rotation is called the slip factor.

In a synchronous motor, the magnetic field is produced by application of direct current through the rotor windings. Therefore, the rotor spins at the same speed as the rotating magnetic field of the stator, and thus the rotor and the stator magnetic fields are synchronous in their speed.

Because of their construction characteristics, the induction motor is basically an inductive load and thus has a lagging power factor, while the synchronous motor can be set so it has a leading power factor (i.e., acts like a capacitor). Therefore, it is important to remember that a synchronous motor can be specified to both provide mechanical power and improve the power factor for a set of induction motors. This option may be more cost-effective than just adding a bank of capacitors.

Three parameters are typically used to characterize an electric motor during full-load operation. These parameters include:

1. The mechanical power output of the motor, P_M. This power can be expressed in kW or horsepower (1 HP = 0.746 kW). The mechanical power is generally the most important parameter in selecting a motor.
2. The energy conversion efficiency of the motor, η_M. This efficiency expresses the mechanical power as a fraction of the real electric power consumed by the motor. Due to various losses (such as friction, core losses due to the alternating of the magnetic field, and resistive losses through the windings), the motor efficiency is always less than 100%. Typical motor efficiencies range from 75 to 95%, depending on the size of the motor.
3. The power factor of the motor, pf_M. As indicated earlier in this chapter, the power factor is a measure of the magnitude of the reactive power needed by the motor.

Using the schematic diagram of Figure 6.5, the real power used by the motor can be calculated as follows:

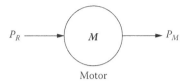

FIGURE 6.5   Definition of the efficiency of a motor.

$$P_R = \frac{1}{\eta_M}.P_M \qquad (6.14)$$

Therefore, the total power and the reactive power needed to operate the motor are, respectively:

$$P_T = \frac{P_R}{pf_M} = \frac{1}{pf_M.\eta_M}.P_M \qquad (6.15)$$

$$P_X = P_R \tan\phi = \frac{1}{\eta_M}.P_M.\tan(\cos^{-1} pf) \qquad (6.16)$$

6.3.3 Energy-Efficient Motors

6.3.3.1 General Description

Based on their efficiency, motors can be classified into two categories: (1) standard efficiency motors and (2) high- or premium efficiency (i.e., energy-efficient) motors. The energy-efficient motors are 2 to 10 percentage points more efficient than standard efficiency motors, depending on the size. Table 6.2 summarizes the average efficiencies for both standard and energy-efficient motors that are currently available commercially. The improved efficiency for the high or premium motors is mainly due to better design and use of better materials to reduce losses. However, this efficiency improvement comes with a higher price of about 10 to 30% more than standard efficiency motors. These higher prices may be the main reason that only one-fifth of the motors sold in the United States are energy efficient.

However, the installation of premium efficiency motors is becoming a common method of improving the overall energy efficiency of buildings. The potential for energy savings from premium efficiency motor retrofits is significant. In the United States alone, there were in 1991 about 125 million operating motors that consumed approximately 55% of the electrical energy generated in the United States (Andreas, 1992). It was estimated that replacing all these motors with premium efficiency models would save approximately 60 Thw of energy per year (Nadel et al., 1991).

To determine the cost-effectiveness of motor retrofits, there are several tools available, including the MotorMaster developed by the Washington State Energy Office (WSEO)

TABLE 6.2 Typical Motor Efficiencies

Motor Mechanical Power Output kW (HP)	Average Nominal Efficiency for Standard Efficiency Motors	Nominal Efficiency for Premium Efficiency Motors
0.75 (1.0)	0.730	0.855
1.12 (1.5)	0.750	0.865
1.50 (2.0)	0.770	0.865
2.25 (3.0)	0.800	0.895
3.73 (5.0)	0.820	0.895
5.60 (7.5)	0.840	0.917
7.46 (10)	0.850	0.917
11.20 (15)	0.860	0.924
14.92 (20)	0.875	0.930
18.65 (25)	0.880	0.936
22.38 (30)	0.885	0.936
29.84 (40)	0.895	0.941
37.30 (50)	0.900	0.945
44.76 (60)	0.905	0.950
55.95 (75)	0.910	0.954
74.60 (100)	0.915	0.954
93.25 (125)	0.920	0.954
111.9 (150)	0.925	0.958
149.2 (200)	0.930	0.962

Source: Adapted from Hoshide, R.K., *Energy Engineering*, 1(1), 6–24, 1994, for standard motors; NEMA, *Energy Management Guide for Selection and Use of Polyphase Motors*, Standard MG-10-1994, National Electrical Manufacturers Association, Rosslyn, VA, 2006, for premium motors.

and currently available from the DOE website (DOE, 2009). These tools have the advantage of providing large databases for cost and performance information for various motor types and sizes.

6.3.3.2 Energy Savings Calculations

A simplified method to calculate the energy savings due to energy-efficient motor replacement is provided by Equation (6.17). Inherent to this method, two assumptions are made: (1) the motor is fully loaded, and (2) the change in motor speed is neglected.

$$\Delta P_R = P_M \cdot \left(\frac{1}{\eta_e} - \frac{1}{\eta_r} \right) \tag{6.17}$$

where P_M is the mechanical power output of the motor, η_e is the design (i.e., full-load) efficiency of the existing motor (e.g., before retrofit), and η_r is the design (i.e., full-load) efficiency of the energy-efficient motor (e.g., after retrofit).

The electric energy savings incurred from the motor replacement is thus

$$\Delta kWh = \Delta P_R . N_h . LF_M \qquad (6.18)$$

where N_h is the number of hours per year during which the motor is operating, and LF_M is the load factor of the motor's operation during 1 year.

Example 6.2 indicates the potential energy use and cost savings and payback period when a standard efficiency 10 HP (7.46 kW) motor is replaced with a premium efficiency motor.

EXAMPLE 6.2

Determine the cost-effectiveness of replacing a 10 HP motor with an efficiency of 85% with a premium efficiency motor with a rated full-load efficiency of 91.70%. Assume:

- The cost of electricity is $0.10/kWh.
- The differential cost of premium vs. standard motor is $300.
- The average load factor of the motor is 0.80.
- The average full-load operating hours of the motor are 5,000 h/year.

Solution: To determine the cost-effectiveness of installing a premium efficiency motor instead of a standard efficiency motor, a simplified economic analysis is used to estimate the simple payback period. The savings in energy use in kWh for the premium efficiency motor can be calculated using Equations (6.17) and (6.18):

$$kWh_{saved} = N_h . HP * 0.746 . LF_M . \left(\frac{1}{\eta_{std}} - \frac{1}{\eta_{eff}} \right)$$

where N_h is the total number of hours (per year) during which the motor is operating at full load (5,000 h/year), HP is the rated motor power output (10 HP), LF_M is the annual average load factor of the motor ($LF_M = 0.80$), and η_{std} and η_{eff} are the respective efficiencies of the standard transformer and the efficient transformer (0.850 and 0.917).

Thus, the energy savings in kWh is calculated as follows:

$$kWh_{saved} = 5000 * 10 * 0.746 * 0.80 * \left(\frac{1}{0.850} - \frac{1}{0.917} \right) = 2565 kWh / yr$$

Therefore, the simple payback period, *SPB*, for investing in the premium efficiency motor is

$$SPB = \frac{\$300}{2565 kWh * \$0.10 / kWh} = 1.2 \, years$$

6.4 Lighting Systems

6.4.1 Introduction

Lighting can account for a significant portion of the energy use in residential buildings. In addition, heat generated by lighting contributes to additional thermal loads that need to be removed by the cooling equipment. Typically, energy retrofits of lighting equipment are very cost-effective, with payback periods of less than 2 years in most applications.

To better understand the retrofit measures that need to be considered in order to improve the energy efficiency of lighting systems, a simple estimation of the total electrical energy use due to lighting is first considered:

$$Kwh_{Lit} = \sum_{j=1}^{J} N_{Lum,j}.WR_{Lum,j}.N_{h,j} \qquad (6.19)$$

where $N_{Lum,j}$ is the number of lighting luminaires of type j in the building to be retrofitted (it should be noted that a luminaire can include several components, such as ballast, electric wiring, housing, and lamps); $WR_{Lum,j}$ is the wattage rating for each luminaire of type j (in this rating the energy use due to both the lamps and ballast should be accounted for); $N_{h,j}$ is the number of hours per year when the luminaires of type j are operating; and J is the number of luminaire types in the building.

It is clear from Equation (6.19) that there are three options to reduce the energy use attributed to lighting systems, as briefly discussed below:

1. Reduce the wattage rating for the luminaires, including both the lighting sources (e.g., lamps) and the power transforming devices (e.g., ballasts); therefore, decrease the term $WR_{Lum,j}$ in Equation (6.19). In the last decade, technological advances such as compact fluorescent lamps and electronic ballasts have increased the energy efficiency of lighting systems.

2. Reduce the time of use of the lighting systems through lighting controls; therefore, decrease the term $N_{h,j}$ in Equation (6.19). Automatic controls have been developed to decrease the use of a lighting system, so illumination is provided only during times when it is actually needed. Energy-efficient lighting controls include the occupancy sensing systems and light dimming controls through the use of daylighting.

3. Reduce the number of luminaires; therefore, decrease the term $N_{Lum,j}$ in Equation (6.19). This goal can be achieved only in cases where delamping is possible due to overillumination.

In this section, only measures related to the general actions described in items 1 and 2 are discussed. To estimate the energy savings due to any retrofit measure for the lighting system, Equation (6.19) can be used. The energy use due to lighting has to be calculated before and after the retrofit, and the difference between the two estimated energy uses represents the energy savings.

6.4.2 Energy-Efficient Lighting Systems

Improvements in the energy efficiency of lighting systems have provided several opportunities to reduce electrical energy use in buildings. In this section, the energy savings calculations for the following technologies are discussed:

- High-efficiency fluorescent lamps
- Compact fluorescent lamps
- Compact halogen lamps
- Electronic ballasts
- LED lighting

First, a brief description is provided for the factors that an auditor should consider to achieve and maintain an acceptable quality and level of comfort for the lighting system. Second, the design and operation concepts are summarized for each available lighting technology. Then, the energy savings that can be expected from retrofitting existing lighting systems using any of the new technologies are estimated and discussed.

Typically, three factors determine the proper level of light for a particular space: age of the occupants, speed and accuracy requirements, and background contrast (depending on the task being performed). It is a common misconception to consider that overlighting a space provides higher visual quality. Indeed, it has been shown that overlighting can actually reduce the illuminance quality and the visual comfort level within a space, in addition to wasting energy. Therefore, it is important when upgrading a lighting system to determine and maintain the adequate illuminance level as recommended by the appropriate authorities.

6.4.2.1 High-Efficiency Fluorescent Lamps

Fluorescent lamps have high efficiency (i.e., light output per watt consumed), diffuse light distribution, and long operating life. A fluorescent lamp consists generally of a glass tube with a pair of electrodes at each end. The tube is filled at very low pressure with a mixture of inert gases (primarily argon) and liquid mercury. When the lamp is turned on, an electric arc is established between the electrodes. The mercury vaporizes and radiates in the ultraviolet spectrum. This ultraviolet radiation excites a phosphorous coating on the inner surface of the tube that emits visible light. High-efficiency fluorescent lamps use a krypton-argon mixture that increases the efficacy output by 10 to 20% from a typical efficacy of 70 lumens/W to about 80 lumens/W. Improvements in phosphorous coating can further increase the efficacy to 100 lumens/W.

It should be mentioned that the handling and disposal of fluorescent lamps is highly controversial due to the fact that mercury inside the lamps can be toxic and hazardous to the environment. A new technology is being tested to replace the mercury with sulfur to generate the radiation that excites the phosphorous coating of the fluorescent lamps. The sulfur lamps are not hazardous and would present an environmental advantage to the mercury-containing fluorescent lamps.

The compact fluorescent lamps (CFLs) are miniaturized fluorescent lamps with a small diameter and shorter length. The compact lamps are less efficient than full-size fluorescent lamps with only 35 to 55 lumens/W. However, they are more energy efficient and have longer life than incandescent lamps. Indeed, CFLs have an average life between 6,000 and 15,000 h, while incandescent lamps have a life of 750 or 1,000 h. Currently, compact fluorescent lamps are being heavily promoted as energy savings alternatives to incandescent lamps, even though they may have some drawbacks. In addition to their high cost, compact fluorescent lamps are cooler and thus provide less pleasing contrast than incandescent lamps.

Among the most common retrofit in lighting systems for residential buildings is the upgrade of the conventional incandescent lamps to more energy-efficient CFLs. For a lighting retrofit, it is recommended that a series of tests be conducted to determine the characteristics of the existing lighting system. For instance, it is important to determine the illuminance level at various locations within the space, especially in working areas such as benches or desks.

6.4.2.2 Compact Halogen Lamps

The compact halogen lamps are adapted for use as direct replacements for standard incandescent lamps. Halogen lamps are more energy efficient, produce whiter light, and last longer than incandescent lamps. Indeed, incandescent lamps typically convert only 15% of their electrical energy input into visible light since 75% is emitted as infrared radiation and 10% is used by the filament as it burns off. In halogen lamps, the filament is encased inside a quartz tube that is contained in a glass bulb. A selective coating on the exterior surface of the quartz tube allows visible radiation to pass through but reflects the infrared radiation back to the filament. This recycled infrared radiation permits the filament to maintain its operating temperatures with 30% less electrical power input.

The halogen lamps can be dimmed and present no power quality or compatibility concerns, as can be the case for the compact fluorescent lamps.

6.4.2.3 Electronic Ballasts

Ballasts are integral parts to fluorescent luminaires since they provide the voltage level required to start the electric arc and regulate the intensity of the arc. Before the development of electronic ballasts in the early 1980s, only magnetic or "core and coil" ballasts were used to operate fluorescent lamps. While the frequency of the electrical current is kept at 60 Hz (in other countries other than the United States, the frequency is set at 50 Hz) by the magnetic ballasts, electronic ballasts use solid-state technology to produce high-frequency (20–60 MHz) current. The use of high-frequency current increases the energy efficiency of the fluorescent luminaires since light is cycling more quickly and appears brighter. When used with high-efficiency lamps (T8, for instance), electronic ballasts can achieve 95 lumens/W as opposed to 70 lumens/w for conventional magnetic ballasts. It should be mentioned, however, that efficient magnetic ballasts can achieve similar lumen/watt ratios as electronic ballasts.

Other advantages that electronic ballasts have relative to their magnetic counterparts include:

- Higher power factor. The power factor of electronic ballasts is typically in the 0.90 to 0.98 range. Meanwhile, the conventional magnetic ballasts have a low power factor (less than 0.80) unless a capacitor is added, as discussed in Section 6.2.

- Less flicker problems. Since the magnetic ballasts operate at 60 Hz current, they cycle the electric arc about 120 times per second. As a result, flicker may be perceptible, especially if the lamp is old during normal operation or when the lamp is dimmed to less than 50% capacity. However, electronic ballasts cycle the electric arc several thousands of times per second and flicker problems are avoided even when the lamps are dimmed to as low as 5% of capacity.
- Less noise problems. The magnetic ballasts use electric coils and generate audible hum, which can increase with age. Such noise is eliminated by the solid-state components of the electronic ballasts.

6.4.2.4 Light-Emitting Diode (LED) Lamps

Light-emitting diodes (LEDs) are the newest lighting technologies available on the market. While they were first introduced as indicator lights, because of advances in their optics and their luminous output, LEDs are now used as lighting fixtures for buildings. Some of the advantages of LEDs include their small size, reliability, flexibility, and energy efficiency. However, LEDs are still expensive as lighting sources for buildings. LEDs are semiconductors, with light created by the flow of electrons in p-n junctions and releasing energy in the form of photons.

LEDs are typically divided into three different size categories: miniature LEDs, mid-range LEDs, and high-power LEDs. The miniature LEDs are very small, usually only a few millimeters in diameter, and are made up of a single semiconductor. Because these lamps are so small, they are not suitable for illuminating spaces. Typical current ratings for miniature LEDs range between 1 and 20 mA. The mid-range LEDs usually have a current rating of 100 mA and are used most often when little luminance levels are desired in specific spaces. High-intensity LEDs are currently suitable for lighting buildings and typically require currents in the range of 100 mA to 1.0 A. With these relatively high currents, overheating can be a concern for LEDs. Thus, they often need a heat sink.

LED lamps are quickly becoming popular because of their energy efficiency. Even though they are still expensive, LED lamps can be cost-effective lighting alternatives for several buildings. Moreover, these lamps offer significant flexibility due to their size and can be placed in a wide range of locations within buildings. Example 6.3 illustrates the cost-effectiveness of LED lamps compared to CFLs in residential buildings.

EXAMPLE 6.3

Consider a home with a total of 10 incandescent lamps rated at 60 W. Determine, the energy savings potential if these lamps are replaced with (1) 13 W CFLs and (2) 7 W LED lamps. The life spans of incandescent lamps, CFLs, and LED lamps are, respectively, 1,300, 8,000, and 30,000 h. Perform the analysis using the life span of the LED lamps. Determine the best lamps to use—CFLs or LED lamps—if the costs of the incandescent lamps, CFLs, and LED lamps are $0.30, $1.50, and $40, respectively. Assume the cost of electricity is $0.10/kWh and perform a simple payback analysis.

Solution:

1. Incandescent lamps: Over the life span of 30,000 h, 230 incandescent lamps are needed. The energy cost for using incandescent lamps over 30,000 h would be

$$kWh = 230 * 60 * 1,300 * \frac{1}{1000} * 0.10 = \$1,794$$

Thus, the total cost for using incandescent lamps is $1,794 + $0.30*230 = $1,863.

2. CFLs: Over the life span of 30,000 h, about 38 CFLs are needed. The energy cost for using CFLs over 30,000 h would be

$$kWh = 38 * 13 * 8,000 * \frac{1}{1000} * 0.10 = \$395$$

Thus, the total cost for using CFL lamps is $395 + $1.50*38 = $452.

3. LED lamps: Over the life span of 30,000 h, about 10 LEDs are needed. The energy cost for using LEDs over 30,000 h would be

$$kWh = 10 * 7 * 30,000 * \frac{1}{1000} * 0.10 = \$210$$

Thus, the total cost for using LED lamps is $210 + $40*10 = $610.

Therefore, CFLs would be the best option to replace the incandescent lamps.

6.4.3 Lighting Controls

As illustrated by Equation (6.19), energy savings can be achieved by not operating the lighting system in cases when illumination becomes unnecessary. The control of the operation of the lighting system can be achieved by several means, including dimming switches, occupancy sensing systems, and through daylighting.

While energy savings can be achieved by manual dimming, the results are typically unpredictable since they depend on the occupant behavior. Scheduled or sensor-based lighting controls provide a more efficient approach to energy savings but can also be affected by the frequent adjustments from occupants. Some of the dimming switches and automatic controls available for lighting systems are briefly discussed below.

6.4.3.1 Dimming Switches

One of the simplest and cost-effective ways to reduce energy use associated with lighting residential buildings is the use of dimming switches. Indeed, these types of switches can save up to 50% of the electrical lighting energy use for a typical residence, compared to the conventional toggle switches. Moreover, they can extend the life span of lighting lamps.

6.4.3.2 Occupancy or Vacancy Sensors

These sensors can be installed in areas with intermittent occupancy in a house, such as a bathroom, a walk-in closet, or a garage. The occupancy sensors save energy by automatically turning off the lights in spaces that are not occupied. For residential applications, occupancy sensors are typically passive infrared (PIR) devices that can simply replace traditional wall switches and can detect motion for spaces of up to 60 m² (600 ft²). Occupancy sensors can save up to 60% of energy use associated with lighting. It should be noted that these infrared sensors operate adequately only if they are in direct line of sight with the occupants.

6.4.3.3 Photosensor-Based Controls

Automatic controls of outdoor lighting can save significant electrical energy use for residential buildings. Timers and motion detection can be used for controlling outdoor lighting. However, photosensors offer an energy-efficient and practical means to control the outdoor lighting while ensuring safety and security around the house. Typically photosensor-based controls can save 20% of energy use associated with outdoor lighting.

6.4.4 Daylighting Harvesting Systems

While in residential buildings there is no need to use special controls to take advantage of the natural light to illuminate indoor spaces inside residential buildings, daylighting harvesting is highly recommended. The illuminance levels available inside rooms are typically expressed as percentages of the illuminance outdoors. These percentages, referred to as demand factors (DFs), are recommended to be 1.5 to 2.5% for visual tasks, 2.5 to 4.0% for moderately difficult tasks, and 4.0 to 8.0% for difficult and prolonged tasks.

Rooms that require good quality lighting throughout the year, such as kitchens, bedrooms, and living rooms, should be placed along the north-south perimeter of the homes. Rooms such as laundry rooms, garages, and closets can be placed in the interior of the homes. In addition to large windows, skylights, and roof monitors, daylighting can be better harvested for rooms with reflective surfaces, especially the ceiling. It is generally recommended that the ceiling surface has a reflectance of at least 0.90 through using high-reflectance paints or tiles.

To estimate the potential annual energy use savings, f_d, associated with daylighting harvesting in reducing the need for electrical lighting in small rooms, the following expression can be used (Krarti et al., 2005):

$$f_d = b\left[1 - \exp\left(-a\tau_w A_w / A_f\right)\right] \tag{6.20}$$

where A_w/A_f is the ratio of the window area over the space floor area (this parameter provides a good indicator of the window size relative to the daylit floor area); a and b are coefficients that depend only on the building location and are given by Table 6.3 for various sites throughout the world; and τ_w is the visible transmittance of the glazing.

Example 6.4 outlines the calculation procedures to estimate the annual energy savings associated with installing daylighting harvesting in a home.

TABLE 6.3 Coefficients *a* and *b* of Equation (6.20) for Various Locations throughout the World

Location	a	b	Location	a	b
Atlanta	19.6	74.3	Casper	19.2	72.7
Chicago	18.4	71.7	Portland	17.8	70.9
Denver	19.4	72.9	Montreal	18.8	69.8
Phoenix	22.3	74.8	Quebec	19.1	70.6
New York City	18.7	67.0	Vancouver	16.9	68.7
Washington, DC	18.7	70.8	Regina	20.0	70.5
Boston	18.7	67.1	Toronto	19.3	70.5
Miami	25.1	74.8	Winnipeg	19.6	70.9
San Francisco	20.6	74.0	Shanghai	19.4	67.3
Seattle	16.6	69.2	K-Lumpur	20.2	72.4
Los Angeles	22.0	74.2	Singapore	23.3	73.7
Madison	18.8	70.0	Cairo	27.0	74.2
Houston	21.6	74.7	Alexandria	36.9	74.7
Fort Worth	19.7	72.9	Tunis	25.2	74.1
Bangor	17.9	70.7	Sao Paulo	29.4	71.2
Dodge City	18.8	72.6	Mexico	28.6	73.6
Nashville	20.0	70.4	Melbourne	20.0	67.7
Oklahoma City	20.2	74.4	Roma	16.0	72.4
Columbus	18.6	72.3	Frankfurt	15.2	69.7
Bismarck	17.9	71.5	Kuwait	22.0	65.3
Minneapolis	18.2	72.0	Riyadh	21.2	72.7
Omaha	18.9	72.3			

EXAMPLE 6.4

Consider a ranch house with 150 m² located in Denver, Colorado, with a total window area of 20 m² with double-pane glazing that has a visible transmittance of 0.55. The home has 20 CFLs, each rated at 13 W. Determine the energy use and cost savings due to daylighting harvesting strategies. Assume that without daylighting harvesting, each CFL is operated 3 h/day when daylighting can be harvested, 7 days/week, 45 weeks/year. The cost of electricity is 0.10/kWh.

Solution: Based on Equation (6.20) using the coefficients *a* = 19.4 and *b* = 72.9, the percent reduction in electrical lighting associated with daylighting harvesting is

$$f_d = 72.9 * \left[1 - \exp\left(-19.4 * 0.55 * 20 / 150 \right) \right] = 55.3\%$$

Therefore, the savings in annual electrical energy use for the home are computed as follows:

$$\Delta kWh = 55.3\% * 20 * 13 * \frac{1}{1000} * 3 * 7 * 45 = 136 kWh / yr$$

Thus, the annual energy cost saving is $13.6/year.

6.5 Electrical Appliances

6.5.1 Typical Energy Use

Appliances account for a significant part of the energy consumption in buildings that used about 41% of electricity generated worldwide in 1990 (IPCC, 1996). Moreover, the operating cost of appliances during their lifetime (typically 10 to 15 years) far exceeds their initial purchase price. However, consumers, especially in the developing countries where no labeling programs for appliances are enacted, do not generally consider energy efficiency and operating cost when making purchases since they are not well informed.

Figure 6.6 indicates that appliances have become more common in U.S. households over the last decade. In particular, microwave ovens, clothes dryers, and dishwashers were available in, respectively, 83, 71, and 50% of U.S. households in 1997. Almost all U.S. households have at least one refrigerator. Figure 6.7 shows the distribution of the capacity of the refrigerators used in the U.S. households. It is clear that the size of the refrigerators used in the U.S. homes has increased over the years.

Average daily energy use profiles as well as the runtime variations for selected electrical appliances are shown in Figures 6.8 through 6.11 for a refrigerator, an oven, a dishwasher, and a clothes dryer (Surles, 2011).

As expected, the refrigerator consumes the most electrical energy during a typical day among the four appliances considered in the figures.

6.5.2 Energy Efficiency Standards

Recognizing the significance and impact of appliances on the national energy requirements, a number of countries have established energy efficiency programs. In particular, some of these programs target improvements of energy efficiency for residential

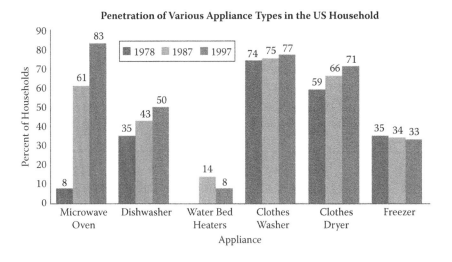

FIGURE 6.6   Penetration levels of selected appliances in U.S. households.

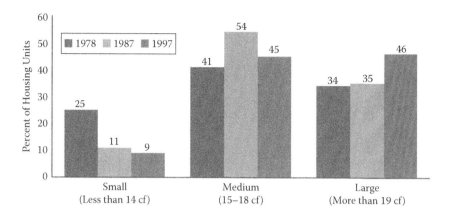

FIGURE 6.7 Size of refrigerators used in U.S. households.

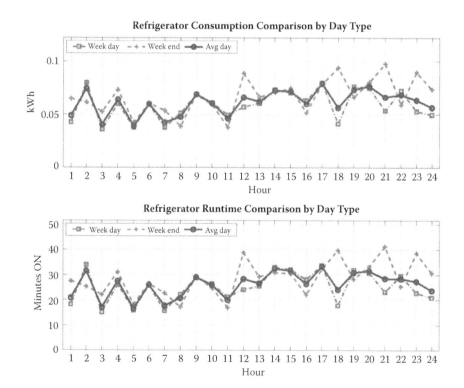

FIGURE 6.8 Average daily profiles for energy use and runtime for a refrigerator.

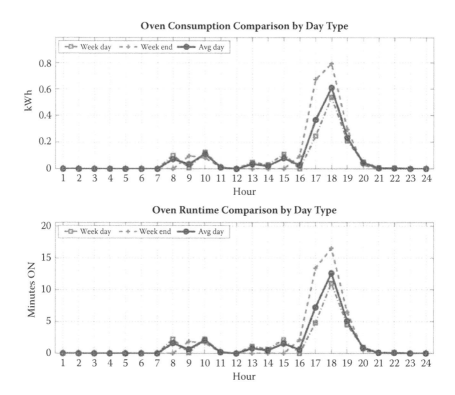

FIGURE 6.9   Average daily profiles for energy use and runtime for an oven.

appliances. Methods to achieve these improvements include energy efficiency standards and labeling programs.

Minimum efficiency standards for residential appliances have been implemented in some countries for a number of residential end uses. The energy savings associated with the implementation of these standards are found to be substantial. For instance, studies have indicated that in the UK, the energy consumption of average new refrigerators and freezers in 1993 was about 60% of the consumption in 1970. Similar improvements have been obtained in Germany (Waide et al., 1997). In the United States, the savings due to the standards are estimated to be about 0.7 EJ per year during the period extending from 1990 to 2010 (1 EJ = 10^{18} J = 1 quadrillion of Btu = 10^{15} Btu).

Energy standards for appliances in the residential sector have been highly cost-effective. In the United States, it is estimated that the average benefit-cost ratios for promoting energy-efficient appliances are about 3.5. In other terms, each U.S. dollar of federal expenditure on implementing the standards is expected to contribute $165 of net present-valued savings to the economy over the period of 1990 to 2010. In addition to energy and cost savings, minimum efficiency standards reduce pollution with significant reduction in carbon emissions. In the period 2000 to 2010, it is expected that energy efficiency standards have resulted in an annual carbon reduction of 4% (corresponding to 9 million metric tons of carbon/year) relative to the 1990 level.

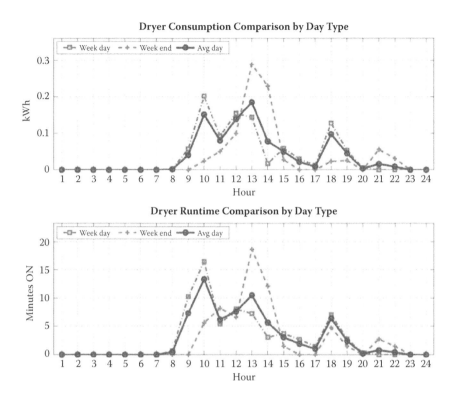

FIGURE 6.10 Average daily profiles for energy use and runtime for a clothes dryer.

Currently, energy efficiency standards are utilized with various degrees of comprehensiveness, enforcement, and adoption in a limited number of countries, as summarized in Table 6.4. However, several other countries are in the process of developing national standards or labeling programs to promote energy-efficient residential appliances.

Minimum efficiency standards have been in use in the United Sates and Canada for more than two decades and cover a wide range of products. In the last few years, standards programs have spread to other countries, like Brazil, China, Korea, Mexico, and the Philippines. Most of these countries have currently one or two products subject to standards, such as the case of European countries, Korea, and Japan. Other countries are in the process of implementing or considering energy efficiency standards for household appliances. Among these countries are Colombia, Denmark, Egypt, Indonesia, Malaysia, Pakistan, Singapore, and Thailand.

Many of the currently existing standards are mandatory and prohibit the manufacture or sale of noncomplying products. However, there are other standards that are voluntary and thus are not mandatory, such as the case of product quality standards established in India.

As indicated in Table 6.4, most countries have established minimum efficiency standards for refrigerators and freezers since this product type has one of the highest

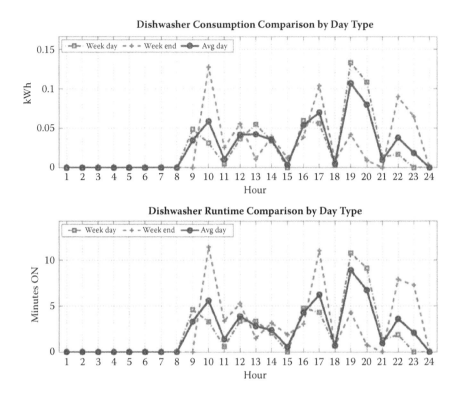

FIGURE 6.11   Average daily profiles for energy use and runtime for a dishwasher.

TABLE 6.4   Status of International Residential Appliance Energy Efficiency Standards

Country/Region	Compliance Status	Products
Australia	Mandatory	R, FR, WH
Brazil	Voluntary	R, FR
Canada	Mandatory	All
China	Mandatory	R, CW, RAC
European Union	Mandatory	R, FR
India	Voluntary	R, RAC, A/C
Japan	Voluntary	A/C
Korea	Mandatory	R, A/C
Mexico	Mandatory	R, FR, RAC
Philippines	Mandatory	A/C
United States	Mandatory	All

Adapted from Turiel, I., *Energy and Buildings,* 26(1), 5, 1997.

Note: Refrigerators (R), freezers (FR), clothes washers (CW), dishwashers (DW), clothes dryers (CD), cooking, water heaters (WH), room air conditioners (RAC), and central air conditioning (A/C).

TABLE 6.5 Maximum Allowable Annual Energy Use (kWh/year) for Refrigerators and Freezers Sold and Manufactured in the United States

Product Category	2001 Standard	2008 Standard
Manual defrost R/FR	248.4 + 8.82 AV	198.72.4 + 7.056 AV
Partial autodefrost R/FR	248.4 + 8.82 AV	198.72.4 + 7.056 AV
Top-mount autodefrost R/FR	276 + 9.8 AV	220.8 + 7.84 AV
Top-mount autodefrost with through-the-door features R/FR	356 + 10.20 AV	284.8 + 8.16 AV
Side-mount autodefrost R/FR	507.5 + 4.91 AV	406 + 3.928 AV
Side-mount autodefrost with through-the-door features R/FR	406 + 10.10 AV	324.8 + 8.08 AV
Bottom-mount autodefrost R/FR	459 + 4.6 AV	367.2 + 3.68 AV
Upright manual FR	258.3 + 7.55 AV	232.47 + 8.08 AV
Upright autodefrost FR	326.1 + 12.43 AV	293.49 + 11.87 AV
Chest RF	143.7 + 9.88 AV	129.33 + 8.892 AV

Source: Energy Star, Energy Star, U.S. Environmental Protection Agency, information provided on the Energy Star website: http://www.energystar.gov, 2009.

Note: R = refrigerators, FR = freezers, AV = adjusted volume = volume of R + 1.63 × volume of FR (for R/FR) = 1.73 × volume of FR (for FR only).

growth rates in terms of both sales value and volume. The existing international energy efficiency standards for refrigerators and freezers set a limit on the energy use over a specific period of time (generally, 1 month or 1 year). This energy use limit may vary depending on the size and configuration of the product. Table 6.5 shows the maximum limits for the allowable annual energy use for U.S. refrigerators and freezers. Two standards are shown in Table 6.5: the standards that are currently effective in 2001 and the updated standards effective in April 2008. It should be noted that models with higher energy efficiencies than those listed in Table 6.5 do exist and are sold in the U.S. market. To keep up with the technology advances, U.S. standards are typically amended or changed periodically. However, any new or amended U.S. standard for energy efficiency has to be based on improvements that are technologically feasible and economically justified. Typically, energy-efficient designs with payback periods of less than 3 years can be incorporated into new U.S. standards.

In addition to standards, labeling programs have been developed to inform consumers about the benefits of energy efficiency. There is a wide range of labels used in various countries to promote energy efficiency for appliances. These labels can be grouped into three categories:

- Efficiency type labels used to allow consumers to compare the performance of different models for a particular product type. For instance, a common label used for refrigerators indicates the energy use and operating cost over a specific period, such as 1 month or 1 year. Other labels, such as labels for clothes washers, show the efficiency expressed in kWh of energy used per pound of clothes washed. Another feature of efficiency labels is the ability to provide consumers with a comparative evaluation of the product models by showing the energy consumption or efficiency of a particular model on a scale of the lowest (or highest) energy use (or efficiency) models.

- Eco-labels provide information on more than one aspect (i.e., energy efficiency) of the product. Other aspects include noise level, waste disposal, and emissions. Green Seal in the United States is an example of an eco-label program that certifies that the products are designed and manufactured in an environmentally responsible manner (Green Seal, 1993). Certification standards have been established for refrigerators, freezers, clothes washers, clothes dryers, dishwashers, and cooktops/ovens.
- Efficiency seals of approval, such as the Energy Star program in the United States, are labels that indicate that a product has met a set of energy efficiency criteria but do not quantify the degree by which the criteria were met. The Energy Star label, established by the U.S. Environmental Protection Agency, indicates, for instance, that a computer monitor is capable of reducing its standby power level when not in use for some time period (Johnson and Zoi, 1992).

In recent years, labeling of appliances has become a popular approach around the world in order to inform consumers about the energy use and energy cost of purchasing different models of the same product. Presently, Australia, United States, and Canada have the most comprehensive and extensive labeling programs. Other countries, such as the European Union, Japan, Korea, Brazil, Philippines, and Thailand, have developed labels for a few products.

In addition to energy efficiency, standards have been developed to improve the performance of some appliances in conserving water. For instance, water-efficient plumbing fixtures and equipment have been developed in the United States to promote water conservation.

The reduction of water use by some household appliances can also increase their energy efficiency. Indeed, a large fraction of the electrical energy used by both clothes washers and dishwashers is attributed to heating the water (85% for clothes washers and 80% for dishwashers). Chapter 8 provides a more detailed discussion on water and energy performance of conventional and energy/water-efficient models currently available for residential clothes washers and dishwashers.

6.5.3 Standby Power Loads

Some electrical devices have a built-in clock as well as battery units that consume energy even when the devices are turned off. Standby power loads, commonly called phantom loads, can be significant in residential homes. Typical values for standby power measured for selected electrical devices are outlined in Table 6.6 (LBNL, 2004). The Energy Star requirements for standby power are also provided in Table 6.6.

To estimate the energy used by phantom loads for any electrical device with a standby power, W_{Stby}, the number of hours when the device is turned off, $N_{h,off}$, should be used:

$$kWh_{Stby} = W_{Stby,j}.N_{h,off} \qquad (6.21)$$

EXAMPLE 6.5
Estimate the annual phantom load for a TV with a set-top box if the TV is operated 2 h/day. Determine the annual cost of the phantom load energy if the electricity cost is 0.10/kWh.

TABLE 6.6 Measured Average Standby Power and Energy Standard Requirements for Selected Electrical Devices in Homes

Electrical Device	Average Standby Power (W)	Energy Star Standby Power (W)
TV (CRT)	3	1
Set-top box, digital cable	17.8	1
Audio mini-system	8.3	1
VCR	4.7	1
DVD player	1.6	1
Desktop computer	2.8	2
Laptop computer	8.9	1
Printer	1.4	1
Fax machine	5.3	2
Copier	1.5	1
Microwave oven	2.9	2

Solution: Based on Table 6.6, the standby power for a TV is 3 W, while that of the set-top box is 17.8 W; thus the W_{Stby} = 20.8 W. Since the TV and its set-top box are operated 2 h/day, it is turned off 22 h/day. Therefore, $N_{h,off}$ = 22 h/day*365 days/year = 8,030 h/year. Using Equation (6.21), the annual energy used by the phantom load associated with the TV and the set-top box is

$$kWh_{Stby} = (20.8 \text{ W}) (8,030 \text{ h/year}) = 167,024 \text{ W/year} = 167 \text{ kWh/year}$$

Thus, the annual cost of the phantom load is $16.7/year based on $0.10/kWh electricity cost. The use of a power strip would save this annual cost.

6.6 Electrical Distribution Systems

6.6.1 Introduction

All electrical systems have to be properly designed and installed in order to provide electrical energy to the utilization equipment as safely, reliably, and economically as possible. Figure 6.12 shows a typical one-line diagram of an electrical system for a residential building. A service entrance feeder is typically used to connect the utility distribution station to the main panel of the building. The main panel includes the protection devices (typically circuit breakers) for the branch circuits serving various electrical devices within the building.

Figure 6.13 illustrates some electrical devices that are served by the main panel through a set of branch circuits. A panel board for an audited house is shown in Figure 6.14.

A typical schedule of the panel is provided in Figure 6.15 for a U.S. home served by a two-phase service entrance (240 V/120 V). The two phases of the panel should

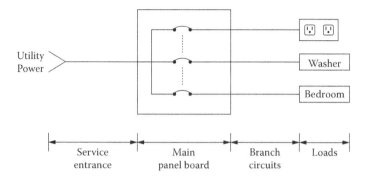

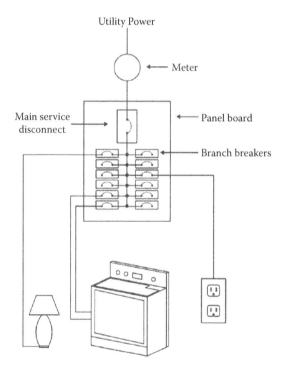

FIGURE 6.12   A schematic one-line diagram for a basic electrical distribution system within a home.

FIGURE 6.13   Common electrical loads served by a distribution panel within a home.

be balanced within the 10% to ensure that the neutral wire does not carry significant currents. The inspection and audit of the distribution system within a building are important tasks before planning a monitoring program to determine the total electrical use or the energy end uses for a dwelling. Current transformers are typically used to measure the current flowing through any electrical device, including the main service entrance, a branch circuit, or a receptacle.

FIGURE 6.14   Layout of the circuit breakers in a house distribution panel.

6.6.2 Electrical Wires

The term *electrical wire* is actually generic and refers typically to both a conductor and a cable. The conductor is the copper or aluminum wire that actually carries electrical current. The cable refers generally to the complete wire assembly, including the conductor, the insulation, and any shielding or protective covering. A cable can have more than one conductor, each with it own insulation.

The size of an electrical conductor represents its cross-sectional area. In the United States, two methods are used to indicate the size of a conductor: the American wire gauge (AWG) for small sizes and thousand of circular mils (MCM) for larger sizes. For the AWG method, the available sizes are from no. 18 to no. 4/0, with the higher number representing the smaller conductor size. For buildings, the smallest size of copper conductor that can be used is no. 14, which is rated for a maximum loading of 15 amp. The AWG size designation soon became inadequate after its implementation in the early 1900s due to the ever-increasing electrical load in buildings. For larger conductors, the cross-sectional area is measured in circular mils. A circular mil corresponds to the area of a circle that has a diameter of 1 mil or 1/1,000th of an inch. For instance, a conductor with a diameter of ½ in. (500 mils) has a circular mil area of 250,000, which is designated by 250 MCM.

To determine the correct size of conductors to be used for feeders and branch circuits in buildings, three criteria need generally to be considered:

- The rating of the continuous current under normal operating conditions. The National Electric Code (NEC, 1996) refers to the continuous current rating as the ampacity of the conductor. The main parameters that affect the ampacity of a conductor include the physical characteristics of the wire, such as its cross-sectional area (or size) and its material, and the conditions under which the wire operates,

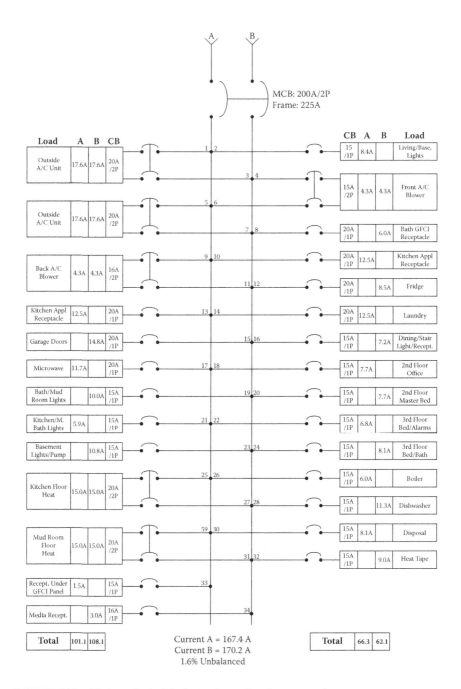

FIGURE 6.15   A balanced schedule for in a house distribution panel.

such as the ambient temperature and the number of conductors installed in the same cable. Table 6.7 indicates the ampacity rating of copper and aluminum conductors with various sizes. Various derating and correction factors may need to be applied to the ampacity of the conductor to select its size.

- The rating of short-circuit current under fault conditions. Indeed, high short-circuit currents can impose significant thermal or magnetic stresses not only on the conductor but also on all the components of the electrical system. The conductor has to withstand these relatively high short-circuit currents since the protective device requires some finite time before detecting and interrupting the fault current.
- The maximum allowable voltage drop across the length of the conductor. Most electrical utilization equipment is sensitive to the voltage applied to them. It is therefore important to reduce the voltage drop that occurs across the feeders and the branch circuits. The NEC recommends a maximum voltage drop of 3% for any one feeder or branch circuit, with a maximum voltage drop from the service entrance to the utilization outlet of 5%.

For more details, refer to Section 220 of the NEC, which covers the design calculations of both feeders and branch circuits (NEC, 2008).

Two conductor materials are commonly used for building electrical systems: copper and aluminum. Because of its highly desirable electrical and mechanical properties, copper is the preferred material used for conductors of insulated cables. Aluminum has some undesirable properties and its use is restricted. Indeed, an oxide film that is not a good conductor can develop on the surface of aluminum and can cause poor electrical contact, especially at the wire connections. It should be noted that aluminum can be considered in cases when cost and weight are important criteria for the selection of conductors. However, it is highly recommended, even in these cases, to use copper conductors for the connections and the equipment terminals to eliminate poor electrical contact.

To protect the conductor, several types of insulation materials are used. The cable (which is the assembly that includes the conductor, insulation, and any other covering) is identified by letter designations, depending on the type of insulation material and the conditions of use. In buildings, the following letter designations are used:

- For the insulation material type: A (asbestos), MI (mineral insulation), R (rubber), SA (silicone asbestos), T (thermoplastic), V (varnished cambric), and X (crosslinked synthetic polymer).
- For the conditions of use: H (heat up to 75°C), HH (heat up to 90°C), UF (suitable for underground), and W (moisture resistant).

Thus, the letter designation THW refers to a cable that has a thermoplastic insulation rated for maximum operating temperature of 75°C and suitable for use in dry as well as wet locations.

Moreover, some types of electrical cables have outer coverings that provide mechanical/corrosion protection, such as lead sheath (L), nylon jacket (N), armored cable (AC), metal-clad cable (MC), and NM (nonmetallic sheath cable).

For a full description and all types of insulated conductors, their letter designations, and their uses, refer to the NEC, article 310 (NEC, 2008).

In general, the electrical cables are housed inside conduits for additional protection and safety. The types of conduits commonly used in buildings are listed below:

- Rigid metal conduit (RMC) can be of either steel or aluminum and has the thickest wall of all types of conduit. Rigid metal conduit is used in hazardous locations such as high exposure to chemicals.
- Intermediate metal conduit (IMC) has a thinner wall than the rigid metal conduit but can be used in the same applications.
- Electrical metallic tubing (EMT) is a metal conduit but with a very thin wall. The NEC restricts the use of EMT to locations where it is not subjected to severe physical damage during installation or after installation.
- Electrical nonmetallic conduit (ENC) is made of nonmetallic material such as fiber or rigid polyvinyl chloride (PVC). Generally, rigid nonmetallic conduit cannot be used where subject to physical damage.
- Electrical nonmetallic tubing (ENT) is a pliable corrugated conduit that can be bent by hand. Electrical nonmetallic tubing can be concealed within walls, floors, and ceilings.
- Flexible conduit can be readily flexed and thus is not affected by vibration. Therefore, a common application of the flexible conduit is for the final connection to motors or recessed lighting fixtures.

It should be noted that the number of electrical conductors that can be installed in any one conduit is restricted to avoid any damage of cables (especially when the cables are pulled through the conduit). The NEC restricts the percentage fill to 40% for three or more conductors. The percentage fill is defined as the fraction of the total cross-sectional area of the conductors, including the insulation, over the cross-sectional area of the inside of the conduit.

When selecting the size of the conductor, the operating costs and not only the initial costs should be considered. As illustrated in Example 6.6, the cost of energy encourages the installation of larger conductors than are required typically by the NEC, especially when smaller size conductors are involved (i.e., no. 14, 12, 10, and 8). Unfortunately, most designers do not consider the operating costs in their design due to several reasons, including interest in lower first costs and uncertainty in electricity prices.

EXAMPLE 6.6

Determine if it is economically feasible to install no. 10 (AWG) copper conductor instead of no. 12 (AWG) on a 100 ft branch circuit that feeds a load of 16 amp. Assume:

- The load is used 10 h/day and 300 days/year.
- The cost of electricity is $0.10/kWh.
- The installed costs of no. 12 and no. 10 conductors are, respectively, $60.00 and $90.00 per 1,000 ft long cable.

TABLE 6.7   Ampacity of Selected Insulated Conductors Used in Buildings

Conductor Size (AWG or MCM)	THW (copper)	THHN (copper)	THW (aluminum)	THHN (aluminum)
18	—	14	—	—
16	—	18	—	—
14	20	25	—	—
12	25	30	20	25
10	35	40	30	35
8	50	55	40	45
6	65	75	50	60
4	85	95	65	75
3	100	110	75	85
2	115	130	90	100
1	130	150	100	115
1/0	150	170	120	135
2/0	175	195	135	150
3/0	200	225	155	175
4/0	230	260	180	205
250	255	290	205	230
300	285	320	230	255
350	310	350	250	280
400	335	380	270	305
500	380	430	310	350
600	420	475	340	385
700	460	520	375	420
750	475	535	385	435
800	490	555	395	450
900	520	585	425	480
1,000	545	615	445	500
1,250	590	665	485	545
1,500	625	705	520	585
1,750	650	735	545	615
2,000	665	750	560	630

Source: Adapted from National Electrical Code, Table 310-16.

Solution: In addition to the electric energy used to meet the load, there is an energy loss in the form of heat generated by the flow of current, I, through the resistance of the conductor, R. The heat loss in watts can be calculated as follows:

$$Watts = R.I^2$$

Using the information by the NEC (Table 6.7), the resistance of both conductors no. 12 and 10 can be determined to be, respectively, 0.193 and 0.121 ohm per 100 ft. Thus, the heat loss for the 400 ft branch circuit if no. 12 conductor is used can be estimated as follows:

$$Watts_{12} = 0.193 * 100 / 100.(16)^2 = 49.4 \qquad W$$

Similarly, the heat loss for the 400 ft branch circuit when no. 10 conductor is used is found to be

$$Watts_{10} = 0.121 * 100 / 100.(16)^2 = 31.0 \qquad W$$

The annual cost of copper losses for both cases can be easily calculated:

$$Cost_{12} = 49.4W * 300days / yr * 10hrs / Day * 1kW / 1000W * \$0.10 / kWh = \$14.8 / yr$$

$$Cost_{10} = 31.0W * 300days / yr * 10hrs / Day * 1kW / 1000W * \$0.10 / kWh = \$9.3 / yr$$

Therefore, if no. 10 is used instead of no. 12, the simple payback period, *SPP*, for the higher initial cost for the branch circuit conductor is

$$SPP = \frac{(\$90 / 1000\,ft - \$60 / 1000\,ft) * 100\,ft}{(\$14.8 - \$9.3)} = 0.54\,yr = 6.5\,months$$

The savings in energy consumption through the use of larger conductors can thus be cost-effective. Moreover, it should be noted that the larger size conductors reduce the voltage drop across the branch circuit, which permits the connected electrical utilization equipment to operate more efficiently. However, the applicable code has to be carefully consulted to determine if larger size conduit is required when larger size conductors are used.

6.7 Summary

In this chapter, an overview is provided for the basic characteristics of electrical systems in residential buildings. In particular, an overview of various electrical equipment, including motors, lighting fixtures, and appliances is outlined. Moreover, several measures specific to electrical systems are described to improve the energy performance of existing electrical installations. Additionally, illustrative examples are presented to evaluate the cost-effectiveness of selected energy efficiency measures specific to residential electrical systems. The measures and the calculation procedure are presented to illustrate the wide range of energy-conserving opportunities that an auditor should address when retrofitting electrical systems for residential buildings.

Problems

6.1 Two motors operate 5,000 h/h at full load. One is 10 HP with an electrical efficiency of 0.75 and a power factor of 0.65, while the other is 7.5 HP with an electrical efficiency of 0.935 and a power factor of 0.85. Determine:
   a. The overall power factor.
   b. The simple payback period of replacing each motor with an energy-efficient motor. Assume that the cost of electricity is $0.12/kWh. The incremental costs for energy-efficient motors are $150 for 10 HP motor and $100 for 7.5 motor.

6.2 Determine the capacitor ratings (in kVAR) to add to a 10 HP motor with an efficiency of 0.85 to increase its power factor from 0.80 to 0.85, 0.90, and 0.95, respectively.

6.3 For an apartment building with a total of 200 incandescent lamps rated at 60 W, determine the energy savings potential if these lamps are replaced with 13 W CFLs. The life spans of incandescent lamps and CFLs are, respectively, 1,300 and 8,000 h. Perform the analysis using the life span of the CFL lamps. Determine the cost-effectiveness of replacing the incandescent lamps with CFLs if the costs of the incandescent lamp and CFL are $0.20 and $1.40, respectively. Assume the cost of electricity is $0.10/kWh. Perform a simple payback period analysis.

6.4 For a house with a total of 20 incandescent lamps rated at 60 W, determine the energy savings potential if these lamps are replaced with (a) 13 W CFLs and (b) 7 W LED lamps. The life spans of incandescent lamps, CFLs, and LED lamps are, respectively, 1,300, 8,000, and 30,000 h. Perform the analysis using the life span of the LED lamps. Determine the best lamps to use, CFLs or LED lamps, if the costs of the incandescent lamps, CFLs, and LED lamps are $0.25, $1.45, and $45, respectively. Assume the cost of electricity is $0.12/kWh and perform a life cycle cost analysis using the proper life cycle period and a 5% discount rate.

6.5 Estimate the annual phantom load for a TV with a set-top box if the TV is operated 5 h/day. Determine the annual cost of the phantom load energy if the electricity cost is 0.12/kWh.

6.6 Estimate the annual phantom load for an audio mini-system operated 3 h/day. Determine the annual cost of the phantom load energy if the electricity cost is 0.10/kWh.

6.7 Consider a ranch house with 100 m² located in Phoenix, Colorado, with a total window area of 15 m² with double-pane glazing that has a visible transmittance of 0.35. The home has 20 incandescent lamps, each rated at 60 W.
   a. Estimate the energy use and cost savings when the incandescent lamps are replaced by 13 W CFLs.
   b. Determine the energy use and cost savings due to daylighting harvesting strategies for both options with the existing incandescent lamps and with CFLs.
   Assume that without daylighting harvesting, each lamp is operated 4 h/day when daylighting can be harvested, 6 days/week, 50 weeks/year. The cost of electricity is 0.12/kWh.

6.8 Determine the energy and cost savings of daylighting harvesting to a thin two-story house building (each floor is 24 ft by 50 ft) with a window-to-floor area ratio of 30%. The building has clear windows (i.e., visible transmittance = 0.78) and is located in Denver, Colorado. The lighting system has a density of 0.5 W/ft² and operates 3,000 h per year. Assume that the cost of electricity is $0.10/kWh.

6.9 Determine if it is economically feasible to install no. 8 (AWG) copper conductor instead of no. 12 (AWG) on a 200 ft branch circuit that feeds a load of 12 amp. Assume:

- The load is used 14 h/day and 320 days/year.
- The cost of electricity is $0.12/kWh.
- The installed costs of no. 12 and no. 8 conductors are, respectively, $60.00 and $105.00 per 1,000 ft long cable.

7

Heating and Cooling Systems Retrofit

Abstract

In this chapter, an overview of commonly used heating and cooling systems for residential buildings is outlined. In addition, retrofit measures as well as control strategies to improve the energy efficiency of heating and cooling systems are provided. Moreover, calculation methods are provided to estimate energy savings due to ventilation controls, temperature settings, and energy-efficient heating and cooling system retrofits.

7.1 Introduction

The heating, ventilating, and air conditioning (HVAC) system maintains and controls temperature and humidity levels to provide an adequate indoor environment for people activity or for processing goods. In several countries, space heating and cooling represent a significant part of the total source and site energy consumed by the residential buildings, as noted in Chapter 1. For instance, space heating and cooling account for 26% of all electricity consumption and 69% of all natural gas consumption in U.S. residential buildings (EIA, 2009). It is therefore important that auditors recognize some of the characteristics of the heating and cooling systems and determine if any retrofits can be recommended to improve their energy efficiency.

There are several types of heating and cooling systems used in residential buildings. Table 7.1 provides a summary of the categories and types of various heating and cooling systems used in the United States. The systems can be used to provide heating only, cooling only, or both heating and cooling. Some systems use forced air to condition spaces, while others utilize water for heating and cooling buildings. Both forced-air and hydronic systems use mechanical devices powered by common energy sources (i.e., natural gas, electricity, or fuel). More innovative systems utilize ambient conditions to heat and cool buildings, such as natural ventilation or ground source heat pumps.

TABLE 7.1   Classification of Heating and Cooling Systems Used for Residential Buildings

Function	Heating Only	Cooling Only	Heating and Cooling
Forced-air systems	Furnaces with ducted air or duct-free air	DX air conditioning Evaporative cooling	Air heat pumps
Hydronic systems	Boilers and baseboard radiators Radiant floors	Radiant ceilings	Radiant walls
Passive/renewable systems	Direct gain systems Trombe walls Sunspaces	Earth air tunnels Natural ventilation	Ground source heat pumps
Others	Electric heaters Wood stoves	Absorption cooling	Thermoactivate foundations

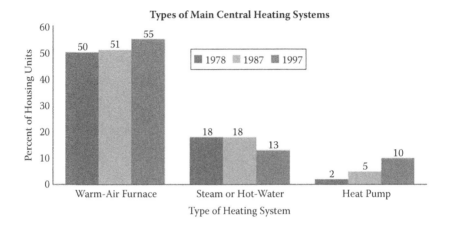

FIGURE 7.1   Penetration levels of heating systems used in U.S. households.

In the United States, most residential buildings are heated by forced-air systems, as indicated in Figure 7.1, which shows the level of penetration of various heating systems in U.S. households over the last three decades. Figure 7.2 presents the energy sources used to operate heating systems in U.S. residential buildings. Over 55% of U.S. households use natural gas for space heating. The use of electricity to provide space heating has been increasing steadily over the last three decades. As indicated by Table 7.1, heating systems include furnaces, boilers using baseboard radiators or radiant floors, electric heaters, air heat pumps, and ground source heat pumps. Table 7.2 lists the range of typical energy efficiencies as well as Energy Star ratings for selected heating systems. Three energy efficiency metrics are used for heating systems: annual fuel utilization efficiency (AFUE) for furnaces and boilers, heating seasonal performance factor (HSPF) for air heat pumps, and coefficient of performance (COP) for ground source heat pumps.

Space cooling is not required in several climate zones and thus is not common for several U.S. households. Indeed, mechanical cooling is mainly used in the southern regions and includes forced-air central systems or unitary window/wall units, as shown

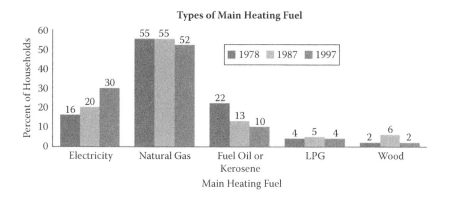

FIGURE 7.2 Common fuel types used for heating U.S. households.

TABLE 7.2 Typical and Energy Star Energy Efficiency Heating Systems

Heating System	Gas Furnace	Oil Furnace	Gas Boiler	Oil Boiler	Air Source Heat Pump	Ground Source Heat Pump
Efficiency metric	AFUE	AFUE	AFUE	AFUE	HSPF	COP
Available systems	78–96%	78–95%	80–99%	80–90%	7.7–10	2.5–3.2
Energy Star rating	90%	83%	85%	85%	8.2 (split systems) 8.0 (packaged units)	3.6 (open loop) 3.3 (closed loop) 3.5 (DX)

in Figure 7.3. In dry climates, evaporative cooling is often used to provide space cooling. Table 7.3 provides the range of typical energy efficiency metrics as well as the Energy Star ratings defined by energy efficiency ratio (EER) or seasonal energy efficiency ratio (SEER) of the cooling systems.

In the following sections, description of various heating and cooling systems are provided along with a basic analysis method and a series of energy efficiency measures suitable for retrofit projects associated with ventilating, heating, and cooling residential buildings.

7.2 Forced-Air Systems

Forced air is the most common method of heating and cooling spaces in the United States. Indeed, 62% of home space heating is done with central warm-air furnaces and 72% of space cooling is done with central air conditioning systems (EIA, 2009). In addition to increasing the energy efficiency of furnaces and air conditioners, minimizing energy consumption of distribution air systems is essential to reduce energy use for space heating and cooling in residential buildings.

Figure 7.4 illustrates the various components of a forced-air system with furnace and direct expansion (DX) cooling. A basic forced-air distribution system includes:

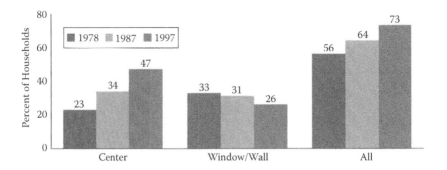

FIGURE 7.3　Penetration levels of A/C systems in U.S. households.

TABLE 7.3　Typical and Energy Star Energy Efficiency Metrics and Values for Cooling Systems

System	Central A/C	Air Source Heat Pump	Ground Source Heat Pump
Efficiency metric	SEER-EER	SEER or EER	EER
Range available	13–21 (SEER) 9–14 (EER)	13–17 (SEER) 9–13.5 (EER)	8.7–20.4 (EER)
Energy Star rating	14 (SEER) 11.5 (EER)	14 (SEER) 11.5 (EER)	3.6 (open loop) 3.3 (closed loop) 3.5 (DX)

- Damper to control the amount of air to be distributed to various spaces. Typically, forced-air systems have an outside air (OA) damper.
- Filter to clear the air from any dirt.
- Cooling coil to condition the supply air to meet the cooling load of the conditioned spaces.
- Fan to move the supply air through the distribution system.
- Humidifiers to add moisture to the supply air in case a humidity control is provided to the conditioned spaces.
- A distribution system (i.e., ducts) where the air is channeled to various locations and spaces.

In some systems, humidifiers and heat recovery devices are added. Humidifiers are used to add moisture to the supply air in case a humidity control is provided to the conditioned spaces. Heat recovery devices are used to preheat the outside air by extracting heat from the exhaust air.

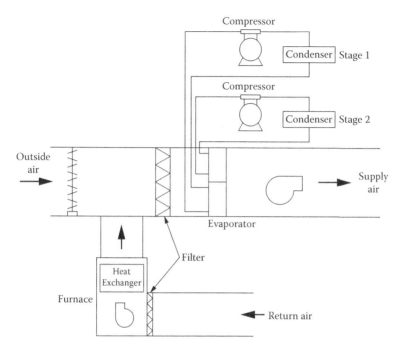

FIGURE 7.4 Typical components for a forced-air system.

In the following sections, selected energy conservation measures are described for various elements of forced-air systems. First, a basic overview of psychrometric analysis of air heating and cooling processes is presented.

7.2.1 Psychrometic Analysis

As indicated in Chapter 3, two principles are generally used to analyze various psychrometric processes, including conservation of mass and conservation of energy. The conservation of mass should be applied to both dry air and water vapor. For a generic process, the moist air can gain or lose heat, Q, and water, m_w, when flowing through a duct by going from state 1 to state 2 as illustrated in Figure 7.5.

By applying mass and energy conservation principles, it can be shown that

Mass balance of dry air: $$m_{a,1} = m_{a,2} \tag{7.1}$$

Mass balance of water vapor: $$m_{a,1}W_1 + m_w = m_{a,2}W_2 \tag{7.2}$$

Energy balance: $$m_{a,1}h_1 + m_w h_w + Q = m_{a,2}h_2 \tag{7.3}$$

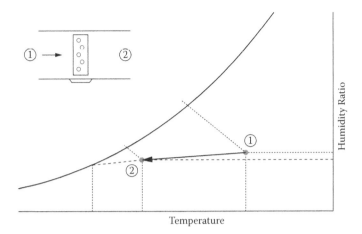

FIGURE 7.5   Cooling process in a psychrometric chart.

In case of a cooling and dehumidification process, moisture and heat have to be removed from the incoming air. The rate of moisture removal through water condensation can be estimated using Equation (7.2) with $m_a = m_{a,1} = m_{a,2}$ using Equation (7.1):

$$m_w = m_a(W_1 - W_2) \qquad (7.4)$$

The rate of heat removal through cooling of moist air using a cooling coil can be determined using Equations (7.3) and (7.4):

$$Q - m_a[(h_1 - h_2) - (W_1 - W_2)h_w] \qquad (7.5)$$

The total heat transfer rate, Q, of Equation (7.5) includes two parts: the sensible heat transfer, Q_s, that is required to reduce the dry bulb temperature from T_1 to T_2, and latent heat transfer, Q_l, which is needed to reduce the specific humidity ratio from W_1 to W_2. These heat transfer rates can be estimated using the psychrometric chart. First, the intersection point a, between the constant dry bulb temperature line ($T = T_1$) and the constant specific humidity ratio line ($W = W_1$) is found. Then, the enthalpy, h_a, for the intersection point is obtained. Finally, the sensible and latent heat transfer rates are calculated as follows:

$$Q_s = m_a(h_a - h_2) \qquad (7.6)$$

$$Q_l = m_a(h_1 - h_a) \qquad (7.7)$$

The ratio of the sensible heat transfer rate, Q_s, and the total heat transfer, $Q_s + Q_l$, is typically referred to as the sensible heat ratio (SHR):

$$SHR = \frac{Q_s}{Q_l + Q_s} \qquad (7.8)$$

A low value of SHR (i.e., SHR < 0.5) indicates that the latent heat transfer is dominant. Example 7.2 illustrates the calculation of SHR for a cooling and dehumidification process of moist air.

Example 7.1 illustrates how to analyze a cooling process and estimate the sensible heat ratio.

EXAMPLE 7.1

Moist air enters a cooling coil at 90°F and 90% relative humidity at a rate of 10,000 cfm. The air is cooled to 57°F and 60% relative humidity. The condensate leaves at 60°F. Find the total cooling coil heat transfer rate as well as the sensible heat ratio of the process.

Solution: Based on the psychrometric chart, we have the following properties for the air entering the cooling deck: h_1 = 41.13 Btu/lba, v_1 = 14.05 ft³/lba, and W_1 = 0.02 lbw/lba. From the psychrometric chart, the properties of the leaving air can be obtained: h_2 = 20.05 Btu/lba, v_2 = 13.15 ft³/lba, and W_2 = 0.006 lbw/lba.

From the steam tables, the enthalpy of liquid water at 60°F is h_w = 28.06 Btu/lbw.

Based on the water mass conservation equation (7.4), the rate of water condensation, m_w, can be determined:

$$m_w = m_a(W_1 - W_2) = \frac{\dot{V}_a}{v_1}(W_1 - W_2) = \frac{2000}{14.05}(0.02 - 0.006) = 1.993 \ lba / min = 119.6 \ lba / hr$$

The total heat transfer rate during the process, Q, is estimated using Equation (7.5):

$$Q = m_a(h_1 - h_2) - m_w h_w = \frac{(2000)(60)}{14.05}(41.13 - 20.05) - (119.6)(28.06) = 176,687 \ Btu / hr$$

In terms of tons of refrigeration, the total heat transfer rate is

$$Q = 14.72 \ tons$$

The sensible and latent heat transfer rates are obtained from Equations (7.6) and (7.7), respectively, using h_a = 26.0 Btu/lba (read from the psychrometric chart):

$$Q_s = m_a(h_a - h_2) = \frac{(2000)(60)}{14.05}(26.0 - 20.05) = 50,819 \ Btu / hr$$

$$Q_l = m_a(h_1 - h_a) = \frac{(2000)(60)}{14.05}(41.13 - 26.0) = 129,224 \ Btu/hr$$

Thus, the sensible heat ratio (SHR) can be calculated as indicated by Equation (7.9):

$$SHR = \frac{Q_s}{Q_s + Q_l} = \frac{50,819}{50,819 + 129,224} = 0.28$$

Thus, the latent load is dominant for this space.

7.2.2 Ventilation

Maintaining acceptable indoor air quality (IAQ) inside buildings is essential for health and indoor comfort. The IAQ level can be assessed by measuring the concentration of indoor pollutants. Several techniques are available to reduce indoor pollutants, including dilution with outdoor air and cleaning of air using filters. Ventilation (controlled) and infiltration (uncontrolled) can help dilute and reduce indoor pollutant concentration and thus improve the IAQ. It should be noted that ventilation and especially infiltration can increase the heating and cooling loads for buildings.

Ventilation airflow requirements in buildings are set by ASHRAE Standard 62, "Ventilation for Acceptable Indoor Air Quality." The typical residential requirement is 0.35 ACH (air changes per hour). That is, the ventilation airflow rate should be 35% of the house volume each hour. For leaky houses, uncontrolled air infiltration is sufficient to provide the required ventilation requirements. However, controlled ventilation may be required for airtight houses. Standard 62 permits the use of a pollutant indicator such as carbon dioxide (CO_2) to assess and control the ventilation rate in order to maintain an acceptable IAQ level within indoor spaces. Typically, an indoor CO_2 concentration of less than 900 ppm is desirable. A well-mixed space model can be used to assess the required ventilation based on the concentration of CO_2 within indoor spaces.

The concentration of a contaminant in a well-mixed space, C_i, is related to the flow and concentration, $\dot{V}_o$ and C_o, of the entering air and the rate of contaminant generation in the space, N_{pol}. Note that the concentration leaving a well-mixed space is assumed to be the room concentration.

$$C_i = C_o + \frac{N_{pol}}{\dot{V}_o} \tag{7.9}$$

If perfect mixing is not achieved within the building, a ventilation effectiveness, E_v, can be defined and estimated as follows:

$$E_v = \frac{1-S}{1-RS} \tag{7.10}$$

where S = fraction of supply air that does not reach conditioned (occupied) space, and R = fraction of recirculated return air.

Example 7.2 shows how the CO_2 level can be used to determine the required ventilation to maintain an acceptable IAQ level within a bedroom in a house.

EXAMPLE 7.2

In a home, the CO_2 concentration in a bedroom was 4,000 ppm. The bedroom size is 120 ft² and room height 8 ft. Find the air change per hour rate required for ventilation if two occupants were in the bedroom. The outdoor CO_2 level is 300 ppm and one person breathes out 0.6 ft³/h CO_2.

Solution: Using Equation (7.9), the ventilation requirement can be estimated:

$$\dot{V}_o = N_{pol}/(C_i - C_o)$$

$$\dot{V}_o = N_{pol}/(C_i - C_o) = (2 \times 0.6 \text{ ft}_{CO2}^3/\text{h})/$$
$$(4,000 \times 10^{-6} \text{ ft}_{CO2}^3/\text{ft}_{air}^3 - 300 \times 10^{-6} \text{ ft}_{CO2}^3/\text{ft}_{air}^3) = 324 \text{ ft}_{air}^3/\text{h}$$

Thus, the required ventilation expressed in air change per hour (ACH) can be determined:

$$\text{ACH} = \dot{V}_o/(\text{room volume}) = 324 \text{ ft}^3/\text{h}/(120 \text{ ft}^2 \times 8 \text{ ft}) = 0.34 \text{ ACH}$$

7.2.3 Air Filters

To remove dust and other unwanted particulates from air supplied to spaces, air filters are typically placed in the air distribution system. There are two main types of air filters commonly used in building HVAC systems:

- Dry type extended surface filters that consist of bats or blankets made up of fibrous materials such as bonded glass fiber, cellulose, wool felt, and synthetic materials
- Viscous impingement filters that include filter media coated with viscous substances (such as oil) to catch particles in the air stream

The efficiency of an air filter to remove particulates from air streams can be measured using standard testing procedures such as ASHRAE Standard 52.1. Table 7.4 outlines typical efficiencies and rated airflow velocities for commonly available air filters. The flat filters are the least efficient, the least expensive, and the most commonly used in residential buildings. The high-efficiency particulate air (HEPA) filters are the most efficient but are generally expensive and are appropriate for clean room applications. To extend the life of the more expensive filters, it is a common practice to add inexpensive prefilters that are placed upstream of the more expensive and effective filters in order to reduce dirt loading.

Air filters affect the pressure pattern across the ducts and can increase the electric power requirement for supply fans. Table 7.4 indicates typical pressure drops created by

TABLE 7.4   Typical Efficiencies and Pressure Drops for Air Filters

Filter Type	Average Efficiency	Rated Air Face Velocity (fpm)	Pressure Drop for Clean Filter (inches of water)	Pressure Drop for Dirty Filter (inches of water)
Flat	85%	500	0.10–0.20	1.00
Pleated	90%	500	0.15–0.40	1.00
Bag	90%	625	0.25–0.40	1.00
HEPA	99.9%	250–500	0.65–1.35	1.00

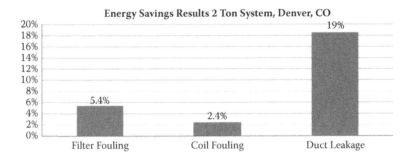

FIGURE 7.6   Typical energy savings associated with cleaning filters and coils and sealing ducts for a home in Denver, Colorado.

clean and dirty air filters. As expected, the pressure drop is higher for dirty air filters. Moreover, the pressure drop increases as the efficiency of the filter increases.

For a typical house in Denver, Colorado, cleaning the filter results in 5.4% energy cost savings, as outlined in Figure 7.6 (Wilson, 2011). Moreover, cleaning the evaporator coil and furnace results in 2.4% energy cost savings. However, reducing supply duct leakage from 12% to 0.1% results in 19% savings.

7.2.4 Duct Leakage

Ducts are generally used to distribute conditioned air to various spaces within buildings. While ducts do not consume energy, they can be a significant source of air leakage and energy waste, decreasing the overall performance of the HVAC system. Several field studies showed that up to 30% of energy input required for a typical HVAC system (including fan electrical power, heating or cooling energy) can be lost in the ducts (Wilson, 2011).

Air leakage in ducts can be measured using a pressurization test similar to the blower door test and tracer gas technique used to estimate air infiltration through the building envelope, as outlined in Chapter 5. Duct leakage can be estimated knowing the pressure in the ducts, P_{in}, and the pressure of the surrounding ducts, P_{out}, using the following equation:

$$\dot{V}_{leak} = C(P_{in} - P_{out})^n \tag{7.11}$$

where C and n are correlation coefficients that are determined based on the pressurization test. If the ducts have relatively large leaks (i.e., holes) in the ducts, the coefficient n is close to 0.5.

7.2.5 Fan Operation

In forced-air distribution systems, fans provide the pressure required to move air through ducts, heating or cooling coils, and filters present within the duct system. Fans generally fall into two categories, as illustrated in Figure 7.7. The centrifugal fan consists of a rotating wheel, generally referred to as an impeller, mounted in the center of a round housing. The impeller is driven by an electric motor through a belt drive. The vane-axial fan includes a cylindrical housing with the impeller mounted inside along the axis of the cylindrical housing. The impeller of an axial fan has blades mounted around a central hub similar to an airplane propeller. Typically, axial fans are more efficient than centrifugal fans but are more expensive since they are difficult to construct.

The performance of a fan depends on the power of the motor, the shape and orientation of the fan blades, and the inlet and outlet duct configurations.

To characterize the operation of a fan, several parameters need to be determined, including the electrical energy input required in kW (or HP), the maximum amount of air it can move in L/s (or cfm) for a total pressure differential (ΔP_T) or a static pressure differential (ΔP_s), and the fan efficiency. A simple relationship exists that allows the calculation of the electrical energy input required for a fan as a function of the amount of airflow, the pressure differential, and its fan efficiency. If the total pressure is used, Equation (7.12a) provides the electrical energy input using metric units.

$$\dot{V}_{leak} = C(P_{in} - P_{out})^n \, kW_{fan} = \frac{\dot{V}_f . \Delta P_T}{\eta_{f,t}} \tag{7.12a}$$

Using English units, the horsepower of the fan can be determined as follows:

$$Hp_{fan} = \frac{\dot{V}_f . \Delta P_T}{6,356 * \eta_{f,t}} \tag{7.12b}$$

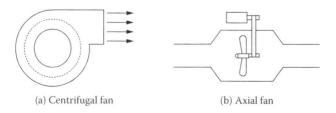

(a) Centrifugal fan   (b) Axial fan

FIGURE 7.7 Fan types commonly used in air distribution systems.

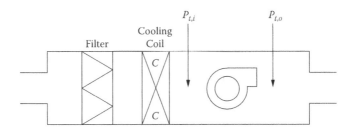

FIGURE 7.8   Measurement of **total** pressure of the fan.

If static pressure is considered instead of total pressure, Equations (7.13a) and (7.13b) should be used. Note that in this case, the static fan efficiency is needed to estimate the fan power rating.

For SI units:

$$kW_{fan} = \frac{\dot{V}_f . \Delta P_s}{\eta_{f,s}}$$ (7.13a)

For IP units:

$$HP_{fan} = \frac{\dot{V}_f . \Delta P_s}{6,356 * \eta_{f,s}}$$ (7.13b)

To measure the total pressure of the fan, a Pitot tube can be used in two locations within the duct that houses the fan, as illustrated in Figure 7.8.

As indicated in Figure 7.6, the total pressures $P_{t,i}$ and $P_{t,o}$ at, respectively, the inlet and outlet of the fan are first measured. Then, the fan total pressure is simply found by taking the difference:

$$\Delta P_T = P_{t,o} - P_{t,i}$$ (7.14)

When any of the fan parameters (such as total pressure ΔP_T, static pressure ΔP_S, or airflow rate $\dot{V}_f$) vary, they are still related through certain laws, often referred to as the fan laws. Fan laws can be very useful to estimate the energy savings from any change in fan operation. These fan laws are summarized in the following three equations:

Equation (7.15) states that the airflow rate, $\dot{V}_f$, is proportional to the fan speed ω:

$$\frac{\dot{V}_{f,1}}{\dot{V}_{f,2}} = \frac{\omega_1}{\omega_2}$$ (7.15)

Equation (7.16) indicates that the fan static pressure, ΔP_S, varies as the fan airflow rate $\dot{V}_f$, or the square of the fan speed ω:

$$\frac{\Delta P_{s,1}}{\Delta P_{s,2}} = \left(\frac{\dot{V}_{f,1}}{\dot{V}_{f,2}}\right)^2 = \left(\frac{\omega_1}{\omega_2}\right)^2$$ (7.16)

Finally Equation (7.17) states that the power used by the fan, kW_{fan}, varies as the cube of the fan airflow rate $\dot{V}_f$, or the square of the fan speed ω:

$$\frac{kW_{fan,1}}{kW_{fan,2}} = \left(\frac{\dot{V}_{f,1}}{\dot{V}_{f,2}}\right)^3 = \left(\frac{\omega_1}{\omega_2}\right)^3 \tag{7.17}$$

Generally, fans are rated based on standard air density. However, when the air density changes, the fan static pressure and the power required to operate the fan would change. Both fan static pressure and fan power vary in direct proportion to the change in air density.

It can be noted from the fan law equation (7.17) that by reducing the amount of air to be moved by the fan, the electrical energy input required is reduced significantly. For instance, a 50% reduction in air volume results in 87.5% reduction in fan energy use. This fact clearly indicates the advantage of using variable air volume fan systems compared to constant volume fans. It should be noted that variable air volume (VAV) systems are not typically considered for residential buildings since most houses are controlled with one thermostat, and thus are considered one thermal zone. However, there is a change in designing heating and cooling systems for residential buildings to include multiple thermal zones, and thus VAV systems and variable speed fans are being introduced to increase energy efficiency and indoor thermal comfort.

Figure 7.9 illustrates the variations of fan static pressure and fan power with flow rates for typical residential system capacities and blower wheel diameters.

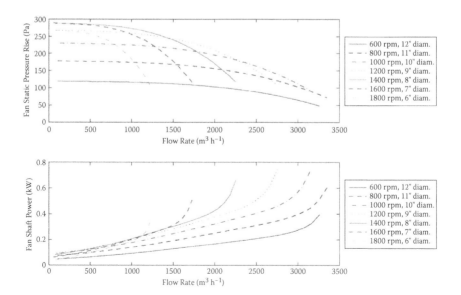

FIGURE 7.9 Fan static pressure and power curves for typical residential systems.

7.2.6 Furnaces

For forced-air systems, furnaces are used to heat air supplied to conditioned spaces through a duct system. Furnaces are basically heat exchangers with typical heat sources including natural gas- or oil-fired burners or electrical resistances. In conventional gas- or oil-fired furnaces, acid combustion gases are exhausted at relatively high temperatures ranging from 250°F (120°C) to 300°F (150°C) to prevent condensation, and thus corrosion of various furnace components. Condensing furnaces are made of stainless steel heat exchangers with extended surface areas to recover heat from the combustion gases, which is allowed to condense. The temperature of the exhausted combustion gases in condensing furnaces is only in the range of 100°F (38°C) to 150°F (65°C).

The seasonal thermal efficiency of furnaces is expressed in terms of the annual fuel utilization efficiency (AFUE). Conventional furnaces use natural gas or oil and have AFUE values varying between 78 and 85% AFUE. Condensing furnaces have AFUE values above 90% AFUE. Replacing an old furnace with a condensing furnace is a common retrofit measure for residential buildings. The energy savings, ΔEU, associated with a furnace retrofit can be estimated using Equation (7.18):

$$\Delta EU = \frac{AFUE_{new} - AFUE_{old}}{AFUE_{new}} . EU_{old} \tag{7.18}$$

where $AFUE_{old}$ and $AFUE_{new}$ are, respectively, the old and new efficiencies of the furnace, and EU_{old} is the energy consumption before retrofitting of the existing furnace.

Example 7.3 illustrates the potential energy savings and the cost-effectiveness associated with replacing a conventional furnace with a condensing furnace in an existing home.

EXAMPLE 7.3

In a home, a gas-fired furnace uses 1,500 therms/year and has an AFUE of 80%. The furnace is replaced by a condensing gas-fired furnace with an AFUE of 95%. Determine the energy use savings and the payback of the heating system retrofit of the existing furnace if the condensing furnace cost is $1,750 and the gas cost $1.5/therm.

Solution: Using Equation (7.18), the gas saved by replacing the existing furnace with the condensing furnace can be estimated:

$$\Delta EU = \frac{0.95 - 0.80}{0.95} . 1500 = 237 \ therms/year$$

Thus, the annual energy cost savings are

$$\Delta EC = 237 * 1.5 = \$355 / yr$$

If the cost of the condensing furnace is $1,750, the simple payback period is

$$SPP = \frac{\$1750}{\$355 / yr} = 5 \ years$$

7.2.7 DX Cooling Systems

Cooling of residential systems is typically achieved using direct expansion or DX air conditioning systems. In DX systems, air is directly cooled by exchanging heat with the refrigerant using the evaporator, as shown in Figure 7.10. Most DX systems, including room or central air conditioners and air-to-air heat pumps, operate as vapor compression cycles. A vapor compression cycle includes a compressor, a condenser, an expansion device, and an evaporator, as illustrated in Figure 7.10.

Note that in the vapor compression cycle of Figure 7.10, chilling (or heat extraction) occurs at the evaporator, while heat rejection is done by the condenser. Both the evaporator and the condenser are heat exchangers. At the evaporator, heat is extracted by the refrigerant from water that is circulated through cooling coils of an air handling unit. At the condenser, heat is extracted from the refrigerant and rejected to the ambient air (for air-cooled condensers) or water (for water-cooled condensers connected to cooling towers).

Generally, the energy efficiency of a cooling system is characterized by its coefficient of performance (COP). The COP is defined as the ratio of the heat extracted divided by the energy input required. In case of an electrically driven cooling system, as represented in Figure 7.10, the COP can be expressed as

$$COP = \frac{Q_{cool}}{W_{comp}}$$
(7.19)

Both Q_{cool} and W_{comp} should be expressed in the same unit (i.e., W or kW), so that the COP has no dimension.

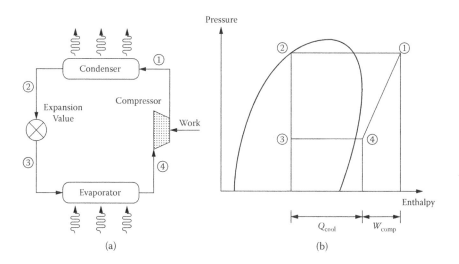

FIGURE 7.10 Typical vapor compression cycle (a) schematic and (b) P-h diagram.

Most manufacturers typically provide the COP of their cooling systems for full-load conditions. The capacity of cooling systems is expressed in kW and is defined in terms of the maximum amount of heat that can be extracted. In addition, the energy efficiency of the electrically powered cooling systems can be expressed in terms of the energy efficiency ratio (EER), which is defined as the ratio of the heat extracted (expressed in Btu/h) over the energy input required (expressed in W). Therefore, the relationship between the EER and the COP is given as follows:

$$EER = 3.413 * COP \qquad (7.20)$$

The definition of the EER provided above is specific to the U.S. HVAC industry. In Europe, the EER is set to be equal to the COP. However, the adopted European standard, EN 814 (Cenelec, 1997), specifies that the term *COP* is to be used for only the heating mode operation of heat pumps. Otherwise, the standard requires the use of the term *EER* to rate the energy efficiency of air conditioners and heat pumps. It should be noted that some manufacturers may use kW/ton as an alternative metric of energy efficiency for cooling systems. All the energy efficiency metrics are related, as indicated by Equation (7.21):

$$EER = 3.413 * COPkW \, / \, ton = {3.516}\big/{COP} = {12}\big/{EER} \qquad (7.21)$$

Energy efficiency of a cooling system varies under part-load conditions. Since cooling systems operate often under part-load conditions throughout the year, other energy efficiency coefficients have been proposed in an attempt to provide a better estimation of the energy performance of the cooling units over a wide range of operating conditions. Currently, the seasonal energy efficiency ratio (SEER) is used as a rating for cooling systems suitable for residential buildings.

To reduce the energy use of cooling systems, the energy efficiency of the equipment has to be improved under both full-load and part-load conditions. In general, the improvement of the energy efficiency of cooling systems can be achieved by one of the following measures:

- Replace the existing cooling systems by others that are more energy efficient.
- Improve the existing operating controls of the cooling systems.
- Use alternative cooling systems such as passive cooling or evaporative cooling systems.

The energy savings calculation from increased energy efficiency of cooling systems can be estimated using the simplified but general expression provided by Equation (7.22):

$$\Delta E_C = . \left(\frac{\dot{Q}_C . N_{h,C} . LF_C}{SEER} \right)_e - . \left(\frac{\dot{Q}_C . N_{h,C} . LF_C}{SEER} \right)_r \qquad (7.22)$$

where the indices *e* and *r* indicate the values of the parameters, respectively, before and after retrofitting the cooling unit; *SEER* is the seasonal efficiency ratio of the cooling unit (when available, the average seasonal COP can be used instead of the SEER); $\dot{Q}_C$ is the rated capacity of the cooling system; $N_{h,C}$ is the number of equivalent cooling full-load cooling hours; and LF_C is the rated load factor and is defined as the ratio of the peak cooling load experienced by the building over the rated capacity of the cooling equipment. This load factor compensates for oversizing of the cooling unit.

It should be noted that the units for both Q_C and SEER have to be consistent; that is, if SEER has no dimension (using the European definition of EER), Q_C has to be expressed in kW. However, if SEER is defined in terms of Btu/h/W (using the U.S. rating), Q_C has to be expressed in Btu/h, or as an alternative, SEER should be divided by 3.413 and Q_C expressed in kW.

When the only effect of the retrofit is improved energy efficiency of the cooling system so that only the SEER is changed due to the retrofit, the calculation of the energy savings can be performed using the following equation:

$$\Delta E_C = \dot{Q}_C . N_{h,C} . LF_C . \left(\frac{1}{SEER_e} - \frac{1}{SEER_r} \right) \tag{7.23}$$

In the following sections, some common energy efficiency measures applicable to cooling systems are described with some calculation examples to estimate energy use and cost savings.

Example 7.4 illustrates a sample of calculations to determine the cost-effectiveness of replacing an existing chiller with a high-energy efficiency chiller.

EXAMPLE 7.4

A home central air conditioner with a capacity of 5 tons and an average seasonal SEER of 10 is to be replaced by a new air conditioner with the same capacity but with a SEER of 17. Determine the simple payback period of the chiller replacement if the cost of electricity is $0.09/kWh and the cost differential of the new chiller is $1,000. Assume that the number of equivalent full-load hours for the air conditioner is 800 per year both before and after the replacement.

Solution: In this example, the energy use savings can be calculated using Equation (7.23) with $SEER_e = 10$, $_{SEERr} = 17$, $N_{h,C} = 800$, $Q_C = $,5 tons*3.516 kW/ton = 17.6 kW, and $LF_C = 1.0$ (it is assumed that the air conditioner is properly sized):

$$\Delta E_C = 17.6 \; kW * 800 \; hrs / yr * 1.0 * \left(\frac{3.413}{10} - \frac{3.413}{17} \right) = 1976 \; kWh / yr$$

Therefore, the simple payback period for investing in a high-efficiency air conditioner rather than a standard system can be estimated as follows:

$$SPP = \frac{\$1,000}{1976 \; kWh / yr * \$0.09 / kWh} = 5.6 \; years$$

A life cycle cost analysis may be required to determine if the investment in a high-energy-efficiency central air conditioner is really warranted.

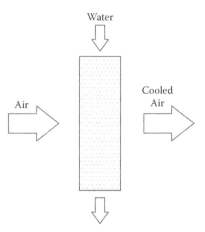

FIGURE 7.11   Basic direct evaporative cooler.

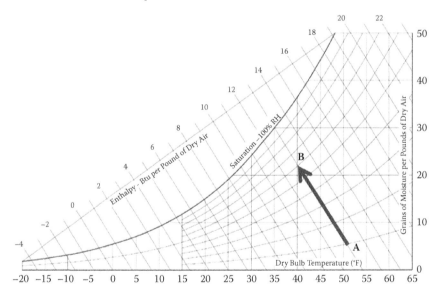

FIGURE 7.12   Direct cooling process in a psychrometric chart.

7.2.8 Evaporative Coolers

Evaporative coolers, also called swamp coolers, provide energy-efficient alternatives to cool houses in dry regions. There are several types of evaporative coolers, including:

Direct evaporative cooler: This type of cooler is often called a swamp cooler and uses a blower to force air through a permeable water-soaked pad, as illustrated in Figure 7.11. As the air passes through the pad, it is filtered, cooled, and humidified. Figure 7.12 illustrates an indirect cooling process in a psychro-metric chart.

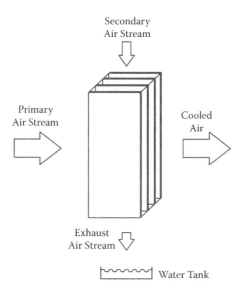

FIGURE 7.13 Basic indirect evaporative cooler.

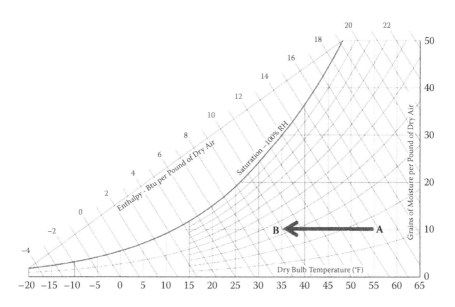

FIGURE 7.14 Indirect cooling process in a psychrometric chart.

Indirect evaporative cooler: A secondary heat exchanger is used to prevent humidity from being added to the air stream that enters the conditioned space, as shown in Figure 7.13. The psychrometric process of an indirect evaporative cooler is presented in Figure 7.14.

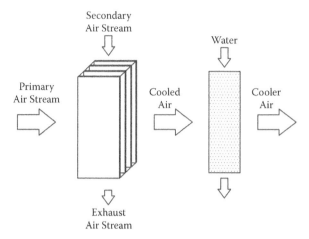

FIGURE 7.15   Basic indirect/direct evaporative cooler.

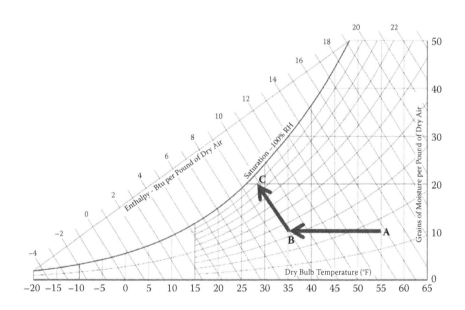

FIGURE 7.16   Indirect/direct cooling process in a psychrometric chart.

Indirect/direct evaporative cooler: Combines both the indirect and direct evaporative cooling. In addition to the water-soaked pad, a secondary heat exchanger is used to prevent humidity from being added to the air stream that enters the conditioned space, as shown in Figure 7.15. The psychrometric process of an indirect evaporative cooler is presented in Figure 7.16.

TABLE 7.5 Wet Bulb Efficiency for Different Types of Evaporative Coolers

Evaporative Cooler Type	$T_{db,in}$	$T_{wb,in}$	$T_{db,out}$	ε_{wb}
Direct	110°F (43.3°C)	72°F (22.2°C)	78°F (25.5°C)	84%
Indirect	100°F (37.8°C)	65°F (18.3°C)	78°F (25.5°C)	63%
Indirect/direct	113°F (45°C)	70°F (21.1°C)	78°F (25.5°C)	81%

Ultracoolers: These coolers are able to cool air below the thermodynamic wet bulb temperature associated with the dry bulb temperature of the outside ambient air. Ultracoolers typically use both direct and indirect processes operating in parallel and in stages to achieve cooler air than a direct or indirect system alone could achieve.

To assess the effectiveness of the evaporative cooling systems, wet bulb efficiency is used to determine how effectively the system cooled the air relative to its wet bulb temperature:

$$\varepsilon_{wb} = \frac{T_{db,in} - T_{db,out}}{T_{db,in} - T_{wb,in}} \tag{7.24}$$

Table 7.5 illustrates the wet bulb efficiency for various evaporative cooling systems, including direct, indirect, and indirect/direct to deliver the same dry bulb temperature to the conditioned space.

It should be noted that evaporative coolers can be used as a stand-alone cooling system in a home or in combination with a refrigeration-based air conditioning system. However, conventional air conditioners should not be operated simultaneously with direct evaporative coolers, because air conditioners dehumidify while evaporative coolers humidify the indoor air.

7.2.9 Other Air Forced Systems

In addition to the systems presented in the previous sections, here are several other types of systems to provide heating and cooling to residential buildings using forced air. Some of these systems are described briefly.

7.2.9.1 Air-to-Air Heat Pumps

Heat pumps can be used for both cooling and heating by simply reversing the refrigeration flow through the unit. The heat sink (or source) for the heat pump can be either air, water, or ground. For commercial and industrial applications, air-to-air heat pumps can have capacities up to 90 kW (or 25 tons), and hydronic heat pumps can have higher cooling capacities. Ground-coupled heat pumps are still small and are mostly suitable for residential applications.

7.2.9.2 Energy Recovery Ventilators

Energy recovery ventilators (ERVs), also known as heat recovery ventilators (HRVs), are used to reclaim energy from exhaust airflows while providing fresh air to the conditioned spaces. HRVs use heat exchangers to heat or cool incoming fresh air to capture up to 80% of the energy that would be otherwise lost from a conditioned exhaust air stream.

7.2.9.3 Desiccant Cooling Systems

Desiccant materials, which absorb moisture, can be dried or regenerated, by adding heat supplied by natural gas, waste heat, or solar radiation. In most systems, a wheel that contains a desiccant turns slowly to extract moisture from incoming air and discharge it to the outdoors. A desiccant system can be combined with a conventional air conditioning system in which the desiccant removes humidity and the air conditioner reduces the air temperature supplied to the conditioned spaces.

7.2.9.4 Ductless Mini-Splits

Mini-split systems can be suitable retrofit options for homes using hydronic heating. Similar to the conventional central DX air conditioners, mini-splits use an outside compressor/condenser and air distribution units. The main difference is that each room or conditioned zone has its dedicated air distribution unit that is connected to the outdoor unit via a conduit carrying the power, refrigerant, and condensate lines.

The main advantage is that by providing dedicated units to each space, it is easier to meet indoor comfort needs of different rooms. By avoiding the use of ductwork, mini-splits also reduce energy losses associated with central forced-air systems. However, mini-split systems can cost 30% more than typical central air conditioners of similar capacities. But, when considering the installation costs and energy savings associated with installing new central ductwork, ductless mini-split systems may be cost-effective.

7.3 Hydronic Systems

Hydronic systems typically utilize water to heat or cool buildings. These systems, while not common in the United States, are popular in several countries, especially Europe and some parts of Asia. Hydronic systems require a piping distribution system in addition to heat sources, such as boilers. To deliver or extract heat from spaces, baseboard radiators, heating/cooling coils, radiant floors, radiant walls, and ceilings are commonly used in hydronic systems. Humidifiers and heat recovery devices are also considered for some systems. Humidifiers are used to add moisture to the supply air in case a humidity control is provided to the conditioned spaces. Heat recovery devices are used to preheat the outside air by extracting heat from the exhaust air. In this section, the energy performance of two hydronic system components is discussed, including hot water boilers and radiant floor heating panels.

7.3.1 Hot Water Boilers

Boilers are used to generate hot water that is circulated through baseboard radiators, fan coils, or radiant floors to heat residential buildings. Boilers have several parts, including an insulated jacket, a burner, a mechanical draft system, tubes and chambers for combustion gas, tubes and chambers for water or steam circulation, and controls (AHSRAE, 2009). There are several factors that influence the energy performance of boilers, including fuel characteristics, firing method, steam pressure, and heating capacity. The main challenge to ensure optimal operating conditions for boilers is to provide the proper excess air for the fuel combustion. It is generally agreed that 10% excess air provides the optimum air-to-fuel ratio for complete combustion. Too much excess air causes higher stack losses and requires more fuel to heat ambient air to stack temperatures. On the other hand, if insufficient air is supplied, incomplete combustion occurs and the flame temperature is reduced.

The general definition of overall boiler thermal efficiency is the ratio of the heat output, E_{out}, over the heat input, E_{in}:

$$\eta_b = \frac{E_{out}}{E_{in}} \tag{7.25}$$

The overall efficiency accounts for combustion efficiency, the stack heat loss, and the heat losses from the outside surfaces of the boiler. The combustion efficiency refers to the effectiveness of the burner in providing the optimum fuel/air ratio for complete fuel combustion. To determine the overall boiler thermal efficiency, some measurements are required. The most common test used for boilers is the flue gas analysis using an Orsat apparatus to determine the percentage by volume of the amount of CO_2, CO, O_2, and N_2 in the combustion gas leaving the stack. Based on the flue gas composition and temperature, some adjustments can be made to tune up the boiler and determine the best air-to-fuel ratio in order to improve the boiler efficiency.

Monographs are available to determine the overall boiler efficiency based on measurement of flue gas composition and temperature. One of these monographs applies to both gas-fired and oil-fired boilers and is reproduced in Figure 7.17 (for IP units) and Figure 7.18 (for SI units), and Example 7.5 illustrates how the monograph can be used to determine the boiler efficiency.

EXAMPLE 7.5

A flue gas analysis of a gas-fired boiler indicates that the CO_2 content is 8% with a gas flue temperature of 204°C (400°F). Determine the overall thermal efficiency of the boiler.

Solution: By reading the monograph of Figure 7.17, the combustion occurs with excess air of 48% and excess O_2 of 7%. The overall boiler thermal efficiency is about 81%.

Similar to furnaces, the long-term performance of boilers is expressed in terms of annual fuel utilization efficiency (AFUE) and accounts for part-load operation. There are several measures by which the boiler efficiency of an existing heating plant can be improved. Among these measures are:

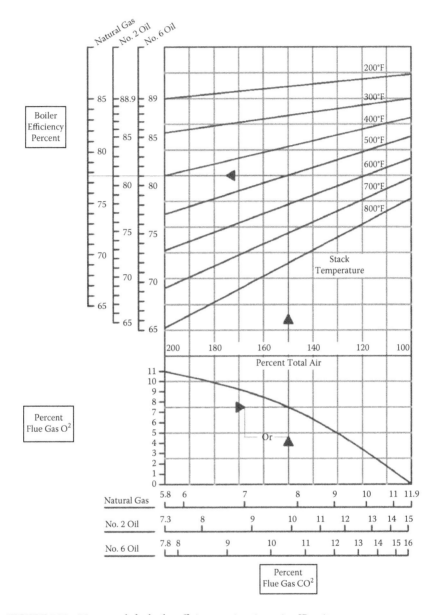

FIGURE 7.17   Monograph for boiler efficiency estimation using IP units.

- Tune-up the existing boiler.
- Replace the existing boiler with a high-efficiency boiler.
- Use modular boilers to heat apartment buildings.

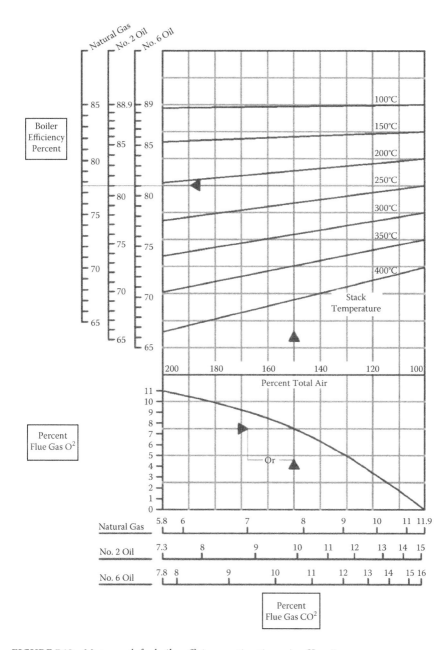

FIGURE 7.18 Monograph for boiler efficiency estimation using SI units.

The net effect of all these measures is some savings in the fuel use by the heating plant. To calculate the savings in fuel use, ΔFU related to the change in the boiler efficiency, the following equation can be used:

$$\Delta FU = \frac{AFUE_{eff} - AFUE_{std}}{AFUE_{eff}} . FU_{std} \tag{7.26}$$

where $AFUE_{std}, AFUE_{eff}$ are, respectively, the old and new seasonal thermal efficiencies of the boiler, and FU_{std} is the fuel consumption before any retrofit of the boiler system.

It is therefore important to obtain both the old and new overall thermal efficiencies of the boiler to estimate the energy savings. The following sections provide a more detailed description of the various boiler improvement measures.

By analyzing the flue gas composition and temperature, the boiler thermal efficiency can be estimated using the monograph provided in Figures 7.17 and 7.18. If it was found that the efficiency is low due to inappropriate excess air, the air-fuel mix can be adjusted and the boiler efficiency improved. To perform this adjustment, some instrumentation is needed (gas flue analyzer and a temperature measurement device). Example 7.6 illustrates how the cost-effectiveness of a boiler retrofit can be evaluated.

EXAMPLE 7.6

An existing boiler with an average AFUE of 75% uses 2,500 therms of natural gas per year. If a condensing boiler with AFUE of 95% is used, determine the annual natural gas use savings. Estimate the payback of installing the condensing boiler if it costs $3,500 and the natural gas cost is $1.75/therm.

Solution: Using Equation (7.26), the annual energy savings in natural gas due to the installation of a condensing boiler can be estimated:

$$\Delta FU = \frac{0.95 - 0.75}{0.95} . 2500 = 526 \; therms \,/\, yr$$

Therefore, the simple payback period (SPP) for replacing the existing boiler with a condensing boiler is estimated to be

$$SPP = \frac{\$3,500}{526 \; therms\,/\,yr * \$01.75\,/\,therm} . \approx 3.8 \, years$$

Other measures that can be considered to increase the overall efficiency of boilers include:

- Install turbulators in the fire tubes to create more turbulence and thus increase the heat transfer between the hot combustion gas and the water. The improvement in boiler efficiency can be determined by measuring the stack flue gas temperature. The stack gas temperature should decrease when the turbulators are installed. As a rule of thumb, a 2.5% increase in the boiler efficiency is expected for each 50°C (90°F) decrease in the stack flue gas temperature.

- Insulate the jacket of the boiler to reduce heat losses. The improvement of the boiler efficiency depends on the surface temperature.
- Install soot blowers to remove boiler tube deposits that reduce heat transfer between the hot combustion gas and the water. The improvement in the boiler efficiency depends on the flue gas temperature.
- Use economizers to transfer energy from stack flue gases to incoming feedwater. As a rule of thumb, a 1% increase in the boiler efficiency is expected for each 5°C (9°F) increase in the feedwater temperature.
- Use air preheaters to transfer energy from stack flue gases to combustion air.

The stack flue gas heat recovery equipment (i.e., air preheaters and economizers) is typically the most cost-effective auxiliary equipment that can be added to improve the overall thermal efficiency of the boiler system.

7.3.2 Radiant Heating Systems

Radiant floor heating systems typically consist of embedded hot water coils in floor slabs of residential and commercial buildings to provide space heating, as illustrated in Figure 7.19. Recently, radiant systems have received renewed interest in the United States due to their inherent advantages compared to conventional all-air heating systems, including low-noise, potential energy savings, uniform temperature distribution within spaces, and superior thermal comfort (ASHRAE, 2008). Experimental studies have demonstrated that radiant heating can achieve 33 to 52% savings when compared to air-to-air heat pump and electric baseboard heating systems, respectively, in residential houses (CADDET, 1999).

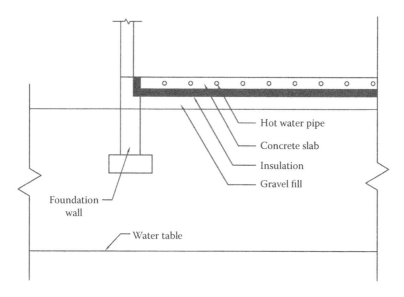

FIGURE 7.19 Cross section of radiant heating floor.

Thermal mass of a heated floor acting as heat storage medium absorbs heat from the embedded heating sources. Heat is then conducted from the heating sources to the panel surfaces. By radiation, the surfaces directly heat objects without heat-transferring media (such as air). However, the room air will be also heated to some degree due to convection heat transfer from the panel surfaces. Therefore, radiant heating systems may require lower energy use due to lower room set-point temperature when compared to conventional forced-air systems. Similar indoor comfort levels can be accomplished with 3 to 4°C (5.5 to 7°F) lower space temperatures. Additionally, infiltration as a part of heating loads is reduced for radiant floor systems due to relatively smaller temperature difference between indoor and outdoor air temperature than air heating systems. Radiant heating systems also provide better thermal comfort to people since the warm panels reduce heat losses through radiation from the human body.

The ASHRAE *Handbook of HVAC Systems and Equipment* (2008) provides useful information to analyze, design, and control radiant floor systems. Similarly, the *Radiant Heating and Cooling Handbook* (Watson and Chapman, 2002) provides a detailed overview of radiant heating systems, including radiant heat transfer, building thermal comfort, design guidelines, and typical control strategies. As an example, the design guidelines of a residential building include the following recommendations:

- Room temperature: 65–72°F (18–22°C)
- Supplied hot water temperature: 95–140°F (35–60°C)
- Floor surface temperature: 75–85°F (24–30°C)
- Drop in water temperature: 15–20°F (8–11°C)
- Maximum length of loop: 200 ft (60 m) in 3/8 in. tube size, 300 ft (90 m) in 1/2 in. tube size
- Tube size: 3/8 in.
- Tube spacing: 4–9 in. (10–23 cm)

In some climates, it may be beneficial to add insulation beneath the heated slab to reduce the heat losses to the ground. Figures 7.20 and 7.21 show typical wintertime temperature profiles for uninsulated and uniformly insulated slab within the ground medium and within the radiant heating slab, where hot water pipes are embedded. Figure 7.20 indicates that lower surface temperature of uninsulated slab is gradually decreased from the center (34°C or 93°F) to the perimeter (2°C or 35.6°F). However, an abrupt change in temperature gradient occurs at the slab edges due to the significant heat transfer between the ground medium and the building foundation walls. At the slab center, heat is lost through a one-dimensional flow path to the water table. Figure 7.21 presents the effect of uniformly adding insulation along the horizontal surface between the slab and the ground. The slab edges still lose a significant amount of heat during wintertime. However, uniformly insulated slab results in a significant reduction of soil temperature changes beneath the slab. The temperature profiles of the ground medium 4 m away from building remain unaffected by the insulation placement.

Controlling the performance of radiant systems can be challenging due to the high thermal mass of both the floor and ground medium. Control strategies for radiant floor heating systems use typically temperature or heat flux modulation techniques. Temperature modulation control sets the supply water temperature to be proportional

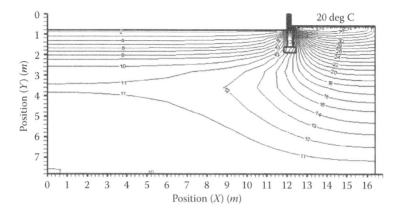

FIGURE 7.20 Soil temperature isotherms for uninsulated slab with embedded hot water pipes during wintertime (Denver, Colorado, midnight).

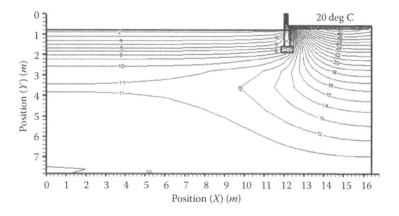

FIGURE 7.21 Soil temperature isotherms for uniformly insulated slab with embedded hot water pipes during wintertime (Denver, Colorado, midnight).

to either outdoor temperature or the difference between a desired set point and room air temperatures. Flux modulation control attempts to ensure that a slab delivers heat to the space at a rate per unit area proportional to the difference between the room air temperature and the slab surface temperature. The heat flux provided to the slab is proportional to the difference between the supply and return water temperatures.

In addition to the mean air temperature (MAT), mean radiant temperature (MRT) and operative temperature (OT) are used for control radiant heating systems. Figure 7.22 illustrates the impact of temperature settings on total heating energy use and the number of hours when thermal comfort levels exceed the acceptable range (i.e., −0.5 £ PMV £ +0.5) for a typical house located in Denver, Colorado (Ihm and Krarti, 2005). As shown in Figure 7.20, total heating energy use is generally proportional to the increasing setpoint temperature for all three control strategies. Depending on the control strategy,

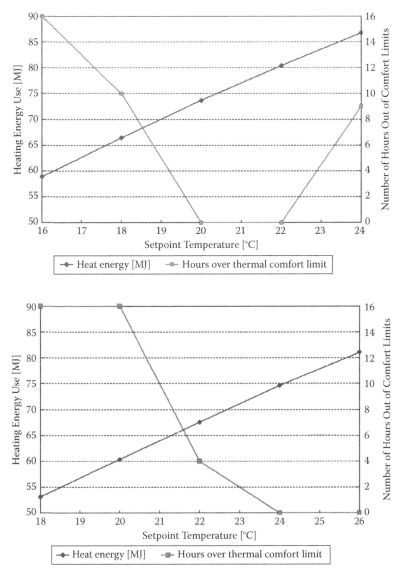

FIGURE 7.22 Impact of temperature set point on energy use of radiant heating system for a house in Denver, Colorado.

lower set-point temperature can save some heating energy use at the expense of a deterioration in the indoor thermal comfort. To further evaluate the performance of control strategies, the set-point temperatures are selected based on both achieving low energy use and satisfying building thermal comfort. Typical set-point temperatures of 20°C (68°F) for MAT-based control, 24°C (75.2°F) for MRT-based control, and 22°C (71.6°F) for OT-based control are recommended to operate radiant heating systems.

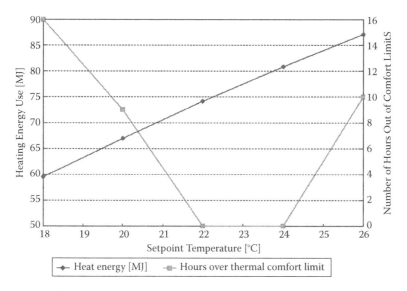

FIGURE 7.22 *(Continued)*

To better operate radiant heating systems, optimal control schemes have been proposed. For instance, Simmonds (1993) used optimal controls to demonstrate that additional energy savings could be achieved if the control is based on maintaining the Fanger's predicted mean vote (PMV) index rather than mean air temperature within a thermal comfort range. A PMV value within ±0.5 is acceptable for thermal comfort according to the ASHRAE comfort standard (2004). Ihm and Krarti (2005) have evaluated optimal control strategies to operate radiant floor heating systems in three locations: Denver, Colorado, Minneapolis, Minnesota, and Albany, New York. They found that compared to the conventional controls using fixed mean air temperature or MAT set point to maintain thermal comfort during occupied periods, the optimal control can reduce the total building energy use by 26.9% for Denver, 11.3% for Minneapolis, and 22.2% for Albany. These results indicated that optimal controls have the potential to save significant energy use relative to conventional MAT-based controls for radiant heating systems, especially for milder climates.

7.3.3 Radiant Cooling Systems

Radiant cooling systems utilize chilled water pipes to distribute cooling energy to various conditioned spaces as alternatives to conventional DX systems that rely on cooled air supplied through ducts. Radiant cooling systems provide direct cooling of occupants by radiative heat transfer from typically cooling ceiling surfaces maintained at temperatures of about 65°F (18.3°C). Heat generated by occupants is absorbed by the radiant cooling surfaces.

There are several systems to provide chilled water in radiant cooling surfaces. Panel systems are the most common of these systems. Aluminum panels that carry tubing can be surface mounted or embedded on floors, walls, or ceilings. Capillary tube systems are also used to provide chilled water through mats of small, closely spaced tubes that

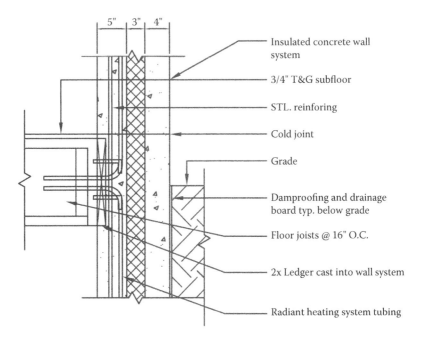

FIGURE 7.23   Cross section of a radiative wall.

are embedded in plastic, gypsum, or plaster on walls and ceilings. Moreover, similar to hydronic radiant heating systems described earlier, concrete layers with embedded tubes can provide the conduit and thermal storage capacity for cooling systems. Figure 7.23 illustrates construction details for a radiative wall that can be used for both cooling and heating homes. In most radiant cooling systems, water is mixed with glycol and cooled by an air-to-water heat pump, a cooling tower, a ground source heat pump, or even well water.

7.4 Ground Source Heat Pumps

In the last decade, the popularity of ground source heat pumps has increased as energy-efficient alternatives to both forced-air and hydronic heating and cooling systems. They typically can meet both heating and cooling needs for homes without significant auxiliary systems. Their main disadvantage is their high installation costs associated with excavation requirements. To reduce this initial cost, ground source heat pumps are being combined and installed within the building foundations. These systems, referred to as thermoactive foundations, are described in this section. First, the conventional ground source heat pumps are presented.

7.4.1 Conventional Ground Source Heat Pumps

Ground source heat pumps (GSHPs) are systems that use geothermal energy for heating and cooling buildings. These systems utilize the ground as a heat source in heating and a heat sink in cooling mode operation. GSHPs can be more energy efficient than the

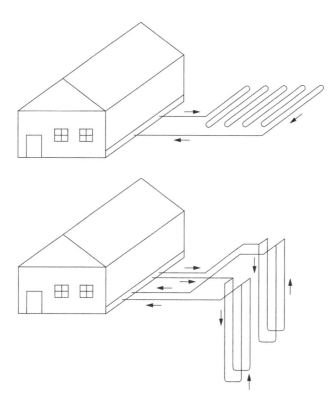

FIGURE 7.24 Typical configurations for ground source heat pump systems.

conventional forced-air or hydronic systems due to higher coefficients of performance and associated improvements in system life cycle and operating costs, reduced greenhouse gas emissions, and improvements in building thermal load profiles (ASHRAE, 2007). A typical GSHP system uses only electricity to power a pump and can be as much as 300 to 400% more energy efficient than a highly efficient furnace. Interest in GSHP is increasing throughout the world with an estimated 550,000 units installed worldwide and with about 66,000 units installed annually, mostly in Europe and North America. About 80% of the units installed worldwide are for residential buildings. With higher energy prices, the upward trend in market growth is expected to remain steady in the United States, Europe, and Asia (EIA, 2008).

GSHPs use either vertical or horizontal (or slinky) loops, depending on the land availability, as illustrated in Figure 7.24. Vertical GSHPs are typically more energy efficient but are more expensive to install. In the United States, the average temperature of the ground below a depth of 5 ft (1.4 m) is on average approximately 50°F (10°C) to 55°F (12.8°C) year-round (Omer, 2006). Conventional borehole heat exchange loops, typically consisting of 5 cm diameter polyvinyl chloride (PVC) tubing, are usually installed to depths up to 60 m below the ground surface (McCartney et al., 2009). The total piping requirements range from 60 to 180 m per cooling ton (approximately 11.5 kW), depending on local soil types, groundwater levels, and temperature profiles (Omer, 2008). In

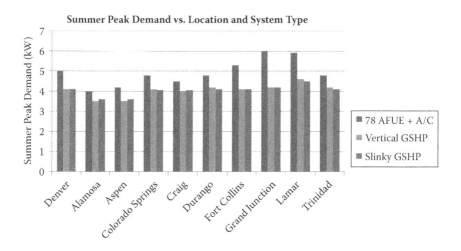

FIGURE 7.25 Summer peak electric demand for the furnace/AC, vertical well GSHP, and slinky GSH systems.

conventional GSHP systems, the heat exchanger tubes are usually filled with a mixture of water and propylene glycol, which permits heat to be extracted from the ground even under subfreezing conditions. A pump is used to cycle the liquid through the closed loop in the ground, while heat is exchanged by a coil in the building heat pump system. The required flow rate through the primary heat exchanger is typically between 1.5 and 3.0 gallons per minute per system cooling ton (0.027 and 0.054 L/s-kW) (Omer 2008).

If properly designed and operated, GSHPs can reduce peak summer electric demand, energy use and cost, and CO_2 emissions. A recent study by Studer and Krarti (2010) has shown that GSHP reduces electricity peak demand, energy use, and CO_2 emissions by 10–30% in a typical Colorado home compared to the conventional heating and cooling system. Figure 7.25 illustrates the peak electrical demand for vertical and horizontal GSHPs when compared to the AC/furnace system commonly utilized in Colorado.

Even with its documented benefits, building designers and owners remain reluctant in installing GSHPs primarily due to the significant initial costs associated with soil excavation and drilling to install the heat exchange loops. As alternatives to GSHPs, thermoactive foundations (also referred to as geothermal foundations, foundation piles, or energy piles), due to their dual purpose as a heat exchanger and a supporting structure for the building, provide a significant cost reduction potential compared to the conventional GSHPs. Ooka et al. (2007) and Sekine et al. (2007) found that a thermoactive foundation system installed for a research facility was approximately 75% less expensive than an equivalent borehole GSHP system. A brief discussion of thermoactive foundations is provided in the following section.

7.4.2 Thermoactive Foundations

Thermoactive foundations offer a distinct advantage over conventional borehole systems since they require less land availability, which can be a major issue for commercial buildings, especially in urban areas. Since concrete has a higher thermal conductivity than soil, thermoactive foundation systems are typically more energy efficient than the conventional GSHPs (Brandl, 2006).

Three main categories of thermoactive foundations have been installed and evaluated mostly in Europe and Asia: (1) cast-in-place concrete pile, (2) precasting concrete pile with a hole in the center to place the heat exchange loops, and (3) steel foundation pile with a blade on the tip of the pile, as depicted in Figure 7.26. Few studies to evaluate the

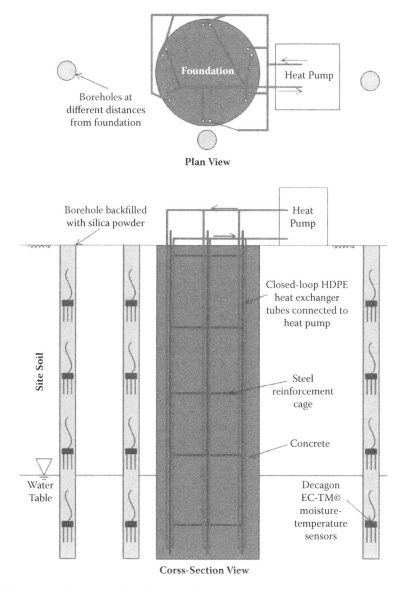

FIGURE 7.26 Thermoactive foundation configured as ground source heat pump.

energy performance of thermoactive foundations have been published. All the evaluated case studies have been mostly in Europe and Japan. No investigation of thermoactive foundations used to heat and cool a building has been found in the literature. Hamada et al. (2007) have carried out a long-term evaluation of a thermoactive foundation installed for a small house in Japan. The space heating operation measurements indicated that the seasonal average temperatures of brine returning from the underground and pile surfaces were 2.4 and 6.8°C, respectively. The average coefficient of performance for space heating was quite high at 3.9, and the seasonal primary energy reduction rate compared with a typical air conditioning system is found to be 23.2%. Brandl (2006) reported that the investment payback period for thermoactive foundations varies between 2 and 10 years, depending on energy prices, building characteristics, ground properties, and climatic conditions. Compared to a conventional system utilizing a gas-fired boiler for heating and an electrical chiller for cooling, Desmedt and Hoes (2007) reported lower energy use but higher initial construction costs for a thermoactive foundation installed for a hospital due to conservative construction regulations in Belgium, but still found that the foundation system would require a payback period of about 10 years.

Unfortunately, no long-term performance data of thermoactive foundations are available to assess the sustainability of these systems' associated with thermomechanical stresses, such as those due to soil freezing-thawing cycles. Indeed, an important issue that may affect soil-foundation interaction in thermoactive foundation elements is the potential of soil freezing during periods of high heating demand (McCartney et al., 2009). Freezing may lead to formation of ice lenses and foundation heave, as well as subsequent settlement after thawing occurs. Moreover, there are no clear design guidelines to properly size thermoactive foundation systems that account for parameters such as soil type and concrete mixture, ground loop conductor type, and water table depth. In addition, measured data of thermal performance of thermoactive foundation systems are limited and are typically not well described.

7.5 Indoor Temperature Controls

The indoor temperature settings during both heating and cooling seasons have significant impacts on the thermal comfort within occupied spaces and on the energy use of the heating and cooling systems. It is therefore important for the auditor to assess the existing indoor air temperature controls within the building to evaluate the potential for reducing energy use or improving indoor thermal comfort without any substantial initial investment. There are four options for adjustments of the indoor temperature setting that can save heating and cooling energy:

1. Eliminating overcooling by increasing the cooling set point during the summer.
2. Eliminating overheating by reducing the heating set point during the winter.
3. Preventing simultaneous heating and cooling operations by separating heating and cooling set points.
4. Reducing heating or cooling requirements during unoccupied hours by setting back the set-point temperature during heating and setting up the set-point temperature (or letting the indoor temperature float) during cooling.

The calculations of the energy use savings for these measures can be estimated based on degree-day methods, as outlined in Chapter 5. Example 7.7 is provided to illustrate the energy use savings due to adjustments in indoor temperature settings.

EXAMPLE 7.7

A home located in Boulder, Colorado, is heated 24 h/day with a set point of 70°F. Determine the annual energy savings associated with setting back the temperature to 60°F between 8:00 a.m. and 5:00 p.m. when the home is unoccupied.

Solution: Using Equation (5.17), the reduction factor f_{DH} in heating degree-days can be estimated for the temperature setback of $\Delta T_{setback} = 10°F$ and the number of hours for the temperature setback, $h_{setback} = 9$ hours. Indeed, for Boulder, Colorado, the correlation coefficients for Equation (5.17) are provided by Table 5.2: $a = 0.2356$, $b = 0.8198$, $m = 0.031$, and $n = 0.010$. Thus, f_{DH} can be calculated:

$$f_{DH} = 0.2356 * \exp(-0.031*10) + 0.8198 * \exp(-0.01*9) = 0.915$$

Therefore, the annual energy savings in natural gas due to the temperature setback is 8.5%.

It should be noted that some of the above-listed measures could actually increase the energy use if they are not adequately implemented. For instance, when the indoor temperature is set lower during the winter, the interior spaces may require more energy since they need to be cooled rather than heated. Similarly, the setting of the indoor temperature higher can lead to an increase in the reheat energy use for the zones with reheat systems.

Instead of hardwired thermostats, wireless thermostats are currently available for residential buildings. Indeed, wireless thermostats with associated receivers provide additional flexibility to control indoor temperature not only in one room but several rooms. Most existing wireless thermostats have one or more remote controllers. The receivers can be wall mounted and connected to the heating or cooling equipment. Multiple thermostats or receivers can be added to allow control of a single heating/cooling system from multiple locations or rooms. Similarly, multiple heating/cooling systems, as the case with several room window units or baseboard heating units, can be controlled from the same wireless thermostat.

7.6 Comparative Analysis of Heating and Cooling Systems

In this section, a comparative analysis of several alternative systems to provide heating and cooling for a residential building is presented. Both passive and active heating and cooling options are considered. The analysis is based on detailed simulation modeling analysis for an apartment building located in Denver, Colorado (Seo and Krarti, 2010),

FIGURE 7.27   Model rendering of apartment building in Denver, Colorado.

as depicted in Figure 7.27. The systems are evaluated to determine their performance in terms of energy efficiency and thermal comfort. A brief description of the modeling approach for each heating and cooling system is provided in the following sections:

System 1: No HVAC system. The building has no heating or cooling system. Energy is used only for systems such as lighting, appliances, and DHW. This case was considered as a reference to assess the level of thermal discomfort within the building if no heating and cooling systems are installed.

System 2: Natural ventilation. With this option, the building is cooled using natural ventilation. No mechanical system is used for heating or cooling. For this building, natural ventilation is considered by opening windows only when outside air temperature is lower than 23°C and indoor and outdoor temperature difference is 1°C over the indoor temperature.

System 3: Earth tube. This option utilizes the heating/cooling energy stored within the ground medium. In this building, an earth tube is used to condition air supplied to each apartment unit in the building. A design flow rate of 0.1 m³/s (212 cfm) is used for each unit with the following earth tube fan control options:

- Cooled/heated air is provided to each unit using fans.
- No fan operation for the period ranging from October 1 to May 30 due to the very small heat extraction capabilities of the earth tube (the outlet air temperature from the tube is lower than the indoor air temperature)

System 4: Radiative walls. Radiative walls, as depicted in Figure 7.23, are used to heat and cool the apartment unit. Three thermostats per unit are used to avoid significant temperature stratification. It is assumed that the heating season starts October 29 and ends April 17, and that the cooling season settings are set for the remainder of the days. To operate the system, operative temperature control is used. Simulation analysis to compare various control strategies has indicated that controls based on operative temperature provide more savings in heating/cooling energy use than controls based on mean-air temperature. Chilled water in this system is supplied through the district cooling system. The pumping energy and site energy for generating the chilled water are accounted for in evaluating this system.

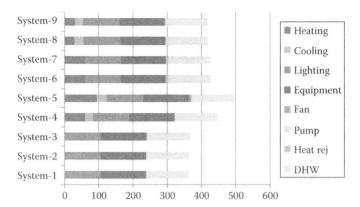

FIGURE 7.28 Summary of annual energy end use for various heating and cooling systems considered for the apartment building.

System 5: DX air conditioner and gas furnace. This system, most commonly used in the United States to condition residential buildings, consists of a gas furnace and an electric air conditioner (AC). A COP of 3.0 for the AC and an efficiency of 80% assumed for this system.

System 6: Radiative wall with cooling tower. This system is a variation of system 4 with an open type cooling tower (condenser) loop directly connected to the radiative wall system loop to reduce cooling energy consumption.

System 7: Radiative wall with evaporative cooler. This system is also a variation of system 4 with an evaporative cooler used to cool the water supplied to the radiant walls.

System 8: Radiative floors. Instead of the radiative walls, radiative floors are utilized to provide both heating and cooling throughout each apartment unit of the building. A radiative floor per floor level for each unit is considered.

System 9: Heating with radiative floor and cooling with radiative wall. This system used two separate water loops: hot water loop to provide heating through the radiative floors as in system 8, and chilled water loop to provide cooling through radiative walls as in system 4.

Figure 7.28 summarizes the energy performance of all the heating and cooling systems described above and the potential energy savings relative to the conventional air conditioning system (i.e., system 5 with gas furnace and DX cooling). Figure 7.29 compares the thermal comfort performance of all the systems expressed in terms of the number of hours where the PMV (Fanger's predicted mean value) is within the −0.5 and 0.5 range for all the apartment units in the building, as recommended by ASHRAE Standard 55 for indoor thermal comfort.

The results provided in Figures 7.28 and 7.29 indicate:

- Without any heating and cooling system, thermal comfort can be achieved 65% of the time. Specifically, the total comfort hours (PMV index within −0.5 and 0.5) are 17,000 for system 1 compared to 26,000 achieved by systems with both heating

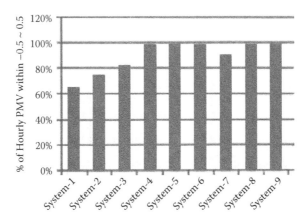

FIGURE 7.29   Summary of annual thermal comfort level for various heating and cooling systems considered for the apartment building.

and cooling capabilities, including system 5 through system 9) out of all 26,280 possible hours (i.e., sum of three floors per unit, i.e., 3*7,760 = 26,280).

- System 3 is only effective during the summer since room temperatures in the winter are typically higher than the air temperatures supplied by the earth tube. Detailed estimation for changeover period (from heating to cooling) is needed to better improve the performance of this system.
- The radiative floor system (system 8) saves almost 50% in heating energy use when compared to the radiative wall system (system 4).
- Evaporative cooling with a cooling tower or an evaporative cooler is effective and is highly recommended for cooling.

7.7 Summary

Significant energy savings can be obtained by improving the energy efficiency of secondary heating and cooling systems in residential buildings. These improvements can be achieved with simple operating and maintenance measures with no or little investment. Better operation and control of heating and cooling systems provides not only energy and cost savings but also improved thermal comfort. For several residential buildings, replacing existing heating or cooling systems with energy-efficient alternatives can be cost-effective.

Problems

7.1   Moist air enters a cooling deck at 90°F and 90% relative humidity at a rate of 10,000 cfm. The air is cooled to 57°F and 60% relative humidity. The condensate leaves at 60°F. Find the total cooling coil heat transfer rate as well as the sensible heat ratio of the process.

7.2   An evaporative cooler is able to cool air by 85% of the difference between the entering air dry and wet bulb temperatures. If the air is at 100°F and 25% relative humidity, what are the outlet conditions (i.e., estimate dry bulb temperature, relative humidity, and humidity ratio)? How much water is evaporated if the airflow is 5,000 cfm? Assume sea level and then 3,000 ft elevation (P_{atm} = 13.2 psia).

7.3 Determine the wet bulb efficiency of a direct evaporative cooler that is able to cool air from 45°C dry bulb and 20% relative humidity to 30°C dry bulb. Show the cooling process in the psychrometric chart.

7.4 Air enters an evaporative cooler, operating at sea level, at a temperature of 90°F and a relative humidity of 15%. The air is cooled to a temperature of 65°F. The evaporative cooler is used to condition a space with a sensible cooling load of 50,000 Btu/h.

    a. Estimate the humidity ratio of the air leaving the evaporative cooler.

    b. Determine the air mass flow rate required to meet the cooling load.

    c. Calculate the water mass flow rate needed to replace the water evaporated in the air.

7.5 During a party, 10 persons have been smoking in a house. Each smoker generates ETS particle matter at an average rate of 7.5 µg/s. By opening windows, an outdoor air intake rate of 300 L/s is introduced. Determine the resulting steady-state indoor air ETS concentration. Compare this concentration to the EPA standard level of 50 µg/m³ exposure to ETS particulate matter. Conclude.

7.6 Consider a ranch house 40 ft by 30 ft with an average wall height of 9 ft with window-to-wall ratio (WWR) = 20%, located in Boston. The house has an average R-value = 13.0 (°F.ft².h/Btu) for the opaque walls (including doors), and UA = 700 Btu/h.°F for the top section of the house (ceiling/roof/end walls with the ceiling R-value = 2.5°F.ft².h/Btu). The windows have an R-value = 1.0 (°F.ft².h/Btu).

    a. If the infiltration rate is estimated to be 200 cfm, determine the BLC of the house neglecting the heat transfer from the slab-on-grade floor.

    b. Determine if the existing gas furnace rated at 60,000 Btu/h is adequate for the house if the design outdoor temperature for Boston is 7°F and the indoor temperature is kept at 70°F.

    c. Determine the cost-effectivess of replacing the existing gas furnace (with a seasonal efficiency of 80%) with a condensing furnace with a seasonal efficiency of 95% at a cost of $2,500. The natural gas cost is $1.75/therm and the heating degree-days for Boston is 5,450°F-days.

7.7 A recent analysis of your gas-fired boiler showed that you have 40% excess combustion air and the flue gas temperature is 500°F.

    a. Determine the efficiency of the existing gas-fired boiler.

    b. How large an annual gas bill is needed to justify the replacement of the existing boiler with a condensing boiler with a seasonal efficiency of 95% at a cost of $4,000.

7.8 An efficiency test of a boiler fired by fuel no. 2 indicated a flue gas temperature of 600°F and excess air of 40%. The annual fuel consumption of the boiler is 5,000 gal/year. The cost of fuel no. 2 is $2.00/gal.

    a. Estimate the efficiency of the boiler as well as the percent in CO_2; O_2 is the flue gas.

    b. Determine the new boiler efficiency and the annual cost savings in fuel use if the percent O_2 in the flue gas is reduced to 3% and the stack temperature is set to 500°F.

    c. Determine the payback period of installing an automatic control system—at a cost of $850—to maintain the same efficiency found in (b) throughout the life of the boiler.

7.9 A combustion efficiency test done on a boiler indicated that the stack temperature is 250°C and the CO_2 content is 10%.

    a. Determine the excess combustion air and the boiler efficiency if the boiler uses fuel no. 2.

    b. Determine the excess combustion air and the boiler efficiency if the boiler uses fuel no. 6.

    c. Determine the excess combustion air and the boiler efficiency if the boiler uses natural gas.

7.10 Consider an old 10-ton air conditioning system operating for 1,500 equivalent full-load hours per year. The system efficiency is rated at SEER = 10. Determine the energy and cost savings if the system is replaced by an air conditioner with SEER = 18. Assume the electricity cost is $0.10/kWh.

7.11 A 5-ton central air conditioner system with an average seasonal COP = 3.5 is operated 2,000 h per year with an average load factor of 70%. This existing air conditioner system needs to be replaced by either air conditioner A or air conditioner B. The manufacturers indicated that the COP for system A is 4.3 and for system B is 4.7.

    a. Estimate the energy cost savings due to replacing the existing air conditioner with system A or system B. The electricity cost is $0.09/kWh.

    b. The cost differential between system B and A is $1,000. Determine if it is cost-effective to replace the existing central air conditioning system with chiller B rather than system A. For this question, a simple payback analysis can be used.

    c. If the discount rate is 5%, determine the electricity price for which it is more cost-effective to replace the existing air conditioner with system A rather than system B. Assume a life cycle of 10 years for both chillers.

7.12 Consider a house with a BLC of 800 Btu/h.°F located in Chicago and heated 24 h/day at 70°F using a gas-fired furnace with AFUE of 80%.

    a. Estimate the energy use savings if the temperature setting is reduced to 68°F.

    b. Determine the reduction in energy use savings if a 10°F temperature setback is implemented from 8:00 a.m. to 5:00 p.m. Express the savings in percent and therms. Assume the balance temperature of the house is 65°F.

    c. Calculate the payback period replacing the existing thermostat with a programmable thermostat at a cost of $250 to implement measures (a) and (b). Do the analysis separately for (a) and (b), and then combine the savings.

8

Water Systems Retrofit

Abstract

This chapter provides common water and energy conservation measures for indoor and outdoor water usage. Since the cost of water can be a significant fraction of the total utility bill, water management should be considered during an energy audit of a residential building. In addition to saving in water usage, energy use can be reduced through water management. In particular, water-saving fixtures and appliances are discussed in this chapter, as well as measures to reduce irrigation water needs and reuse wastewater. Moreover, energy-efficient domestic hot water heaters and the use of solar water heating systems are presented.

8.1 Introduction

In recent years, the cost of water usage has increased significantly and represents an important portion of the total utility bills, especially for residential buildings. In some western U.S. cities where the population growth has been high, the cost of water has increased by more than 400% during the past 10 years. In the future, it is expected that the cost of water will increase at higher rates than the cost of energy. In the United States, energy used for domestic hot water accounts for 15% of residential energy consumption, the third largest energy end use in homes after space heating (47%) and lighting and appliances (24%), as indicated by a survey conducted for residential energy consumption (EIA, 2009). Therefore, it is worthwhile to explore potential savings in water use and water heating expenditures during a building energy audit.

Under the Energy Policy Act of 1992, the U.S. government has recognized the need for water management and requires federal agencies to implement any water conservation measure with a payback period of 10 years or less. The Federal Energy Management Program (FEMP) has been established to help federal agencies identify and implement cost-effective conservation measures to improve both energy and water efficiency in federal facilities. The technical assistance offered by FEMP includes development of water conservation plans, training information resources, and software tools. In particular, FEMP has developed WATERGY, computer software that can be used to estimate potential water and associated energy savings for buildings.

There are several conservation strategies that can be considered for water usage and domestic hot water energy use in residential buildings. These strategies can be grouped into four main categories:

1. Indoor water management with the use of water-efficient plumbing systems (such as low-flow showerheads and water-efficient dishwashers and washing machines)
2. Outdoor water management associated with irrigation and landscaping (including the use of low-flow sprinkler heads, irrigation control systems, and xeriscape)
3. Recycling of water usage by installing processing systems that reuse water
4. Use of energy-efficient water heaters and solar water heating systems

Some of the proven water use and water heating conservation technologies and techniques are described in the following sections.

8.2 Indoor Water Management

The use of water-conserving fixtures and appliances constitutes one of the most common methods of water conservation, particularly in residential buildings. In general, retrofit of toilets, showerheads, and faucets with water-efficient fixtures can be performed with little or no change in lifestyle for the building occupants. Similarly, water-saving appliances such as dishwashers and clothes washers can provide an effective method to reduce indoor water usage in buildings. Another common and generally cost-effective method to conserve water is repairing leaks. It is estimated that up to 10% of water is wasted due to leaks (DeMonsabert, 1996). In addition to water savings, energy use reduction can be achieved when the water has to be heated, as in the case of domestic hot water applications (i.e., showering and hand washing). In the following sections, selected water-conserving technologies are described with illustrative examples to showcase the potential of water and associated energy savings due to implementation of these technologies.

8.2.1 Water-Efficient Plumbing Fixtures

Water, distributed through plumbing systems within buildings, is used for a variety of purposes, such as hand washing, showering, and toilet flushing. In recent years, water-efficient plumbing fixtures and equipment have been developed to promote water conservation. Table 8.1 summarizes a typical U.S. household end use of water with and without conservation. The average U.S. home can reduce inside water usage by about 32% by installing water-efficient fixtures and appliances and by reducing leaks.

In this section, some of the proven water-efficient products are briefly presented with some calculation examples that illustrate how to estimate the cost-effectiveness of installing water-efficient fixtures. As a general rule, it is recommended to test the performance of water-efficient products to ensure user satisfaction before any retrofit or replacement projects.

TABLE 8.1 Average U.S. Household Indoor Water End Use with and without Conservation

End Use	Without Conservation (gal/capita/day)	With Conservation (gal/capita/day)	Savings (gal/capita/day)
Toilets	20.1 (27.7%)	9.6 (19.3%)	10.5 (52%)
Clothes washers	15.1 (20.9%)	10.6 (21.4%)	4.5 (30%)
Showers	12.6 (17.3%)	10.0 (20.1%)	2.6 (21%)
Faucets	11.1 (15.3%)	10.8 (21.9%)	0.3 (2%)
Leaks	10.0 (13.8%)	5.0 (10.1%)	5.0 (50%)
Other domestic	1.5 (2.1%)	1.5 (3.1%)	0 (0%)
Baths	1.2 (1.6%)	1.2 (2.4%)	0 (0%)
Dish washers	1.0 (1.3%)	1.0 (2.0%)	0 (0%)
Total	72.5 (100%)	49.6 (100%)	22.9 (32%)

Adapted from AWWA, *Water Use Inside the Home,* report of American Water Works Research Foundation, 1999.

8.2.1.1 Water-Saving Showerheads

The water flow rate from showerheads depends on the actual inlet water pressure. In accordance with the Energy Policy Act of 1992, the showerhead flow rates are reported at an inlet water pressure of 80 psi. The water flow rate is about 4.0 gpm (gallons per minute) for older showerheads, and is 2.2 gpm for newer showerheads. The best available water-efficient showerheads have flow rates as low as 1.5 gpm. In addition to savings in water usage, water-efficient showerheads provide savings in heating energy cost. The calculation procedure for the energy use savings due to reduction in the water volume to be heated is presented in Section 8.22 and is illustrated in Example 8.1.

8.2.1.2 Water-Saving Toilets

Typical existing toilets have a flush rate of 3.5 gpf (gallons per flush). After 1996, toilets manufactured in the United States are required to have flush rates of at least 1.6 gpf. Since 2003, high-efficiency toilets with water usage of less than 1.3 gpf are available. Therefore, significant water savings can be achieved by retrofitting existing toilets, especially when they become leaky. Leaks in both flush valve and gravity tank toilets are common and are often invisible. The use of dye tablet testing helps the detection of toilet water leaks.

8.2.1.3 Water-Saving Faucets

To reduce water usage for hand washing, low-flow and self-closing faucets can be used. Low-flow faucets have aerators that add air to the water spray to lower the flow rate. High-efficiency aerators can reduce the water flow rates from 4 gpm to 0.5 gpm. Flow rates as low as 0.5 gpm are adequate for hand washing in bathrooms. Self-closing faucets are metered and are off automatically after a specified time (typically 10 s), or when the user moves away from the bathroom sink (as detected by a sensor placed on the faucet). The water flow rates of self-closing faucets can be as low as 0.25 gpc (gallons per cycle).

EXAMPLE 8.1

Determine the annual energy, water, and cost savings associated with replacing an existing showerhead (having a water flow rate of 2.5 gpm) with a low-flow showerhead (1.6 gpm). An electric water heater is used for domestic hot water heating with an efficiency of 90%. The temperature of the showerhead water is 110°F. The inlet water temperature for the heater is 55°F. The showerhead use is 10 min per shower, 2 showers per day, and 300 days per year. Assume that the electricity price is $0.09/kWh and that the combined water and wastewater cost is $5/1,000 gal.

Solution: The annual savings in water usage due to replacing a showerhead using 2.5 gpm by another using only 1.6 gpm can be estimated as follows:

$$\Delta m = 2 * [(2.5 - 1.6)gpm] * 10 \, min/day * 300 days/yr = 5,400 gal/yr$$

The energy savings incurred from the reduction in the hot water usage can be estimated as indicated below:

$$\Delta E = 5,400 gal/yr * 8.33 Btu/gal. \,°F[(110 - 55) \,°F]/0.90 = 2.749 * 10^6 Btu/yr$$

Therefore, the annual cost savings in energy use and in water use are, respectively:

$$\Delta Cost = [(2.749x106 Btu/yr)/(3.413 Btu/Wh)] * kW/1000W * \$0.09/kWh$$

$$= \$72.50/yr$$

and

$$\Delta Cost = 5,400 gal/yr * \$5.0/1000 gal = \$27/yr$$

Thus, the total annual savings incurred from the water-efficient showerhead are $99.50. These savings make the use of water-saving showerheads a cost-effective measure.

8.2.1.4 Repair Water Leaks

It is important to repair leaks in water fixtures even if these leaks consist of few water drips per minute. Over a long period of time, the amount of water wasted from these drips can be significant, as indicated in Table 8.2. The daily, monthly, and annual water wasted due to leaks can be estimated using Table 8.2 by simply counting the number of drips in 1 min from the leaky fixture. It should be noted that a leak of 300 drips per minute (i.e., 5 drips per second) corresponds to steady water flow. Example 8.2 illustrates the cost-effectiveness of replacing a leaky water fixture.

TABLE 8.2 Volumes of Water Wasted from Small Leaks

Number of Drips per Minute	Water Wasted per Day (gal/day)	Water Wasted per Month (gal/month)	Water Wasted per Year (gal/year)
1	0.14	4.3	52.6
5	0.72	21.6	262.8
10	1.44	43.2	525.6
20	2.88	86.4	1,051.2
50	7.20	216.0	2,628.0
100	14.40	432.0	5,256.0
200	28.80	864.0	10,512.0
300	43.20	1,296.0	15,768.0

EXAMPLE 8.2

Determine the annual water use and cost savings when repairing a leaky fixture that caused 100 drips per minute when the fixture was shut off. Estimate the payback of a new fixture that can be installed at a cost of $65. The water costs $6/1,000 gal.

Solution: The annual savings in water usage due to replacing the leaky fixture and avoiding the 100 drips per minute is about 5,256 gal, as indicated in Table 8.2. Thus, the annual cost savings in water use is

$$\Delta Cost = 5,256 gal / yr * \$6.0 / 1000 gal = \$31.5 / yr$$

Thus, the simple payback period for the leaky fixture replacement is

$$SPP = \$65 / (\$31.5 / yr) = 2.1 yrs$$

Therefore, the replacement of the leaky fixture is a cost-effective measure.

8.2.1.5 Water/Energy-Efficient Appliances

In addition to water-efficient fixtures, water can be saved in residential buildings by using water-efficient appliances such as clothes washers and dishwashers. The reduction of water needed to clean dishes or clothes can actually increase the energy efficiency of the household appliances. Indeed, a large fraction of the electrical energy used by both clothes washers and dishwashers is attributed to heating the water (85% for clothes washers and 80% for dishwashers). Typical water and energy performance of conventional and efficient models available for residential clothes washers and dishwashers is summarized in Table 8.3. Example 8.3 provides an estimation of the potential water and energy savings due to the use of efficient clothes washers.

EXAMPLE 8.3

Estimate the annual cost savings incurred by replacing an existing clothes washer (having 2.65 ft³ tub volume) with a water/energy efficiency appliance that has an energy factor of 2.50 ft³/kWh and uses 42 gal per load. The washer is operated based on 400 cycles (loads) per year. The water is heated using an electric heater. Assume that the electricity cost is $0.09/kWh and that water/sewer cost is $5/1,000 gal.

Solution: Using the information provided in Table 15.3, the water savings due to using an efficient clothes washer is 13.0 gal per load. Based on 400 loads per year, the annual water savings is 5,200 gal, which amounts to a cost savings of $26.

The annual electrical energy savings per load can be calculated using the energy factors as indicated in Table 8.3:

$$\Delta E = 2.65\, f^3 t\,/\,load * [(1/1.18 - 1/2.50)kWh\,/\,ft^3] * 400.loads\,/\,yr = 474.3\ kWh\,/\,yr$$

Thus, the annual electrical energy cost savings is $43.

Therefore, the total annual cost savings achieved by using an energy/water-efficient clothes washer is $69. The cost-effectiveness of the efficient clothes washer depends on the cost differential in purchasing price (between the efficient and conventional clothes washer models). For instance, if the cost differential is $200, the payback period for using an efficient clothes washer can be estimated to be

$$SPP = \frac{\$200}{\$69} = 2.9\ years$$

In the case illustrated by this example, the use of a water/energy-saving clothes washer is cost-effective.

TABLE 8.3   Water and Energy Efficiencies of Residential Dishwashers and Clothes Washers

Performance	Dishwashers	Clothes Washers
Older Models		
Water use (gal/cycle)	14.0	55.0
Energy factor[a]	0.46	1.18
Energy Star Models		
Water use (gal/cycle)	6.5	7.5
Energy factor[a]	0.46	1.8

[a] Energy factor is a measure of the energy efficiency of the appliance. For dishwashers, the energy factor is the number of full wash cycles per kWh. For clothes washers it is the volume of clothes washed (in ft³) per kWh per cycle. The water use for Energy Star clothes washers is expressed in gallons used per cubic foot of volume of clothes washed per cycle.

8.2.2 Domestic Hot Water Usage

In most buildings, hot water is used for hand washing and showering. To heat the water, electric or gas boilers or heaters are generally used. The energy input, E_w, required to heat the water can be estimated using a basic heat balance equation as expressed by Equation (8.1):

$$E_w = m_w c_{w,p}(T_{w,in} - T_{w,out})/\eta_{WH} \tag{8.1}$$

where m_w is the mass of the water to be heated. The hot water requirements depend on the building type. ASHRAE *Applications* handbook (2007) provides typical hot water use for various building types. For households, Table 8.4 provides three different draw profiles representing low, medium, and high use. For apartment buildings, the domestic hot water use depends on the size of the apartments, as shown in Table 8.5. $C_{w,p}$ is the specific heat of water, $T_{w,in}$ is the water temperature entering the heater (typically, this temperature is close to the deep ground temperature, i.e., well temperature), $T_{w,out}$ is the water temperature delivered by the heater and depends on the end use of the hot water, and η_{WH} is the efficiency of the water heater.

TABLE 8.4 Daily Draw Volumes for Various Fixtures and Appliances for Three Different Draw Profiles in gal/day (L/day)

Fixture/Appliance	Low Use	Medium Use	High Use
Bath 1	2.92 (11.1)	3.76 (14.2)	4.67 (17.7)
Bath 2	1.44 (5.45)	1.23 (4.66)	0.98 (3.71)
Clothes washer	12.3 (46.6)	15.1 (57.2)	16.7 (63.2)
Dishwasher	4.12 (15.6)	4.98 (18.9)	5.71 (21.6)
Kitchen sink	11.3 (42.8)	13.2 (50.5)	15.8 (59.8)
Sink 2	1.63 (6.17)	1.74 (6.59)	2.18 (8.25)
Sink 3	1.53 (5.79)	2.11 (7.99)	2.14 (8.10)
Sink 4	1.61 (6.09)	2.14 (8.10)	2.41 (9.12)
Shower 1	14.4 (54.5)	15.1 (57.2)	18.2 (68.9)
Shower 2	3.71 (14.0)	5.56 (21.0)	6.19 (23.4)
Total	55.0 (208)	64.9 (246)	75.0 (284)

TABLE 8.5 Typical Draw Volumes for Apartment Buildings

Number of Apartment Units (N)	Peak Hourly gal (L)/unit	Maximum Daily gal (L)/unit	Average Daily gal (L)/unit
$N \leq 20$	12.0 (45.5)	80.0 (303.2)	42.0 (159.2)
$20 < N \leq 50$	10.0 (37.9)	73.0 (276.7)	40.0 (151.6)
$50 < N \leq 75$	8.5 (32.2)	66.0 (250.0)	38.0 (144.0)
$75 < N \leq 200$	7.0 (26.5)	60.0 (227.4)	37.0 (140.2)
$N > 200$	5.0 (19.0)	51.5 (195.0)	35.0 (132.7)

TABLE 8.6   Typical Hot Water Temperature
for Common Residential Applications

Applications	Hot Water Temperatures
Dishwashers	140–160°F (60–71°C)
Showers	105–120°F (41–49°C)
Faucet flows	80–120°F (27–49°C)
Clothes washers	78–93°F (26–34°C)

Based on Equation (8.1) there are four approaches to reduce the energy required to heat the water:

1. Reduction of the amount of water to be heated (i.e., m_w). This approach can be achieved using water-saving fixtures and appliances as discussed in Section 8.2.1. Further reduction in domestic hot water use can be realized through changes in water use habits of building occupants.
2. Reduction in the delivery temperature. The desired delivery temperature depends mostly on the end use of the hot water, as outlined in Table 8.6 for residential buildings. It should be noted that a decrease of the water temperature occurs through the distribution systems (i.e., hot water pipes). The magnitude of this temperature decrease depends on several factors, such as the length and the insulation level of the piping system. For hand washing, the delivery temperature is typically about 120°F (49°C). For dish washing, the delivery temperature can be as high as 160°F (71°C). However, booster heaters are recommended for use in areas where high delivery temperatures are needed, such as dishwashers. A detailed analysis of water/energy-efficient appliances can be found in Koomey et al. (1994).
3. Increase the temperature of the water entering the water heater using heat recovery devices, including drain-water heat recovery systems. When hot water goes down the drain, it carries away with it energy, typically 80–90% of the energy used to heat water in a home. To capture this energy, drain-water heat recovery systems can be used to preheat cold water entering the water heater. These drain-water heat recovery systems can be used effectively with all types of water heaters, especially with tankless and solar water heaters. In particular, they can be used to recover heat from hot water used in showers, bathtubs, sinks, dishwashers, and clothes washers. The payback period for installing a drain-water heat recovery system ranges from 2.5 to 7.0 years, depending on how often the system is used.
4. Increase in the overall efficiency of the hot water heater. This efficiency depends on the fuel used to operate the heater and on factors such as the insulation level of the hot water storage tank. The National Appliance Energy Conservation Act (NAECA, 2004) defined the minimum acceptable efficiencies for various types and sizes of water heaters, as indicated in Table 8.7. The energy factor (EF) is a measure of the water heater efficiency and is defined as the ratio of the energy content of the heated water to the total daily energy used by the water heater.

It should be noted that thermal losses are common in pipes connecting water heating systems to various end-use fixtures in residential buildings. The magnitude of these

TABLE 8.7 Minimum NAECA Efficiency Standards for Water Heaters

Water Heater Fuel Type	Minimum Acceptable Energy Factor (EF)
Electric—storage	0.97 (0.00132 × rated storage volume in gallons)
Electric—instantaneous	0.93 (0.0013 × rated storage volume in gallons)
Gas—storage	0.67 (0.0019 × rated storage volume in gallons)
Gas—instantaneous	0.62 (0.0019 × rated storage volume in gallons)
Oil	0.59 (0.0019 × rated storage volume in gallons)

distribution losses depends on several factors, including the layout of the distribution system and the draw profile for hot water use. Example 8.4 presents a simplified method to assess the cost-effectiveness of reducing the distribution losses associated with water heating systems.

EXAMPLE 8.4

a. Estimate the annual energy cost of domestic hot water heating for a home with a medium draw profile of 250 L/day. The average service water temperature is 50°C. Due to the length of the distribution system, an average of a 10°C decrease in water temperature occurs before the hot water reaches the various fixtures. The heater is an electric water heater with an efficiency of 95%. The cost of electricity is $0.09/kWh. Assume that the entering water temperature is 12°C and that the home is occupied 325 days per year.
b. Determine the cost savings if the losses in hot water temperature in the distribution system are virtually eliminated by insulating the pipes.

Solution:

a. Using Equation (8.1), the annual fuel use by the water heater can be estimated. First, the water mass can be determined:

$$m_w = 250L / day * 325days / yr * 1kg / L = 81.250 * 10^3 kg / yr$$

Therefore, the energy needed to heat the water from 12°C to 60°C (= 50°C + 10°C) is

$$kWh = 81.250 * 10^3 kg / yr * 4.81kJ / kg\ ^\circ C * [(60-12)^\circ C)] / (0.95 * 3.6 * 10^3 kJ / kWh)$$

or

$$kWh = 5,485kWh / yr$$

Thus, the annual fuel energy cost for water heating is

$$Cost = [5,485kWh / yr] * \$0.09 / kWh = \$494 / yr$$

b. When the thermal piping system losses are eliminated (so that the actual hot water temperature from the heater is reduced from 60°C to 50°C), a reduction in the annual cost of heating the water results and is estimated to be

$$\Delta Cost = [1-(50-12)/(60-12)]*\$494/yr = \$103/yr$$

It should be noted that the losses through the distribution systems are generally referred to as parasite losses and can represent a significant fraction of the energy use requirements for water heating. An economic evaluation analysis should be conducted to determine the cost-effectiveness of insulating the piping system.

A more detailed discussion of the energy efficiency of domestic hot water heaters, water distribution systems, as well as solar water heating systems is provided in Section 8.5.

8.3 Domestic Hot Water Systems

8.3.1 Types of Water Heaters

The overall energy efficiency of a water heater is characterized by the energy factor (EF), defined as the ratio of the heat delivered as hot water to the total energy use, Q_{in}, by the water heater, as indicated by Equation (8.2):

$$EF = m_w c_{w,p}(T_{w,in} - T_{w,out})/Q_{in} \tag{8.2}$$

Figure 8.1 illustrates the values parameters used in the definition of EF. Typically EF is determined through a testing procedure that calls for 64 gal (or 243 L) to be drawn in

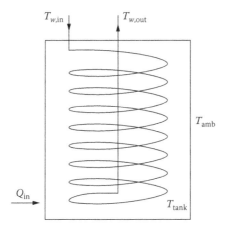

FIGURE 8.1   Simplified model of a storage tank water heater.

six equal amounts over 24 hours with $T_{w,in}$ = 58°F (14°C), T_{amb} = 67.5°F (19.5°C), and T_{tank} = 135°F (57°C).

In several countries, the energy efficiency standards for water heaters are defined using limits on acceptable EF values, as shown in Table 8.7 for the United States.

While the U.S. residential water heater market is currently dominated by storage type gas and electric water heaters, there are other alternatives for domestic water heating, including tankless, condensing, solar, and heat pump water heater models. A brief overview for the various types of water heaters suitable for residential buildings is provided below:

1. Storage water heaters: These systems are the standard water heaters in the United States and use typically electricity, natural gas, oil, or propane, depending on fuel availability and cost. The storage capacity of these heaters typically varies from 20 to 80 gal (i.e., 75.7 to 302.8 L), and they are able to supply high flow rates of hot water, although only for limited periods of time. However, storage water heaters are subject to standby as well as distribution heat losses.

2. Tankless water heaters: These units, also referred to as instantaneous or demand water heaters, are equipped with a gas burner or electric element that automatically ignites when a fixture is turned on and hot water is delivered on demand. Therefore, tankless heaters have lower standby heat losses than storage water heaters. Gas-powered tankless heaters provide typically higher hot water output than electric units. However, these units have relatively low flow rates and thus are more suitable for households with low demands. The tankless heaters are popular in several countries but are not commonly used in the United States. Indeed, tankless water heaters represent only 1% of the U.S. market since they are unable to meet large loads. However, due to recent developments and designs, tankless water heaters are becoming more suitable to U.S. households.

3. Heat pump water heaters (HPWHs): These systems provide a more efficient alternative to traditional electric water heaters. With a HPWH, heat is moved from a heat source (such as ambient air, ground, or another water tank) into the water using typically a refrigeration cycle. The most widely available type of residential HPWH is an air source HPWH. A typical HPWH unit has a coefficient of performance (COP) between 2 and 3, but can reach 4 under ideal conditions with high-temperature heat sources and suitable refrigerant types (Hashimoto, 2006; Zhang et al., 2007). The performance of a HPWH unit depends strongly on the external air temperatures and thus changes seasonally.

4. Condensing water heaters: These heaters are more energy efficient than traditional storage type gas-fired water heaters. Since condensing heaters are made up with materials resistant to corrosion, their flue gas can condense and provide more energy to the water through a heat exchanger. Typically, condensing boilers can have energy factors (EFs) up to 85% and annual fuel utilization efficiencies (AFUEs) of over 90% (Pescatore et al., 2004). Due to the need for a corrosion-resistant heat exchanger, a condensing water heater can be significantly more expensive than a traditional noncondensing water heater. However, some studies have shown that a condensing water heater combined with a condensing furnace could be a cost-effective option due to higher energy savings potential.

1. Solar water heaters: These systems can typically provide more than 50% of the energy required by a household for water heating (DOE, 2009). A backup water heating system is needed to supply the remaining domestic hot water needs. Currently, there are over 100 different models of solar water heaters. However, solar water heaters currently make up less than 1% of the market in the United States due to their high initial costs. Worldwide, there is 127.8 GW of solar water heaters installed as of 2006 (Weiss et al., 2008). In particular, China currently has the largest installed base of solar water heaters, with a total of 108 km^2 of collector area installed (over 60% of the total installed globally). However, the majority of the solar waters installed in China use evacuated tube collectors, which are typically less efficient for residential water heating applications (Han et al., 2010). There are several types of collectors that can be used for solar water heating, including flat plate collectors, evacuated tube collectors, and integrated collector storage (ICS) systems. Solar collectors in the United States are rated by the Solar Rating and Certification Corporation (SRCC), which tests different collector designs under standardized conditions to characterize the thermal performance of the collector (SRCC, 1994).

2. Combisystems: The solar combisystems utilize solar collectors to meet both space heating and domestic hot water thermal load requirements. These systems are typically suitable for residential buildings. Compared to conventional solar water heater (SWH) systems for domestic hot water only, combisystems require significant capital investments and their design can be complex. However, combisystems can improve the solar collector's utilization independent of occupant hot water use. Although solar combisystems are available in the United States, the bulk of solar combisystem implementation in the last 20 years has occurred in Europe. In fact, the market share of solar collectors installed in solar combisystem applications as compared to SWH applications in several northern European countries is about 50% (Weiss, 2003). Solar combisystems typically reduce auxiliary space heating energy consumption by 10 to 60%, depending on the size and type of system installed. Even higher solar fractions can be achieved with large seasonal thermal storage. Seasonal storage systems store summer solar heat in a large stratified tank for later use in the winter.

Based on the results of a recent study using typical U.S. costs of electricity ($0.095/kWh), natural gas ($1.40/therm), and oil ($2.40/gal), life cycle costs of various water heaters are estimated using a discount rate of 5%, as illustrated in Table 8.6. It should be noted that the life expectancy of water heaters depends on several factors, including water chemistry and maintenance level. As indicated in Table 8.8, both tankless and heat pump water heaters can be competitive with storage type water heaters over a life span of 13 years.

Recent studies have indicated that tankless and storage tank heaters react differently to draw profiles. Indeed, while draw profiles do not affect sensibly the energy factor of a storage type water heater, they do impact significantly the tankless unit energy. Specifically, both experimental and simulation analyses have shown that the length of

TABLE 8.8 Typical Efficiency, Installed, and Operating Costs for Various Water Heaters

Water Heater Type	Efficiency (EF)	Installed Cost	Yearly Energy Cost	Life (years)	LCC (based on 13 years)
Conventional gas storage	0.6	$850	$350	13	$4,562
High-efficiency gas storage	0.65	$1,025	$323	13	$4,451
Condensing gas storage	0.86	$2,000	$244	13	$4,588
Conventional oil-fired storage	0.55	$1,400	$654	8	$8,337
Typical efficiency electric storage	0.9	$750	$463	13	$5,661
High-efficiency electric storage	0.95	$820	$439	13	$5,476
Tankless gas (no pilot) 4	0.8	$1,600	$262	13	$4,379
Electric heat pump water heater	2.2	$1,660	$190	13	$3,675
Solar with electric backup	1.2	$4,800	$175	13	$6,656

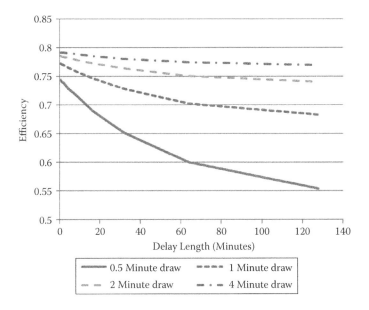

FIGURE 8.2 Impact of draw profile on the energy efficiency of tankless units.

delay between draws has a dramatic impact on the energy efficiency of tankless units, with longer delays allowing the heat exchanger to reach ambient temperature and dramatically decrease efficiency. Figure 8.2 illustrates the impact on the energy efficiency of a tankless unit of the draw profile characterized by the duration of the draws and the time delay between two consecutive draws (Grant et al., 2011). The results in Figure 8.6 clearly indicate that the heater efficiency is sensitive to delay length if the draws are short. However, once the draw reaches 4 min, even a 2 h delay between each draw can only reduce the unit efficiency by 2% relative to its the steady-state efficiency with a continuous draw.

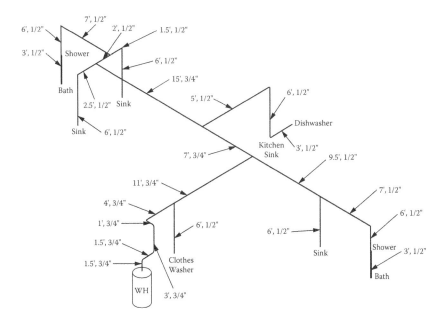

FIGURE 8.3   A schematic view of a typical DHW distribution system in a U.S. home.

8.3.2 Distribution Systems

In most hot water distribution systems for residential buildings, thermal losses are common in pipes connecting the water heater to various end-use fixtures. The magnitude of these losses depends on several factors, such as the location and layout of the distribution system as well as the hot water use profile. The HWSIM program is currently the most detailed domestic hot water (DHW) distribution system modeling tool available in the United States because it can model a full week of various draw profiles, several different types of pipes, ambient temperatures, and main temperatures. In particular, HWSIM can be used to analyze the cost-effectiveness of energy-saving measures specific to DHW distribution systems.

Figure 8.3 illustrates a trunk-and-branch configuration for a DHW distribution system in a single-family home in the United States. The distribution system is made up of a network of copper pipes with different diameters connecting the storage tank to various fixtures and appliances. The thermal losses associated to the DHW distribution system, and thus the potential savings from any improvements, depend on several factors: draw profiles, climate zones, water heater locations, and piping insulation levels.

For a prototypical U.S. home, there are significant variations in the DHW distribution thermal losses, depending on climates and usage profiles, as shown in Figures 8.4 and 8.5. In addition to five U.S. climates, three DHW use profiles of Table 8.3 are considered in the results illustrated in Figures 8.4 and 8.5. As shown in Figure 8.4, the distribution system losses represent typically 10 to 20% of the total energy used by the water heating system. Figure 8.5 indicates that the distribution losses in the case of a high-use

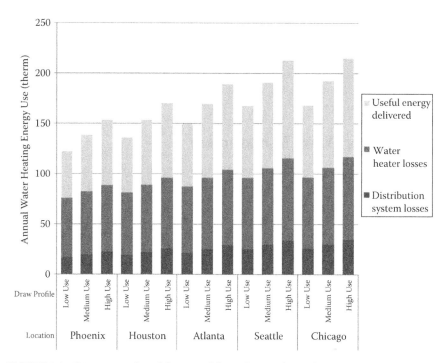

FIGURE 8.4 Disaggregated useful energy delivered, water heater losses, and distribution system losses for a typical U.S. home.

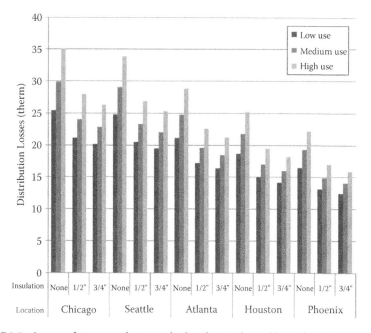

FIGURE 8.5 Impact of piping insulation on the distribution thermal losses for a typical U.S. home.

home in a cold climate are twice as large as those obtained for a low-use home in a hot climate. In most climates, an average difference of 30% in the distribution losses can be obtained between the low use and high use. Adding insulation to the distribution system can reduce the distribution losses by as much as 20%, but the additional reduction in distribution losses from increasing the insulation thickness from ½ in. (12.7 mm) to ¾ in. (19.05 mm) is fairly small.

8.4 Outdoor Water Management

Outdoor water conservation includes mostly innovative strategies for landscaping and irrigation of lawns and trees. Specifically, water savings can be achieved by reducing the overwatering of lawns using adequate irrigation control systems, or by replacing all or part of a landscape with less water-dependent components, such as rocks and indigenous vegetation, a method known as xeriscaping. Other areas to conserve water outdoor use include swimming pools and HVAC equipment, such as evaporative cooling systems.

8.4.1 Irrigation and Landscaping

In addition to its esthetics, vegetation consisting of trees, shrubs, or turfgrass can have a positive impact on the energy use in buildings by reducing cooling loads, especially in hot and arid climates. However, with water costs rising, it is important to reduce the irrigation cost for a vegetated landscape. The amount of water necessary for irrigation depends on several factors, including the type of plant and the climate conditions.

In a study conducted in a residential neighborhood consisting of 228 single-family homes near Boulder, Colorado, Mayer (1995) found that 78% of the total water used in the test neighborhood during the summer is attributed to lawn irrigation. Therefore, outdoor water management provides a significant potential to reduce water use for buildings. Some of the practical recommendations to reduce irrigation water use include:

- Water lawns and plants only when needed. The installation of tensiometers to sense the soil moisture content helps to determine when to water.
- Install automatic irrigation systems that provide water during early mornings or late evenings to reduce evaporation.
- Use a drip system to water plants.
- Add mulch and water-retaining organic matter to conserve soil moisture.
- Install windbreaks and fences to protect the plants against winds and reduce evapotranspiration.
- Install rain gutters and collect water from downspouts to irrigate lawns and garden plants.
- Select trees, shrubs, and groundcover based on their adaptability to the local soil and climate.
- The amount of irrigation water use is generally difficult to determine exactly and depends on the type of vegetation and the local precipitation. Typically, the water needed by a plant is directly related to its potential evapotranspiration (ET) rate. The ET rate of a plant measures the amount of water released through evaporation

TABLE 8.9 Annual Normal Precipitation and ET Rates for Selected U.S. Locations

Location	ET for Turfgrass (inches)	ET for Common Trees (inches)	Precipitation (inches)
Phoenix, Arizona	48.10	30.05	3.77
Austin, Texas	38.86	24.29	21.88
San Francisco, California	26.90	16.82	3.17
Boulder/Denver, Colorado	26.96	16.85	13.92
Boston, Massachusetts	24.13	15.08	22.38

Note: The values are provided for a growing season assumed to extend from April to October for all locations.

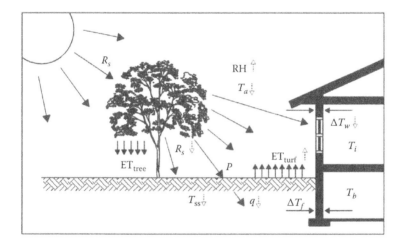

FIGURE 8.6 Effects of ET and other factors on the local climate around a typical house.

and transpiration of moisture from the leaves. Typical values for annual normal precipitation and ET rates in selected U.S. climates are listed in Table 8.9. Figure 8.6 illustrates various factors that may affect the ET and the local climate around a typical residential building (Conchilla and Krarti, 2002). The maximum possible ET rate can be estimated using one of several calculation procedures, including the Penman method (Periera, 1996) based on climate driving forces (i.e., solar radiation, wind speed, ambient temperature).

A well-designed landscaping, while consuming water, can actually save energy to heat and cool a building. Indeed, some computer simulation studies indicated that trees well positioned around a house can save up to 50% in energy use for cooling. Table 8.10 provides the magnitude of energy savings in the cooling loads for a typical residential dwelling surrounded by a variable number of trees in selected U.S. locations based on a simulation analysis (Conchilla, 1999). The placement configurations of the trees are illustrated in Figure 8.7. The savings in building cooling energy use attributed to vegetation is the effect of the cooler microclimate that trees and soil cover create due to

TABLE 8.10   Percent Savings of Annual Cooling Loads
for a Single-Family Home for Various Vegetation Types

Location	Turfgrass	4 Trees	8 Trees	12 Trees
Phoenix, Arizona	1.2%	4.5%	8.0%	13.1%
Austin, Texas	2.9%	4.8%	9.2%	14.7%
San Francisco, California	18.7%	27.2%	40.3%	57.0%
Denver, Colorado	3.4%	10.4%	18.7%	30.7%
Boston, Massachusetts	8.9%	8.6%	19.7%	34.8%

Note: These savings are estimated relative to a bare soil with no
groundcover and no trees.

(a)   (b)

(c)   (d)

FIGURE 8.7   Number and location of trees around a house considered in the analysis performed
by Conchilla (1999).

shading and evapotranspiration (ET) effects. A number of studies (Huang, 1987; Taha
1997; Akbari, 1993; Akbari et al., 1997; Conchilla, 1999) showed that summertime air
temperatures can be 2 to 6°F (1 to 3°C) cooler in tree-shaded areas than in treeless locations. During the winter, the trees provide windbreaks to shield the buildings from the
wind effects, such as air infiltration, and thus reduce heating energy use.

8.4.2 Wastewater Reuse

The reuse options for building applications are generally limited to graywater use and
rainwater harvesting. However, the available options depend on the type and location of
the building, and the legal regulations applicable to water reuse. Sewage water treatment
facilities are other options that are available but require large investments that are too
costly for individual buildings to consider.

Graywater is a form of wastewater with a lesser quality than potable water but higher quality than black water (which is water that contains significant concentrations of organic waste). Sources of black water include water that is used for flushing toilets, washing in the kitchen sink, and dish washing. Graywater comes from other sources, such as washing machines, baths, and showers, and is suitable for reuse in toilet flushing. In addition, graywater can be used instead of potable water to supply some of the irrigation needs of a typical domestic dwelling landscaped with vegetation. It is believed that graywater can be actually beneficial for plants since it often includes nitrogen and phosphorus, which are plant nutrients. However, graywater may also contain sodium and chloride, which can be harmful to some plants. Therefore, it is important to chemically analyze the content of graywater before it is used to irrigate the vegetation around the building.

Several graywater recycling systems are available, ranging from simple and low-cost systems to sophisticated and high-cost systems. For instance, a small water storage tank can be easily connected to a washing machine to recycle the rinse water from one load to be used in the wash cycle for the next load. The most effective systems include settling tanks and sand filters for treatment of the graywater.

Another method to conserve water used for irrigation is rainwater harvesting, especially in areas where rainfall is scarce. The harvesting of rainwater is suitable for both large and small landscapes and can be easily planned in the design of a landscape. There is a wide range of harvesting systems to collect and distribute water. Simple systems consist of catchment areas and distribution systems. The catchment areas are places from which water can be harvested, such as sloped roofs. The distribution systems, such as gutters and downspouts, help direct water to landscape holding areas, which can consist of planted areas with edges to retain water. More sophisticated water harvesting systems include tanks that can store water between rainfall events or periods. These systems result in larger water savings but require higher construction costs and are generally more suitable for large landscapes, such parks, schools, and commercial buildings.

8.5 Summary

This chapter outlined some measures to reduce water usage in buildings. In particular, landscaping and wastewater reuse, water-saving plumbing fixtures, and heat recovery systems can provide substantial water and energy reduction opportunities. Solar domestic hot water heaters are in most climates a cost-effective alternative to heat water for residential buildings. The auditor should perform water use analysis and evaluate any potential water management measures for the building.

Problems

8.1 An apartment building has 200 units, and each apartment unit uses 37 gal of hot water per day for 300 days each year. The temperature of the water as it enters the heater is 55°F (an annual average). The water must be heated to 150°F in order to compensate for a 20°F temperature drop during storage and distribution, and still be delivered at the tap at 130°F. The hot water is generated by an oil-fired boiler (using fuel 2) with an annual efficiency of 0.70. The cost of fuel is $1.50 per gallon. Determine:

a. The fuel cost savings if the delivery temperature is reduced to 90°F.

b. The fuel cost savings if hot water average usage is reduced by two-thirds.

8.2 Determine the best insulation thickness, using simple payback period analysis, to insulate a hot water storage tank with a capacity of 500 gal. The cost of insulation is as follows: $2.60/ft² for 1 in. fiberglass insulation, $2.95/ft² for 2 in. fiberglass insulation, and $3.60/ft² for 3 in. fiberglass insulation. The average ambient temperature surrounding the tank is 65°F. Assume appropriate dimensions for the 500-gallon tank.

8.3 Estimate the annual cost savings incurred by replacing an existing clothes washer (having a 3.00 ft³ tub volume) with a water/energy efficiency appliance that has an energy factor of 2.25 ft³/kWh and uses 50 gal per load. The washer is operated based on 350 cycles (loads) per year. The water is heated using an electric heater. Assume that the electricity cost is $0.10/kWh and that the water/sewer cost is $7/1,000 gal.

8.4 Using monthly calculation based on weather conditions of Denver, Colorado, determine the annual evaporative losses from a 100 ft² outdoor swimming pool with water temperature kept at 80°F throughout the year. Determine the cost-effectiveness of a cover if the pool is operated on average 10 hours per day and 250 days per year. The cost of cover is $1.50/ft². Assume a gas-fired boiler is used to heat the water at a cost of $2/gal and a boiler efficiency of 85%.

9

Net-Zero Energy Retrofits

Abstract

In this chapter, energy efficiency measures are combined with renewable energy technologies to retrofit existing homes to be low-energy or zero-energy buildings. In particular, optimization methodologies are introduced to determine the best combination of energy efficiency strategies and solar thermal and photovoltaic (PV) systems to reduce the reliance of residential buildings on primary energy resources. Both source energy use and life cycle cost are accounted for in the optimization procedure to retrofit existing homes to zero-energy buildings. In addition to the optimization techniques, general guidelines are provided to select the most cost-effective energy efficiency measures suitable for existing residential buildings. In particular, a set of measures as well as PV system sizes are provided for typical homes located in representative U.S. climates.

9.1 Introduction

In the United States, homes built before 2000 represent 92% of the U.S. housing stock (DOE, 2005). Similar conditions are encountered by several countries, as outlined in Chapter 1, with aging and energy-inefficient housing stock. Unfortunately, most of the existing retrofit and weatherization programs are limited in scope and impact, with only a few predefined retrofit measures typically considered, and even fewer measures are implemented and supported through incentive programs. Common evaluated and implemented energy efficiency measures in residential buildings include attic insulation, programmable thermostat, weather stripping, and cooling/heating system replacement. In order to reduce residential energy use significantly, novel approaches to retrofit programs should be considered to combine both a comprehensive list of energy efficiency strategies and renewable energy technologies. Even with current technologies, there is a significant potential to improve cost-effectively the energy efficiency of existing buildings. A recent study by the World Business Council for Sustainable Development (WBCSD, 2009) found that several energy efficiency projects are feasible with today's energy costs. Specifically, the study found that at oil prices of $60 per barrel, investments in existing building energy efficiency technologies can reduce related energy use and carbon footprints by 40% with five discounted payback years.

Comprehensive retrofit programs that encompass several energy efficiency measures require more detailed and sophisticated analysis tools that only recently have been applied to improve the energy performance of existing buildings. For instance, concepts such as low-energy or zero-energy buildings have been applied traditionally to design new buildings and often utilize detailed simulation analysis tools coupled with optimization techniques. These same concepts are only now being extended to existing buildings to select the most cost-effective energy efficiency measures that minimize initial costs and maximize energy savings. In particular, optimization of energy efficiency measures suitable for existing buildings entails evaluating various retrofit options until a set of measures is identified that achieves minimum cost or maximum energy savings. While exhaustive enumeration can be performed, in which every possible combination of possible energy efficiency measures is evaluated, optimization techniques are often utilized to minimize the number of combinations that need to be analyzed.

While the concept of zero net energy (ZNE) buildings has been mostly applied to new construction, it can be considered for retrofit projects, including residential buildings. Typically, zero net energy is defined in terms of either site energy or resource energy. Site energy consists of energy produced and consumed at the building site. Source or primary energy includes site energy as well as energy used to generate, transmit, and distribute this site energy. Therefore, source energy provides a better indicator of the energy use of buildings and their impact on the environment and the society, and thus is better suited for zero net energy building analysis. Indeed, an analysis based on source energy effectively allows different fuel types, such as electricity and natural gas commonly used in buildings, to be encompassed together.

It is generally understood that ZNE buildings produce as much energy as they consume on-site annually. These buildings typically include aggressive energy efficiency measures and active solar water heating systems. Moreover, ZNE buildings employ grid-tied, net-metered renewable energy generation technologies, generally photovoltaic (PV) systems, to produce electricity. Effectively, ZNE buildings use the grid as "battery storage" to reduce required generation system capacity.

Over the last two decades, few tools have been developed for selecting optimal packages of energy efficiency measures for residential buildings. Most of these tools are for designing new homes and are based on the sequential search technique. The sequential search methodology was first applied for optimizing the energy efficiency level of the residential building sector (Meier, 1982). Then, the methodology was implemented to a spreadsheet-based analysis tool and to the EnergyGauge-Pro software (FSEC, 2007). Recently, the sequential search methodology has been used to find a path to design zero-energy homes using a building energy simulation-based computer program referred to as the building energy optimization, or BEopt tool (Christensen et al., 2004). Starting from a reference building model, BEopt identifies a cost-optimal set of building energy options needed to achieve a variety of energy savings levels, including the zero net energy (ZNE) option (i.e., 100% energy savings). The building energy options are chosen from predefined or custom efficiency and renewable energy

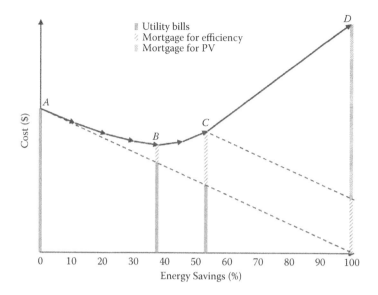

FIGURE 9.1 BEopt Pareto graph of the optimal path to ZNE.

measures. Energy savings for each measure are calculated relative to a reference building on a source energy basis.

Originally, BEopt was used primarily to optimize design options for new homes. It can now be used for optimizing the energy retrofit options for existing residential buildings, even though automatic calibration capabilities are not available in the current version of the tool. Chapter 10 discusses common calibration approaches suitable for detailed energy simulation models, including automatic calibration algorithms. A conceptual plot of the cost/energy savings Pareto graph from BEopt is illustrated in Figure 9.1. At the starting point of the optimization (point A), only utility bills comprise the building's energy-related cost, as no efficiency upgrades have been included to cause an increase in mortgage, which includes the energy use cost and the capital cost for implementing energy efficiency or renewable systems. As energy efficiency measures are introduced into the building, incremental mortgage costs increase and utility bills decrease until the marginal cost of saved energy equals the cost of utility power. At this point, the curve reaches a minimum and the global cost-optimum point is reached (point B). Additional efficiency measures with marginal costs more expensive than the cost of fuel are introduced until the marginal cost of saved energy equals the marginal cost of producing PV energy (point C). At this point, PV capacity is added until all source energy use is offset (point D). For roofs where there is a limit to the number of PV panels that can fit on a building's roof, additional energy efficiency measures are employed after the PV system selection until zero net energy is reached. These additional energy efficiency measures may have a marginal cost of saved energy larger than that of the produced energy from PV. If all energy efficiency measures in the optimization search space are exhausted before zero net energy is attained, the optimization stops at its maximum possible energy savings level.

Other optimization techniques and algorithms have been considered to optimize building energy use (Wright et al., 2002; Wetter, 2004; Tuhus and Krarti, 2009). Some of these optimization techniques are briefly presented in Section 9.2.

9.2 Building Retrofit Optimization Approaches

Inherently, building energy retrofits consider several energy efficiency measures and thus are multivariate. The energy efficiency measures encompass a wide range of options, including building envelope insulation, HVAC (heating, ventilation, and air conditioning), equipment, appliances, lighting, water heating, and possibly renewable energy systems. To accurately select the best package for the energy efficiency measures, all the possible combinations of the measures would need to be evaluated. This approach, often referred to as the brute force optimization approach, can be time-consuming and in several cases not needed to select the optimal package. Instead, some optimization techniques and methodologies can be considered. Several optimization tools are now available and can be applied to new design retrofit of buildings. For instance, GenOpt is a generic optimization program that can be connected with an external building energy simulation program in order to minimize a cost function. It includes several optimization methodologies suitable for continuous, discrete, univariate, or multivariate optimization problems (Wetter, 2004). In this section, a set of optimization techniques are briefly presented. These techniques can handle multivariate and discrete optimizations and have been successfully applied to select the optimal package of energy efficiency measures for residential buildings (Tuhus-Dubrow and Krarti, 2009).

9.2.1 Sequential Search

The sequential search technique, used in BEopt, is a direct search method that identifies the building energy option that will best decrease the cost function at each iterative step. It begins by simulating a user-defined reference building model. The reference model is typically a calibrated energy simulation model for an audited building. A simulation is performed for each potential energy efficiency measure individually, and the most cost-effective measure is chosen and used in the building model description for the next point along the path. There are a number of discrete options for a given energy efficiency measure category, such as glazing type, heating system, and ceiling insulation. The most cost-effective option is defined as the one that gives the largest reduction in annual costs for the smallest reduction in source energy use. Annual costs are a combination of mortgage costs (which increase as more expensive energy-efficient options are included) and utility costs. The process is repeated, ultimately defining a path from the reference building to the minimum cost point, and then to a zero net energy building.

Without modifications, this simple algorithm would not reliably find the correct least-cost path, due to the problem of interactive effects between different options. Three special cases have been identified for the sequential search technique as implemented in BEopt: invest/divest, large steps, and positive interactions (Christensen at al., 2004). The invest/divest case is a result of negative interactive effects. BEopt attempts to account for this by evaluating the removal of options in the current building design, which may result in a more cost-optimal point. For example, a highly efficient HVAC system may have been selected as the most cost-effective option at an early point in the process. Later on, however, the improvement of the building envelope may cause the efficient HVAC option to not be cost optimal, so it is removed from the building design.

The large steps case is another example of negative interaction among options. There may be a large energy-saving option that is available at a current point, but is less cost-effective than another option that does not save as much energy. The latter option is chosen, and then the most cost-effective option is again chosen at that second point, which results in a third point. However, it is possible that the original large energy-saving option available at the first point is more cost optimal than the third point. In order to solve this problem, the sequential search keeps track of points from previous iterations and compares them to the current point. If a previous point is more cost optimal, it replaces the current point.

A positive interaction occurs if two options are more cost-effective when present together than they would be separately. An example could be the presence of both large south-facing windows and thermal mass for passive solar heating. The sequential search will only find these positive interactions if one of the options is first selected individually. This inability to always identify synergistic options is a potential deficiency with the sequential search method.

9.2.2 Genetic Algorithm

The genetic algorithm (GA) uses the evolutionary concept of natural selection to converge on an optimal solution over many generations. This technique differs from traditional optimization methods in a number of areas. First, rather than working with one potential solution at a time, the technique works with a set of solutions called a population. This ensures a global approach to the optimization and helps the GA avoid getting stuck in local minima, which can be a problem with other methods. Second, the GA works with encodings of the parameters, not the parameters themselves. Parameters are traditionally encoded as binary strings, although other methods are possible. Finally, GAs use probabilistic methods for determining the parameter values in each successive iteration, rather than deterministic rules. This means that each time a GA is run, the path toward convergence is different, and the end result may be different as well.

Each individual in the population represents a different solution to the problem. Every option for each parameter has a corresponding binary representation, and the parameters are concatenated to form the complete binary string. A new generation is formed at the end of each iteration of the algorithm, consisting of a new population, and this process is repeated until satisfactory convergence criteria are reached, or the maximum number of generations is reached. The algorithm uses only three operators to produce a new population for the next generation: selection, crossover, and mutation.

There are a number of different ways to handle selection. One method is to rank the population in ascending order by fitness value (after the cost function is evaluated for each individual), and assign probabilities for selection based on each individual's rank. This is called rank weighting. A virtual roulette wheel is spun (by generating a random number between 0 and 1) to determine the members in the new population selected for reproduction.

Once the population for reproduction is selected, the individuals are paired off and mated using the crossover process. A cross-point is selected at random for each pairing, and two new individuals are created by joining the first part of the first string with the second part of the second string, and vice versa. Mutation is the last step in the formation of the population for the next generation, and involves flipping a bit at random in the population from a 0 to a 1, or vice versa. Mutation is intended to prevent the GA from converging prematurely and helps to maintain a global search. The mutation rate is set at the beginning of the algorithm. Finally, this mutated population becomes the population of the next generation, and the process is repeated until convergence is reached.

9.2.3 Particle Swarm Optimization

PSO shares many similarities with genetic algorithms. Like GAs, the technique works with a set of solutions called a population. Each potential solution is called a *particle*. Instead of using evolutionary methods, however, the PSO is based on the social behavior of bird flocks or fish schools. Each particle is characterized by a velocity with which it explores the cost function. The velocity and position of each particle are updated after each successive iteration of the algorithm. The particle velocity and position are governed by Equations (9.1) and (9.2):

$$v^{new} = v^{old} + c_1 r_1 (p^{localbest} - p^{old}) + c_2 r_2 (p^{globalbest} - p^{old}) \tag{9.1}$$

$$p^{new} = p^{old} + v^{new} \tag{9.2}$$

where v = particle velocity, p = particle position, r_1, r_2 = independent uniform random numbers between 0 and 1, c_1 = cognitive acceleration constant, c_2 = social acceleration constant, $p^{localbest}$ = best local solution (best particle in current population), and $p^{globalbest}$ = best global solution (best particle so far in all generations).

The two acceleration constants are usually numbers between 0 and 4. The particle swarm optimization technique has become popular for the same reasons as the GA, in that it is easy to implement with relatively few parameters to adjust.

9.2.4 Applications

Optimization techniques have been used in a wide range of applications to improve the energy performance of buildings. Indeed, optimization-based analysis approaches have been widely used to improve the control and operation of buildings (Huang and Lam, 1997; Wright et al., 2002; Fong et al. 2006). More recently, optimization techniques have been proposed to select building envelope and heating and cooling system design features (Caldas and Norford, 2003; Christensen et al., 2004; Wanga et al., 2005; Tuhus-Dubrow and Krarti, 2010; Bichiou and Krarti, 2011). In particular, Tuhus-Dubrow and Krarti (2010) developed a simulation environment using a genetic algorithm optimization technique to select the best combinations of several building envelope features in order to optimize energy consumption and life cycle costs. Figure 9.2 illustrates the

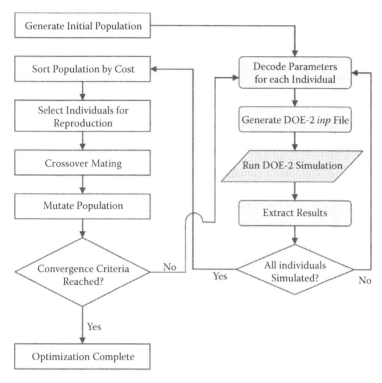

FIGURE 9.2 Flowchart diagram for the developed simulation/optimization tool.

flowchart of the optimization simulation environment used to select building envelope features for a typical U.S. home. Table 9.1 shows optimal building orientation, shape, and building envelope characteristics selected to minimize the life cycle cost (LCC) value obtained using the GA optimization technique. As indicated by the results of Table 9.1, insulating the walls and the attic as well as reducing the air infiltration are selected for all the climates. Foundation insulation is not selected for hot climates such as Phoenix, Arizona, and Miami, Florida. However, increasing the mass level within the home is not selected for any climate due to its high implementation cost.

Table 9.2 summarizes the results of selecting the heating and cooling system and its efficiency for a typical home located in five U.S. climates (Bichiou and Krarti, 2011). The selection of the heating and cooling system is based on minimizing the LCC without a budget constraint. Several heating and cooling systems were considered in the optimization analysis, including a ground source heat pump (GSHP) with vertical borehole wells or horizontal loops, air conditioner, gas furnace, electric heating, and evaporative cooling. Figure 9.3 presents an example of a Pareto optimal plot that shows the energy cost savings and LCC values for selecting both building envelope features and heating and cooling system type and settings for a two-story home located in Chicago. The Pareto plot is obtained using the sequential search method. The optimal solution found by the sequential search method coincides with those obtained using the GA and PSO optimization techniques, as illustrated by Figure 9.3.

TABLE 9.1   Specifications for a Residential Building Envelope Optimization for Five U.S. Climates

Climate	Boulder	Phoenix	Chicago	Miami	San Francisco
Minimum found (LCC in $)	$38,104	$36,569	$41,450	$38,267	$46,967
Azimuth (degrees)	270	315	270	337.5	0
Shape	Rectangle	Rectangle	Rectangle	Rectangle	Rectangle
Aspect ratio	1.0	1.0	1.0	1.0	1.0
Wall construction	R21, 2 × 6	R21, 2 × 6	R21, 2 × 6	R21, 2 × 6	R21, 2 × 6
Ceiling insulation	30	30	30	30	30
Foundation level	4 ft R5 perimeter, R5 gap	Uninsulated	4 ft R15 perimeter, R5 gap	Uninsulated	4 ft R5 perimeter, R5 gap
Glazing type	Low e very high SHGC argon	Low e low SHGC argon	Low e standard SHGC argon	Low e low SHGC argon	Low e very high SHGC argon
Infiltration	Typical	Tight	Typical	Tight	Typical
Mass level	Light	Light	Light	Light	Light

TABLE 9.2   Optimized Selection of HVAC System Features for Five U.S. Sites

	Boulder	Chicago	Miami	Phoenix	San Francisco
Algorithm	Sequential search	Sequential search	Sequential search	Sequential search	Particle swarm optimization
System efficiency	98	—	80	98	98
HVAC system type	GSHP vertical rectangle 6 × 3	GSHP vertical 10 pipes	AC with electric resistance	Evaporative cooler with furnace	AC with furnace
Annual utility cost ($)	$940	$1,154	$1,313	$1,118	$1,441
Life cycle cost ($)	$45,258	$43,886	38,924	$38,036	$41,181

Alspector and Krarti (2009) have used GA-based optimization to both automatically calibrate a detailed building energy simulation model for an audited house using monthly utility data and select a package of energy efficiency measures subject to a budget constraint. Table 9.4 illustrates LCC-based optimization results to select energy retrofit measures for a residential building located in Denver, Colorado, when three budget levels are imposed: $1,000, $5,000, and $10,000. As shown in Table 9.3, more aggressive energy efficiency measures are selected when the retrofit budget increases. It is interesting to note as the budget increases and more retrofit measures are implemented, the payback period of the retrofit package is reduced.

9.3 Renewable Energy Systems

Throughout this book, various energy efficiency measures have been presented and evaluated. In this section, selected renewable energy systems are briefly described to meet thermal loads as well as to generate electrical energy needs for retrofitted residential

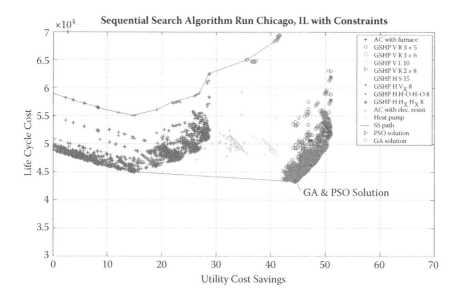

FIGURE 9.3 Pareto graph of LCC as a function of energy cost reduction for a house in Chicago.

buildings. Specifically, three solar energy systems are discussed: passive solar heating systems, hot water systems, and photovoltaic (PV) systems.

9.3.1 Passive Solar Heating Systems

Using existing building envelope components, passive solar heating strategies can be considered when retrofitting residential buildings. Indeed, a passive solar heating system typically requires windows for solar energy aperture and walls and floors to capture and store solar energy. Two passive solar systems can be suitable for retrofit applications, including direct gain and sunspace configurations, as illustrated in Figure 9.4. As shown in Figure 9.4(a), by increasing the size of the window or adding more thermal mass, solar radiation heat can be effectively stored in the building and distributed uniformly to all spaces. Similarly, heat can be captured and released with some time delay to the house when a sunspace or greenhouse space is present or added, as illustrated in Figure 9.4(b).

Several analysis methods are available to assess the potential energy savings associated with passive solar heating systems. In addition rules of thumb, simplified evaluation methods can be used, such as the load collector ratio (LCR) method (Balcomb et al., 1980). Moreover, most detailed building energy simulation programs presented in Chapter 3 can be used to model the performance of the solar passive heating systems.

9.3.2 Solar Thermal Collectors

For residential buildings, solar thermal collectors can be used for domestic water heating, space heating, or combination domestic water and space heating. In this section, solar domestic hot water systems as well as combisystems are briefly described.

TABLE 9.3 Optimized Selection of Energy Retrofit Measures for a House in Denver, Colorado, Based on Three Budget Constraint Levels

	Base Construction	Measure Construction	Electricity Savings (kWh)	Gas Savings (MMBtu)	Utility Savings ($)	Incremental Cost ($)	Simple Payback Period (years)
			Case 2				
			Budget $1,000				
Slab insulation	Uninsulated	4 ft R10 perimeter R5 gap	−43	12	116.10	697	6.0
All measures	N/A	N/A	−43	12	116.10	697	6.0
			Budget $5,000				
Slab insulation	Uninsulated	4 ft R5 perimeter R5 gap	−33	11	107.01	499	4.7
Wall insulation	Brick	R15—2 × 4 frame	180	55	566.31	978	1.7
Window type	Single-paned clear	Low E high SHGC argon	255	19	213.10	2,812	13.2
Furnace upgrade	82	86	0	2	20.00	652	32.6
All measures	N/A	N/A	608	100	1,055.10	4,941	4.7
			Budget $10,000				
Slab insulation	Uninsulated	4 ft R5 perimeter R5 gap	−33	11	107.01	499	4.7
Wall insulation	Brick	R19—2 × 6 + 1 infoam	461	86	901.77	2,588	2.9
Window type	Single-paned clear	Low E std. SHGC argon	353	17	201.98	2,812	13.9
All measures	N/A	N/A	994	123	1,320.10	5,899	4.5

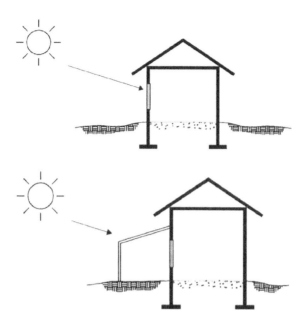

FIGURE 9.4 Passive solar heating systems suitable for retrofit applications.

9.4 Solar Domestic Hot Water Systems

Solar domestic hot water (SDHW) systems are often cost-effective to consider for residential buildings. There are several types of SDHW systems, and they are divided into two primary categories: direct and indirect. In direct or open-loop systems, water is directly heated through the solar collectors. However, heat exchangers are used to heat water for indirect or closed-loop systems. Moreover, SDHW systems can use passive (i.e., natural convection) or active strategies (i.e., pump) to circulate water from the collector to the tank. Thermosiphon and integrated collector systems (ICSs) are examples of direct and passive SDHW systems. Drainback systems are examples of indirect and active SDHW. Figure 9.5 shows a typical drainback SDHW system suitable for residential buildings located in cold climates. The main feature of a drainback system is the fail-safe setup used to ensure that the collector loop system, including the collector and the pipes, would not freeze by removing water from the loop when the system is not collecting solar heat.

The thermal efficiency of a SDHW system is closely related to the thermal efficiency of its collector efficiency, typically expressed as a function of the inlet fluid (i.e., water for direct systems) temperature, ambient air temperature, and solar radiation falling on the collector:

$$\eta_C = F_r(\tau\alpha)_n - F_r U_l * (T_{inlet} - T_{ambient})/G_{net} \tag{9.3}$$

η_C = collector efficiency, $F_r(\tau\alpha)_n$ = collector optical gain coefficient, $F_r U_l$ = collector heat loss coefficient, T_{inlet} = fluid temperature at inlet of collector, $T_{ambient}$ = ambient air

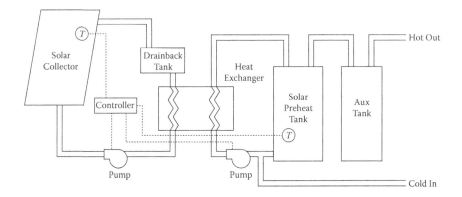

FIGURE 9.5   Typical components of a drainback SDHW system.

TABLE 9.4   Typical Collector Parameters

Collector Type	$F_r(ta)_n$	F_rU_l (W/m²)
Selective—metal-glass	.779	4.77
Nonselective—metal-glass	.768	7.245
Polymer—selective	.779	4.77
Polymer—nonselective	.739	8.216
Polymer—unglazed	$.88 - .029{*}v_{wind}$	$10.24 + 4.69{*}v_{wind}$

temperature, and G_{net} = net incident radiation (includes short-wave solar and net sky infrared radiation).

Table 9.4 provides typical parameter values for selected SDHW collector types, including a selective and nonselective metal-glass collector as well as a number of polymer collectors: glazed, with selective coating; glazed standard; and unglazed.

9.4.1 Solar Combisystems

Solar combisystems utilize solar collectors to meet thermal loads associated to both space heating and domestic hot water. These systems are typically considered for detached single-family homes, groups of family homes, and multifamily buildings. As compared to conventional solar domestic hot water (SDHW) systems, combisystems have the advantage of their high solar collector's utilization independent of occupant hot water use. However, these systems require a relatively large capital investment and their design can be intricate. The market share of solar collectors installed in solar combisystem applications, as compared to SWH applications, is about 50% for several northern European countries, including Austria, Switzerland, Denmark, and Norway (Weiss, 2003). Solar combisystems first appeared in both the United States and Europe in the mid-1970s. In the United States, federal tax credits allowed the technology to be cost-effective. During the 1980s, the solar water heater/space heater market in the United States began to fade due to the end of tax credits. Beginning in the 1990s, European installers and solar companies began offering simpler and cheaper solar combisystems, but these systems were often custom-built (Weiss, 2003). Today, limited commercialized systems exist in both the U.S. and Europe. As a result, most combisystems are still custom-designed and built.

Solar combisystems consist of five components: a solar collector loop, a storage tank, a control system, an auxiliary heater, and a heat distribution system. Combisystems typically use a stratified sensible heat storage tank ranging from 100 to 1,000 gallons and are coupled to up to 400 ft^2 of solar collectors. The heat distribution system consists of a hydronic heating loop serving radiant floors, baseboard heaters, or air handler heating coils that are directly or indirectly heated by the storage tank. Solar combisystems typically reduce auxiliary space heating energy usage by 10 to 60%, depending on the size and type of system installed. Even higher solar fractions can be achieved with large seasonal thermal storage. Seasonal storage systems store summer solar heat in a large stratified tank for later use in the winter. These systems typically have a storage volume between 5,000 and 20,000 gallons (Weiss, 2003). Detailed simulation programs such as TRNSYS can be used to model the performance of combisystems.

9.4.2 PV Systems

Most PV systems installed for residential buildings are used to generate electricity that can be either used directly in the house (stand-alone systems) or sold to the grid (grid-connected systems). Recently, there is an interest in using hybrid PV systems to generate both electricity and heat through photovoltaic/thermal (PV/T) collectors. A brief overview of grid-connected PV systems and PV/T collectors is provided in this section.

9.4.2.1 Grid-Connected PV Systems

Figure 9.6 illustrates the basic design for a grid-connected PV system suitable for a residential building. A PV system typically includes a PV panel made up of several arrays to generate electricity from solar radiation, an inverter to convert direct current (DC)

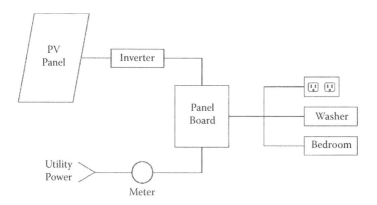

FIGURE 9.6 Typical grid-connected PV system for a house.

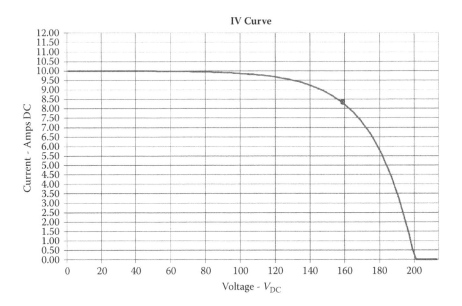

FIGURE 9.7   Typical I-V curve for a PV panel.

generated by the PV panel to alternate current (AC) commonly provided to the house by the utility company, and a set of regulators and controllers and protection devices.

Flat-plate PV panels are commonly installed in residential buildings with fixed mounting systems. However, other PV panels exist, such as one- or two-axis tracking systems and concentrating systems. More recently, flexible shingle type PV systems can be used as both PV panels and roofing shingle materials. The efficiency of the PV panel is typically by its I-V curve, as presented in Figure 9.7 with the highlighted maximum power point (MPP), where the generated electrical power is maximized, as indicated in Figure 9.8. It should be noted that the two characteristic points are often given as part of the manufacturers' specifications of PV panels, including the short-circuit current or I_{sc} (obtained when $V = 0$) and the open circuit voltage or V_{oc} (obtained when $I = 0$). Table 9.5 presents typical PV module characteristics, including rated power, short-circuit current, open-circuit voltage, and efficiency.

As noted in Table 9.5, the conversion efficiency of commercially available PV modules is in the 10–13% range under rated conditions. It should be noted that the thin-film PV module is being used for building integrated PV applications (such as shingle-PV systems) and their efficiency is improving. A simple analysis of PV systems is to utilize the efficiencies for all of its components. In particular, for a typical PV system with a PV panel, inverter, and MPP tracker, the overall system efficiency can be estimated as follows:

$$\eta_{PV,system} = \eta_{PV,Module} * \eta_{INV} * \eta_{MPPT} * \eta_{VR} \tag{9.4}$$

where $\eta_{PV,Module}$ is the PV panel efficiency (refer to Table 9.5 for typical efficiency values), η_{INV} is the inverter efficiency (typically over 95%), η_{MPPT} is the maximum power point tracker efficiency (typically over 95%), and η_{VR} is the voltage regulator efficiency (typically over 95%).

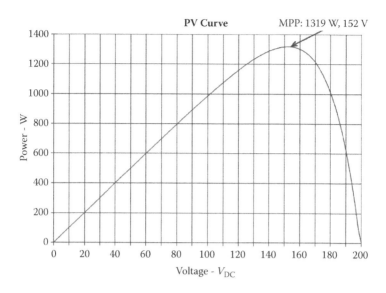

FIGURE 9.8   Variation of electrical power generated by a PV panel.

TABLE 9.5   Typical PV Module Characteristics

PV Cell Type	Rated Power	Short Circuit Current, I_{sc}	Open Circuit Voltage, V_{oc}	Efficiency
Single crystal	175 W	5.55 A	44.4 V	13.5%
Multi crystal	75 W	4.75 A	21.8 V	11.6%
Thin film	40 W	2.68 A	23.3 V	9.4%

Then the electrical energy generated, E_{gen} by the PV system, can be estimated on an hourly basis:

$$E_{gen} = \eta_{PV,system} * A_{PV} * G_{net} \qquad (9.5)$$

where A_{PV} is the area of the PV panel, and G_{net} is the net incident radiation hitting the PV panel and depends on the tilt angle and tracking system used for the PV system.

Several tools are available to estimate the generated electrical energy generated by PV systems, such PVWatts, TRNSYS, and PV-Chart.

9.4.2.2 Photovoltaic/thermal Collectors

Hybrid photovoltaic/thermal (PV/T) collectors convert solar energy into electricity and heat. A typical residential PV/T collector is analogous to a flat-plate solar thermal collector, except a photovoltaic (PV) panel is attached on the top of the metallic absorber plate. The PV cells are composed of semiconductor materials that convert high-energy photons of incident solar radiation into electricity. The lower-energy photons are absorbed by the PV panel and generate heat within the cells. This heat is rejected to the ambient conditions while also being removed from behind the PV cells by a heat transfer fluid. If extracted, heat from the PV panels can offer multiple benefits. The heat can be used to meet various thermal loads while improving the PV electrical efficiency due to its lower operating temperature. Both air and fluids are used to recover heat from the PV systems. The amount of thermal energy and its quality are important factors when considering the value of the output heat. The performance of a PV/T system is characterized by both the amount of electrical energy and the useful heat produced. Currently, there are only very limited PV/T systems commercially available.

9.5 Near-Optimal Retrofit Analysis Methodology

9.5.1 General Methodology Description

Instead of using an optimization technique to find the best package energy and renewable energy measures to retrofit an existing building to zero net energy, a sequential search analysis procedure similar to the BEopt approach can be utilized. The basic sequential search process involves the following steps:

Step 1: Start by the evaluation of different types of energy efficiency measures (e.g., wall insulation addition, window replacement, and HVAC equipment efficiency improvement) to determine the most cost-effective measure in terms of marginal cost. The baseline building model for this first iteration is a calibrated model using either utility data or monitored data as described in Chapter 10. Using a Pareto graph similar to that in Figure 9.9, the life cycle cost is shown in the *y*-axis and the percent savings of source energy is provided in the *x*-axis for each measure. The measure with the steepest negative slope (i.e., ratio cost over percent energy savings) is selected as shown in Figure 9.9.

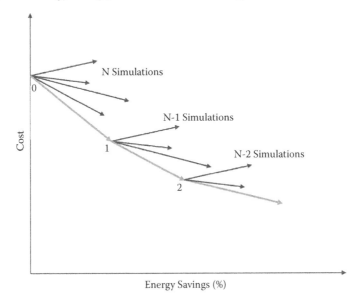

FIGURE 9.9 Sequential analysis procedure for selecting energy efficiency measures.

Step 2: The chosen measure in step 1 is then removed from the potential energy efficiency measures for the second iteration analysis. In this step, the new baseline building model consists of the calibrated model modified to include the selected measure after the first iteration outlined in step 1. The remaining energy efficiency measures are then simulated using the new baseline building model. Again, the measure with the most effective marginal cost is selected after this iteration, as outlined in Figure 9.9.

Step 3: Repeat the same procedure described for step 2 until an optimal point is reached and the slope of LCC over percent energy savings becomes positive, as illustrated in Figure 9.10. The positive slope indicates that a reduction in building energy use results in increasing the LCC due to significant implementation costs. It should be noted that now the measure with the lowest positive slope is selected.

Step 4: After reaching an optimal point, the most cost-effective efficiency measure's marginal cost is compared to the cost of photovoltaic (PV) energy. At the point where further improving the building has a higher marginal cost, PV is employed until zero net source energy is achieved, as illustrated in Figure 9.9.

Step 5: A final check is needed to ensure that the size of the required PV system can be installed based on the constraints of the available spaces (for instance, the house roof area). If no sufficient space is available to install the required PV system, then additional energy efficiency measures need to be selected until 100% source energy savings are achieved. In this case, the selection process is similar to that described for step 4 without the need to compare with the marginal cost of the PV system.

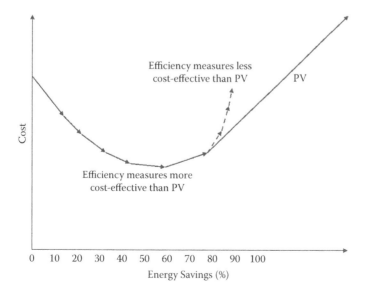

FIGURE 9.10 Sequential selection procedure for the path to ZNE.

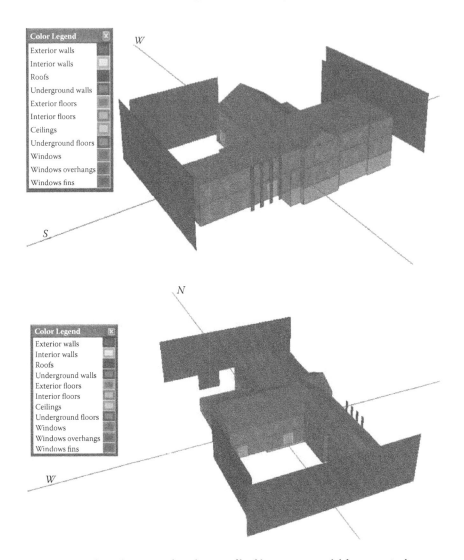

FIGURE 9.11   Three-dimensional renderings of building energy model for a sorority house.

9.5.2 Case Study for ZNE Retrofit

To illustrate the sequential selection procedure, an audit of a large sorority house is carried out. A calibrated building model is developed using a detailed energy simulation program as outlined in Figure 9.11. Note that the building is surrounded by other buildings that are modeled as exterior shading elements. A list of energy efficiency measures is evaluated individually. Table 9.6 summarizes the list of building envelope, lighting, and HVAC system retrofit measures considered for the audit analysis. Only the results for the energy and economic impacts of individual measures are provided in Table 9.6. Using the LCC analysis, all the measures can be cost-effective, except adding wall

TABLE 9.6 Energy and Economical Analysis of Energy Efficiency Measures

EEM	Only individual energy efficiency measures are evaluated.							
	Electricity Savings (kWh/year)	Natural Gas Savings (therm/year)	Annual Costs ($/year)	Total Cost Savings ($/year)	Installed Cost ($)	Simple Payback (years)	Life Cycle Costs[a] (years)
Thermostat management	407	498	$15,629	$350	$0	0.00	$194,772
Reduce infiltration rates	50	366	$15,738	$241	$1,200	4.98	$197,330
Install additional roof insulation	251	686	$15,511	$468	$4,360	9.32	$197,661
Install wall insulation	398	1,772	$14,798	$1,181	$19,155	16.22	$203,571
Install basement insulation	196	206	$15,822	$157	$7,642	48.68	$204,819
Replace all of the incandescent bulbs with compact fluorescent (CFL) bulbs	13,636	−356	$15,108	$871	$272	0.31	$189,150
Replace all of the incandescent flood lights with compact fluorescent (CFL) bulbs	3,585	−23	$15,702	$277	$157	0.57	$195,959
Retrofit T-12 lighting with T-8 lighting and program start electronic ballasts	3,260	−51	$15,749	$230	$450	1.96	$196,497
Install occupancy sensors in the bathrooms, laundry room, and living rooms	3,239	−62	$15,757	$222	$1,285	5.79	$196,589
Baseline	0	0	$15,979	0	0	—	$199,134

[a] LCC analysis assumes a life cycle of 20 years and a discount rate of 5%.

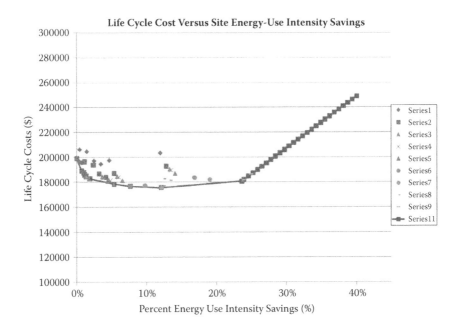

FIGURE 9.12   Life cycle costs vs. percent primary energy savings.

insulation and foundation insulation, as indicated by Table 9.6, by comparing the LCC values obtained for all the energy efficiency measures and the baseline building model.

Using the sequential selection procedure described in Section 9.4.1, an optimal set of energy conservation measures is determined. The results of the sequential selection procedure are summarized in the Pareto graph of Figure 9.12. The optimal package with the lowest LCC value consists of all of the energy efficiency measures listed in Table 9.6, except for the basement insulation measure. The optimal package achieves an annual energy use reduction of 23.63% and costs $26,879 to implement, while reducing the annual electric bills by $3,638 per year. In addition, the optimal package can reduce life cycle costs from $199,133 to $180,675 over a 20-year period with a 5% discount rate. By adding a PV system in the roof, further reduction in source energy use can be achieved. In particular, for a neutral LCC point (as indicated by the dashed line in Figure 9.11), a 13.5 kW PV system needs to be installed. For this PV size, a reduction of 28% in the building energy use can be achieved, but an additional $40,500 to the total installed cost is needed. A zero net energy retrofit could be achieved with the addition of 50 kW of PV panels. However, this PV system would not be cost-effective and would be too large to install on the available roof area of the house.

9.6 Guidelines for U.S. Home Retrofits

In this section, the optimal packages for energy retrofitting existing U.S. single-family detached homes are described as well as the packages needed to achieve cost-neutral

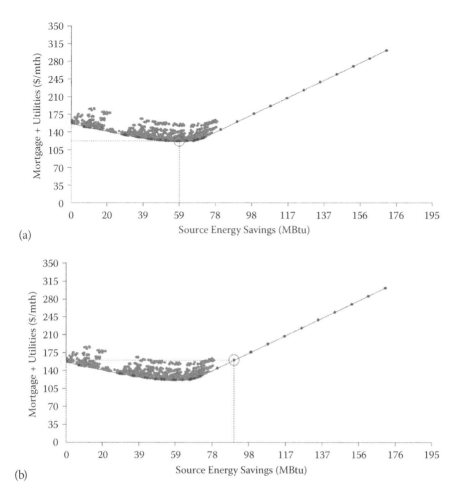

FIGURE 9.13 Selection procedure toward zero net energy building: (a) optimal cost and (b) neutral cost retrofits.

retrofits. Figure 9.13 illustrates both the minimum cost and neutral cost retrofits using the sequential search technique of BEopt (Christensen et al., 2004). The benefits of the minimum cost retrofit are clear: since the energy retrofit measures at this point in the Pareto graph of Figure 9.13(a), the annual energy costs are significantly reduced (due to energy use savings) with rather small implementation costs. The neutral cost point provides an indicator of how much energy use savings can be achieved through efficiency and renewable energy technologies while maintaining the same life cycle cost associated with the existing building without any retrofit.

Using the characteristics of existing U.S. homes as described in Appendix A, a detailed analysis has shown that the minimum cost and neutral cost retrofits can achieve significant energy use savings (Albertsen et al., 2011). Figures 9.14 and 9.15 summarize the energy use savings and the implementation costs incurred by minimum cost retrofits for

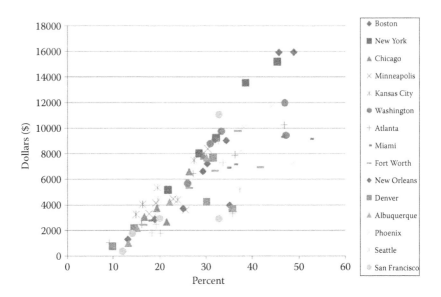

FIGURE 9.14 Implementation costs and source energy use savings for minimum cost retrofits for selected U.S. sites.

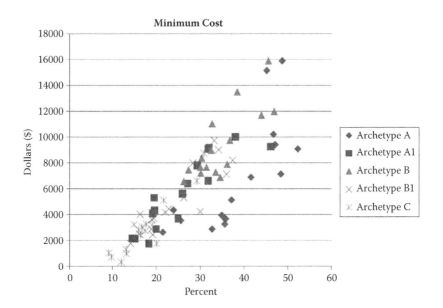

FIGURE 9.15 Implementation costs and source energy use savings for minimum cost retrofits for selected U.S. house vintages.

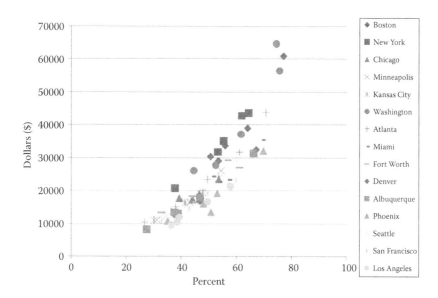

FIGURE 9.16 Implementation costs and source energy use savings for neutral cost retrofits for selected U.S. sites.

various climates and house vintages, respectively. Chapter 1 describes in detail the characteristics of various archetypes for existing U.S. homes. As indicated in Figures 9.14 and 9.15, energy use savings of up to 50% can be achieved especially for archetype A and B homes located in the East (i.e., New York, Boston, and Washington, DC). To achieve these significant savings, relatively high implementation costs are needed with about $12,000 to $16,000 per retrofitted house. Similarly, Figures 9.16 and 9.17 provide the energy use savings and the implementation costs incurred by neutral cost retrofits for various climates and house vintages, respectively. Even more energy use savings can be achieved for the neutral cost retrofits, with savings of up to 80%. As indicated in Figures 9.16 and 9.17, archetype A and B homes located on the East Coast provide the best opportunities for neutral cost retrofits but require significantly high implementation costs with about $40,000 to $65,000 per retrofitted house.

Tables 9.7 and 9.8 summarize the energy efficiency measures that were found to be the most cost-effective for various house vintages and U.S. climates (Albertsen et al., 2011) for, respectively, minimum cost and neutral cost retrofits. For most of the vintages and climates, improving the insulation values and replacing lighting, HVAC equipment, and appliances are the most effective energy efficiency measures. In particular, the minimum cost package of retrofit measures includes typically low-cost and easy-to-implement alternatives, such as ceiling insulation, lighting replacement, and some appliance and equipment replacement. The neutral cost retrofit package has almost twice the energy savings, as outlined in Table 9.8, with more comprehensive retrofit options, such as adding insulation to the walls, foundation, and ceiling, as well as replacing all appliances except the cooking range. In addition, HVAC equipment replacement is recommended for the neutral cost retrofit in almost all the U.S. regions except in the

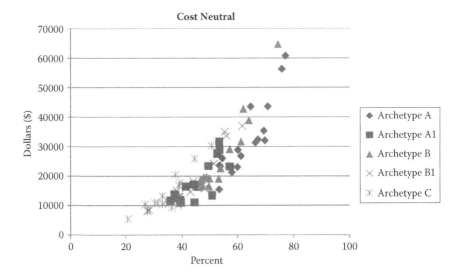

FIGURE 9.17   Implementation costs and source energy use savings for neutral cost retrofits for selected U.S. house vintages.

hottest climates, where it is not cost-effective to replace the furnace. Finally, replacing the hot water heater is always cost-effective in the cost-neutral retrofit but not in the minimum cost retrofit.

9.7 Summary

In this chapter, a sequential selection methodology is introduced to select an optimal package of energy efficiency measures that minimize life cycle cost for retrofitting existing residential buildings. The methodology is based on a sequential search optimization technique and can be utilized to find the path toward zero net energy retrofit of buildings. In addition to the minimum cost and zero net energy retrofits, the methodology can provide insights on how to retrofit buildings while maintaining the same life cycle costs as the existing preretrofit conditions. Based on a detailed energy simulation analysis, it is found that up to 50% of energy use savings can be obtained when minimum life cycle cost retrofits are considered for single-family U.S. homes. Even higher energy savings can be achieved when neural life cycle cost retrofits are implemented. In the case of minimum cost retrofits, renewable energy systems are suitable for installation, including solar domestic hot water and PV systems.

TABLE 9.7 Energy Efficiency Measures for Minimum-Cost Retrofits for U.S. Climate Zones

	Zone 1					Zone 2					Zone 3					Zone 4					Zone 5				
	A	A1	B	B1	C	A	A1	B	B1	C	A	A1	B	B1	C	A	A1	B	B1	C	A	A1	B	B1	C
Walls			■				■	■					■					■					■		
Interzonal walls		■		■				■	■				■	■				■	■		■	■	■		
Ceiling insulation		■											■				■				■				
Foundation				■					■	■				■											
Interzonal foundation										■															
Window type																					■				
Infiltration						■																■			
Refrigerator																					■				
Cooking range																									
Dishwasher																									
Clothes washer	■	■	■	■						■		■	■				■	■			■	■	■		
Clothes dryer	■	■	■	■			■	■		■		■	■		■		■	■	■	■	■	■	■		
Hardwired lighting	■	■	■				■	■				■	■				■	■			■	■	■		
Plug-in lighting	■	■	■				■	■				■	■					■			■	■	■		
Air conditioner																									
Furnace																									
Water heater																									
Solar DHW																									
PV system																									

TABLE 9.8   Energy Efficiency Measures for Neutral-Cost Retrofits for U.S. Climate Zones

	Zone 1					Zone 2					Zone 3					Zone 4					Zone 5				
	A	A1	B	B1	C	A	A1	B	B1	C	A	A1	B	B1	C	A	A1	B	B1	C	A	A1	B	B1	C
Walls																									
Interzonal walls																									
Ceiling insulation																									
Foundation																									
Interzonal foundation																									
Window type																									
Infiltration																									
Refrigerator																									
Cooking range																									
Dishwasher																									
Clothes washer																									
Clothes dryer																									
Hardwired lighting																									
Plug-in lighting																									
Air conditioner																									
Furnace																									
Water heater																									
Solar DHW																									
PV system																									

10

Methods for Estimating Energy Savings

Abstract

In this chapter, an overview of methods that can be used to estimate energy and cost savings from implementing energy conservation measures is presented. These methods are especially used in energy projects financed through performance contracting and are often referred to as measurement and verification (M&V) tools. The methods and their applications are briefly described with some examples to illustrate how savings are estimated. In addition, a case study is presented to apply selected M&V methods to estimate energy savings from retrofits of 30 homes.

10.1 Introduction

After a building energy audit, a set of energy conservation measures (ECMs) is typically recommended. Unfortunately, several of the ECMs that are cost-effective are often not implemented due to a number of factors. The most common reason for not implementing ECMs is the lack of internal funding sources (available to owners or managers of the buildings). Indeed, energy projects have to compete for limited funds against other projects that are perceived to have more visible impacts, such as improvement in esthetics or extension of floor areas.

Over the last decades, there have been several mechanisms to fully or partially fund energy retrofit projects of existing buildings, including tax credits, weatherization programs, and performance contracting (Krarti, 2010). An important feature of these funding mechanisms is the need for a proven protocol for measuring and verifying energy cost savings. An acceptable measurement and verification protocol should provide a systematic approach to ensure that cost and energy savings have indeed incurred from the implementation of the energy retrofit projects.

The predicted energy savings for energy projects based on an energy audit analysis can be significantly different from the actual savings measured after implementation of the energy conservation retrofits. For instance, Greely et al. (1990) found in a study of over 1,700 commercial building energy retrofits that a small fraction (about 16%) of the energy projects have predicted savings within 20% of the measured results. Therefore, accepted and flexible methods to measure and verify savings are needed to encourage investments in building energy efficiency.

Direct "measurements" of energy savings from energy efficiency retrofits or operational changes are almost impossible to perform since several factors can affect energy use, such as weather conditions, levels of occupancy, and HVAC operating procedures. For instance, Eto (1988) found that during abnormally cold and warm weather years, energy consumption for a building can be, respectively, 28% higher and 26% lower than the average weather year energy use. Thus, energy savings cannot be easily obtained by merely comparing the building energy consumption before (pre) and after (post) retrofit periods.

Over the last two decades, several measurement and verification (M&V) protocols have been developed and applied with various degrees of success. Among the methods proposed for the measurement of energy savings are those proposed by the National Association of Energy Service Companies (NAESCO, 1993), the Federal Energy Management Program (FEMP, 1992, 2000, 2008), the American Society of Heating Refrigeration and Air Conditioning Engineers (ASHRAE, 1997, 2002), the Texas LoanSTAR program (Reddy et al., 1994), and the North American Energy Measurement and Verification Protocol (NEMVP) sponsored by the Department of Energy (DOE) and later updated and renamed the International Performance Measurement and Verification Protocol (IPMVP, 1997, 2002, 2007).

In this chapter, general procedures and methods for measuring and verifying energy savings from retrofit projects are presented. Some of these methods are illustrated with calculation examples or with applications reported in the literature.

10.2 General Procedure

To estimate the energy savings incurred by an energy retrofit project, it is important to first identify the implementation period of the project, that is, the construction phase where the building is subject to operational or physical changes due to the retrofit. Figure 10.1 illustrates an example of the variation of the building electrical energy use in a fan that has been retrofitted from constant volume to a variable air volume. The time-series plot of the building energy use clearly indicated the duration of the construction period, the end of the preretrofit period, and the start of the postretrofit period. The duration of the construction period depends on the nature of the retrofit project and can range from a few hours to several months. Figure 10.2 shows the monthly variation of natural gas energy use associated with preretrofit, implementation, and postretrofit periods for a residence that has been retrofit with weather stripping and a new furnace.

The general procedure for estimating the actual energy savings, ΔE_{actual}, from a retrofit energy project is based on the calculation of the difference between the preretrofit

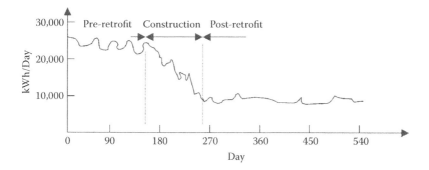

FIGURE 10.1 Daily variation of a building energy consumption showing preretrofit, construction, and postretrofit periods.

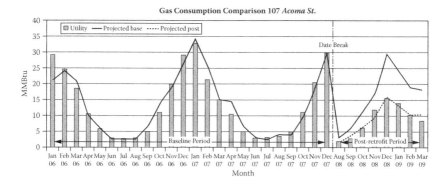

FIGURE 10.2 Monthly variation of a building energy consumption showing preretrofit, construction, and postretrofit periods for a residence.

energy consumption predicted from a model and the postretrofit energy consumption obtained directly from measurement (Kissock et al., 1998):

$$\Delta E_{actual} = \sum_{j=1}^{N} \Delta E_j = \sum_{j=1}^{N} \left(\tilde{E}_{pre,j} - E_{post,j} \right) \tag{10.1}$$

where N is the number of postretrofit measurements (for instance, during one year, $N = 365$ daily data can be used); $\tilde{E}_{pre,j}$ is the energy use predicted from a preretrofit model of the building based on weather and operating conditions observed during the postretrofit period; and $E_{post,j}$ is the energy used by the building measured during the postretrofit period.

Therefore, it is important to develop a preretrofit energy use model for the building before estimating the retrofit energy savings. This preretrofit model helps determine the baseline building energy use knowing the weather and operating conditions during the postretrofit period.

In some instances, the energy savings estimated using Equation (10.1) may not be representative of average or typical energy savings from the retrofit project. For instance, the measured energy use during the postretrofit period may coincide with abnormal weather conditions, and thus may lead to retrofit energy savings that are not representative of average weather conditions. In this case, a postretrofit energy use model for the building can be considered instead of measured data to estimate normalized energy savings that would occur under typical weather and operating conditions. The normalized energy savings, ΔE_{norm}, can be calculated as follows:

$$\Delta E_{norm} = \sum_{j=1}^{N} \Delta \tilde{E}_j = \sum_{j=1}^{N} \left(\tilde{E}_{pre,j} - \tilde{E}_{post,j} \right) \tag{10.2}$$

$\tilde{E}_{pre,j}$ is the energy use predicted from a preretrofit model of the building using normalized weather and operating conditions, $E_{post,j}$ is the energy use predicted from a postretrofit model of the building using normalized weather and operating conditions.

It should be noted that the energy savings calculation procedures expressed by Equations (10.1) and (10.2) can be applied to subsystems of a building (such as lighting systems, appliances, motors, and air conditioners) as well as to the entire building. In recent years, several techniques have been proposed to estimate various energy end uses of a building. Some of these approaches are discussed in Section 10.3.

There are several approaches to estimate the energy savings from energy retrofits. These methods range from simplified engineering approaches to detailed simulation and measurement techniques. For specific projects, the method to be used in M&V of energy savings depends on the desired depth of the verification, the accuracy level of the estimation, and the accepted cost of the total M&V project. In general, the cost of the M&V procedure depends on the metering equipment needed to obtain detailed data on the energy consumption of the building and its end uses. The installation cost of the metering equipment can be significant and can represent more than 10% of the total energy retrofit project costs. A number of suggestions have been proposed to reduce the cost of metering equipment, including:

- Use of statistical sampling techniques to reduce the metering requirements, such as the case for air conditioner retrofit projects where only a selected number of air conditioner circuits are metered. A complete metering of the entire lighting circuits within a building, while feasible, can be very costly.
- End-use metering for specific systems directly affected by the retrofit project. For a lighting retrofit project, for example, metering may be needed only for the lighting circuits.
- Limited metering requirements with stipulated calculation procedures to verify savings. For instance, the energy savings incurred from a replacement of motors can be estimated using simplified engineering methods that may require only the metering of operation hours.

Some of the most commonly used methods for verifying savings are briefly presented in the following section.

10.3 Energy Savings Estimation Models

10.3.1 Simplified Engineering Methods

These methods are commonly used in projects that include retrofits of weather stripping, lighting systems, or motors. The calculation methods can be based on simplified assumptions and models. Example 10.1 provides an illustration of a simplified engineering method that one can use to verify savings in retrofitting incandescent lamps to CFLs for a home. Refer to Chapter 6 for further details about the calculation procedures to estimate the energy and cost savings for lighting retrofit projects.

EXAMPLE 10.1

A house has 40 incandescent lamps; each is rated at 60 W. The homeowner decided to replace all the incandescent lamps with 3 W CFL fixtures. On average, each lamp is operated 2 h per day, 7 days per week, 52 weeks per year. Estimate the annual energy savings of this retrofit. Determine the cost savings due to the lighting retrofit if the electricity cost is $0.09/kWh.

Solution: The reduction in the electrical energy input for all the incandescent lamps is first determined:

$$\Delta kW_{light} = (60 - 23) * 40 = 1,480\,W$$

The total number of hours for operating the lighting system during 1 year can be estimated as follows:

$$\Delta t_{light} = 2\,hrs\,/\,day * 7\,days\,/\,week * 52\,weeks\,/\,year = 728\,hrs\,/\,yr$$

Thus, the annual energy savings due to the lighting retrofit is

$$\Delta kWh_{light} = 1.48\,kW * 728\,hrs\,/\,yr = 1,077\,kWh\,/\,yr$$

Therefore, the annual cost savings that resulted from the stipulated savings is

$$\Delta C_{light} = \$0.09\,/\,kWh * 1,077\,kWh\,/\,yr = \$97\,/\,yr$$

10.3.2 Regression Analysis Models

Using metered data, models for building energy use for pre- and postretrofit periods can be established using regression analysis. The first developed regression model for building energy use estimation has been an application of the variable-base degree-day (VBDD) method (refer to Chapters 3 and 5 for a more detailed description of the VBDD method). Indeed, Fels (1986) proposed the Princeton Scorekeeping Method (PRISM) to correlate the monthly utility bills and the outdoor temperatures to estimate the energy use for heating and cooling and estimate any energy savings for residential retrofit measures. In recent years, general regression approaches have been proposed to establish baseline models for building energy use, which can be used to estimate retrofit savings. In particular, steady-state single-variable regression models have been developed and applied successfully to predict energy use for residential buildings. Moreover, detailed simulation energy models have been used to predict energy use of existing buildings and their associated retrofits. A brief description of the single-variable regression analysis approaches is provided in this section.

The single-variable regression models constitute the basis of the main M&V procedure adopted by the International Performance Measurement and Verification Protocol (IPMVP, 1997, 2002, 2007). A simple linear correlation is assumed to exist between the building energy use and one independent variable. The ambient temperature is typically selected as the independent variable, especially to predict residential building heating and cooling energy use. Degree-days with properly selected balance temperature can be used as another option for the independent variable.

Ambient-temperature-based regression models have been shown to predict building energy use with an acceptable level of accuracy even for daily data sets (Kissock et al., 1992; Katipamula et al., 1994; Kissock and Fels, 1995) and can be used to estimate energy savings (Claridge et al., 1991; Fels and Keating 1993). Four basic functional forms of the single-variable regression models have been proposed for measuring energy savings in residential buildings. The selection of the function form depends on the application and the building characteristics. Figure 10.3 illustrates the four basic functional forms commonly used for ambient temperature linear regression models. The regression models, also called change-point or segmented-linear models, combine both search methods and least-squares regression techniques to obtain the best-fit correlation coefficients. Each change-point regression model is characterized by the number of the correlation coefficients. Therefore, the two-parameter model has two correlation coefficients (β_0 and β_1) and assumes a simple linear correlation between building energy use and ambient temperature. Table 10.1 summarizes the mathematical expressions of four change-point models and their applications. In general, the change-point regression models are more suitable for predicting heating rather than cooling energy use. Indeed, these regression models assume steady-state conditions and are insensitive to the building dynamic effects, solar effects, and nonlinear HVAC system controls, such as on-off schedules.

Figure 10.4 provides an example of a three-parameter model for gas usage of a home in Colorado using the ambient temperature linear regression method. Figure 10.5 illustrates for the same home a two-parameter model using a heating degree-day linear regression model.

Other types of single-variable regression models have been applied to predict the energy use of heating and cooling equipment, such as pumps, fans, furnaces, and air conditioners. For instance, Phelan et al. (1996) used linear and quadratic regression models to obtain correlation between the electrical energy used by fans and pumps and the fluid mass flow rate.

Figure 10.6 shows the results of change-point regression analyses for several residences located throughout the United States using monthly utility data. Tables 10.2 and

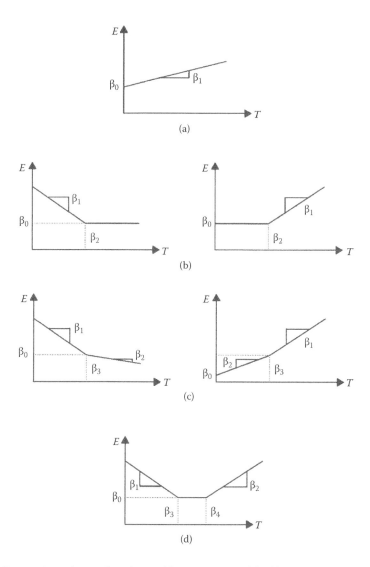

FIGURE 10.3 Basic forms of single-variable regression models: (a) two-parameter model, (b) three-parameter models, (c) four-parameter models, and (d) five-parameter model.

10.3 summarize the basic features for the residences and the coefficients for change-point regression models that best describe the energy use of each house.

10.3.3 Detailed Computer Simulation Models

Detailed computer simulation programs can be used to develop baseline models for building energy use. A brief discussion of the existing computer programs that can be applied to building energy simulation is provided in Chapter 3. The main feature

TABLE 10.1   Mathematical Expressions and Applications of Change-Point Regression Models

Model Type	Mathematical Expression	Applications
Two-parameter (2-P)	$E = \beta_0 + \beta_1.T$	Buildings with constant air volume systems and simple controls
Three-parameter (3-P)	Heating: $E = \beta_0 + \beta_1.(\beta_2 - T)^+$ Cooling: $E = \beta_0 + \beta_1.(T - \beta_2)^+$	Buildings with envelope-driven heating or cooling loads (most residential buildings follow this model)
Four-parameter (4-P)	Heating: $E = \beta_0 + \beta_1.(\beta_3 - T)^+ - \beta_2.(T - \beta_3)^+$ Cooling: $E = \beta_0 - \beta_1.(\beta_3 - T)^+ + \beta_2.(T - \beta_3)^+$	Buildings with variable air volume systems or high latent loads; also, buildings with nonlinear control features (such as economizer cycles and hot deck reset schedules)
Five-parameter (5-P)	$E = \beta_0 + \beta_1.(\beta_3 - T)^+ + \beta_2.(T - \beta_4)^+$	Buildings with systems that use the same energy source for both heating and cooling (i.e., heat pumps, electric heating and cooling systems)

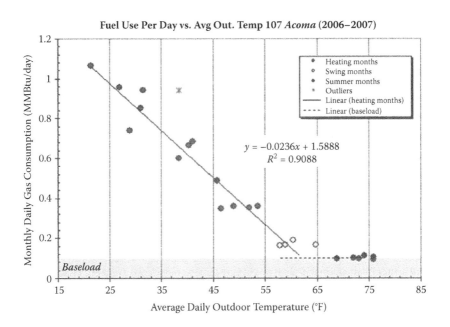

FIGURE 10.4   Three-parameter regression model based on outdoor ambient temperature for a home in Colorado.

specific to the application of computer simulation models to estimate retrofit energy savings is the calibration procedure to match the baseline model results with measured data (Hewett, 1986; Yoon et al., 2003; Zhu, 2006). Typically, the calibration procedure of a computer simulation model can be time-consuming and require significant efforts,

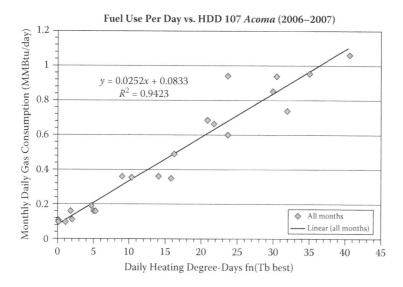

FIGURE 10.5 Two-parameter regression model based on heating degree-day for the same home considered in Figure 10.4.

TABLE 10.2 Characteristics of Houses Used in the Change-Point Regression Analysis

House	Location	State	Occupants	Cooling System	Heating System
1	Newton	Massachusetts	5	Central AC	Central gas heat
2	Anchorage	Alaska	2	None	Baseboard heat
3	Newton	Massachusetts	3	Central AC	Central gas heat
4	Newton	Massachusetts	4	Central AC	Central electric heat
5	Aberdeen	Maryland	2	Central AC	Central gas heat
6	Anchorage	Alaska	2	None	Central gas heat
7	Baltimore	Maryland	2	Central AC	Central electric heat
8	Hopkinton	Massachusetts	5	In-window/in-wall AC	Baseboard heat

TABLE 10.3 Values of Coefficients of the Change-Point Regression Models

House	Slope Heating (kWh/day °F)	T_b Heating (°F)	T_b Cooling (°F)	Slope Cooling (kWh/day °F)	Baseload (kWh/day)
1	0.00	27.0	76.3	0.00	16.4
2	−0.16	40.7	40.7	−0.15	11.8
3	−0.10	60.4	60.4	0.62	6.9
4	−2.75	60.6	76.4	0.00	18.4
5	0.00	29.9	63.1	1.84	18.9
6	−0.36	31.1	31.1	−0.15	23.0
7	−2.88	58.9	74.6	2.49	37.7
8	−0.21	57.6	65.6	0.68	25.8

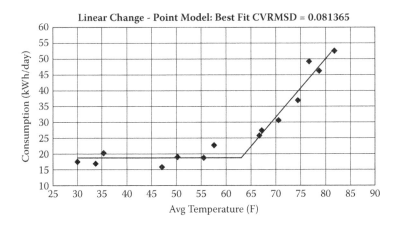

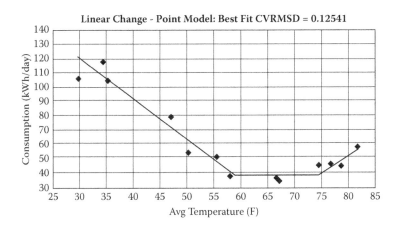

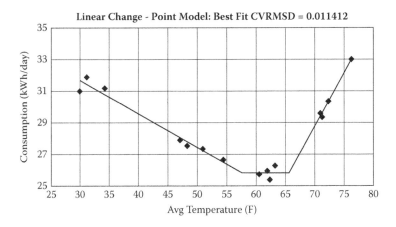

FIGURE 10.6   Change-point regression models for eight homes.

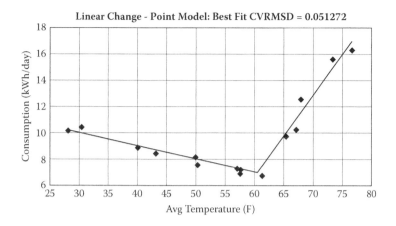

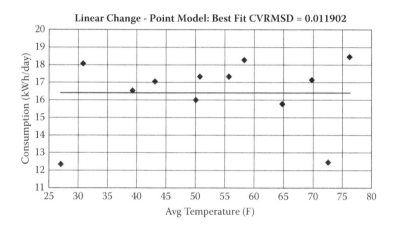

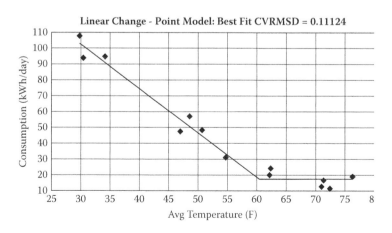

FIGURE 10.6 *(Continued)*

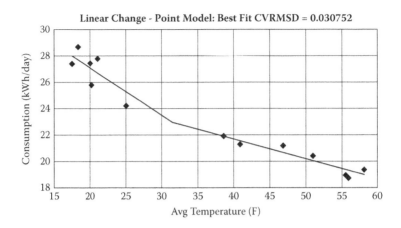

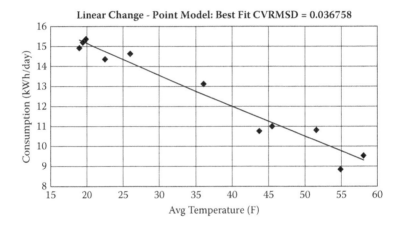

FIGURE 10.6 *(Continued)*

especially when daily or hourly measured data are used. Unfortunately, there is no consensus on a general calibration procedure that can be considered for any building type. To date, calibration of building simulation programs is rather an art form that relies on user knowledge and expertise. However, several authors have proposed graphical and statistical methods to aid in the calibration process, and specifically to compare the simulated results with measured data. These comparative methods can then be used to adjust either manually or automatically certain input values until the simulated results match the measured data according to predefined accuracy level. Figure 10.7 illustrates a typical calibration procedure for building energy simulation models.

Calibration of computer simulation models is generally time-consuming, especially for hourly predictions. Two methods have been used to compare the simulated results to the measured data and help reduce the efforts involved in the calibration procedure:

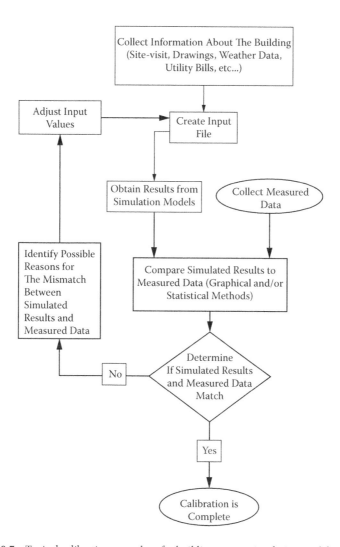

FIGURE 10.7 Typical calibration procedure for building energy simulation models.

- Graphical techniques to efficiently view and compare the simulated results and the measured data. Several graphical packages have been utilized or developed for energy computer simulation calibration, including two-dimensional time-series plots and *x-y* plots (for monthly and daily data), three-dimensional surface plots (for hourly data), and box-whisker-mean (BWM) plots (for weekly data). Haberl and Abbas (1998a, 1998b) and Haberl and Bou-Saada (1998) provide a detailed discussion of the most commonly used graphical tools to calibrate energy computer simulation models.

- Statistical indicators to evaluate the goodness of fit of the simulated results compared to the measured data. Several indicators have been considered, including mean difference, mean bias error (MBE), root mean square error (RMSE), and coefficient of variance (CV). A discussion of these indicators can be found in Haberl and Bou-Saada (1998). To obtain these indicators, classical statistical tools can be used, such as SAS (1989).

More recently, some attempts have been presented to make the calibration process more automatic using optimization techniques (Alspector, 2008; Slusher, 2010) using utility data. The calibration process can be reduced to minimizing the objective function presented by Equation (10.3), representing the average percent error (APE) in the monthly primary energy usage. In the United States, a significant fraction of buildings use two energy sources, typically electricity and natural gas.

$$APE = \sqrt{\sum_{m=Jan}^{Dec} \left[\frac{E_{m,util} - E_{m,sim}}{E_{yr,util}} \right]^2 + \left[\frac{G_{m,util} - G_{m,sim}}{G_{yr,util}} \right]^2} \qquad (10.3)$$

where $E_{m,util}$ = monthly electric utility consumption, $E_{m,sim}$ = monthly simulated electric consumption, $E_{yr,util}$ = annual electric utility consumption, $G_{m,util}$ = monthly natural gas utility consumption, $G_{m,sim}$ = monthly simulated natural gas consumption, and $G_{yr,util}$ = annual natural gas utility consumption.

Common optimization techniques considered in the automatic calibration approaches include the genetic algorithm (GA), the particle swarm optimization (PSO), and the sequential search technique (Tuhus-Dubrow and Krarti, 2009). An example of the results of calibration using the GA optimization is illustrated in Figure 10.8 for a

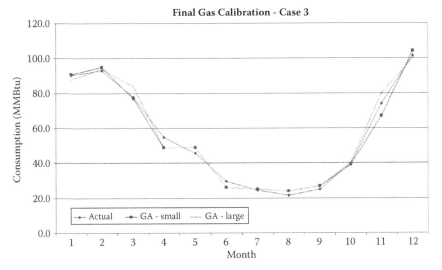

FIGURE 10.8   Comparison of the calibration results obtained for large- and small-parameter sets for monthly natural gas usage.

residence using both large- and small-parameter sets to carry out the calibration procedure (Alspector and Krarti, 2010).

Automatic calibration tools have been developed based on various techniques, including optimization techniques such as the brute force approach or genetic algorithm. Figure 10.9 illustrates the calibration results obtained from one tool developed by Slusher and Krarti (2011).

10.3.4 Calibration Criteria

ASHRAE Guideline 14 (ASHRAE, 2002) provides guidelines for ensuring reliable measurement and verification of energy savings from any implemented ECMs. The procedures outlined in the guideline are applicable to commercial, residential, and industrial applications to determine savings with reasonable accuracy. Savings are determined through comparison of pre- and postretrofit data with modeled or predicted data. Guideline 14 provides calculation of energy savings procedures for three different methods. The first method defined by Guideline 14 is the whole building approach, which uses whole building metered energy consumption data normalized for external factors, such as weather for the pre- and postretrofit period to determine savings. The second method is the retrofit isolation approach, which uses measured end-use consumption data for pre- and postretrofit periods to determine savings. The third method is the whole building calibrated simulation approach, which defines procedures for creating calibrated building energy simulation models for the determination of savings.

ASHRAE Guideline 14 (2002) recommends a set of statistical indicators as calibration tolerances, including monthly errors, annual error, and coefficient of variation of the root mean square of monthly errors. The error between the simulated energy consumption and the measured energy consumption (utility data) is analyzed on a monthly basis:

$$ERR_{month}(\%) = \left[\frac{(M-S)_{month}}{M_{month}} \right] \times 100 \qquad (10.4)$$

where the variable M is the measured energy consumption for the month in kJ, kWh, or Btu, and S is the predicted energy consumption obtained from the simulation model. The annual error is defined as

$$ERR_{year}(\%) = \sum_{month} \left[\frac{ERR_{month}}{N_{month}} \right] \qquad (10.5)$$

As indicated by Equation (10.5), the annual error is an average of the monthly error over the analysis year and may not accurately represent the true error due to sign value differences; therefore it is typically useful to calculate and provide the coefficient of variation. The coefficient of variation is found by first calculating the root mean square monthly errors using the following equation:

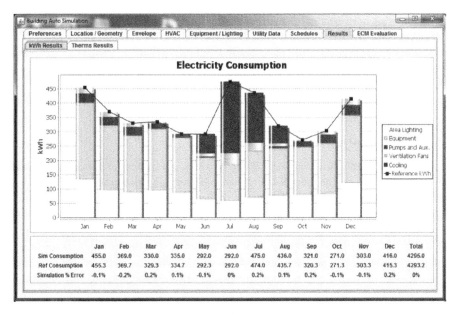

(a)

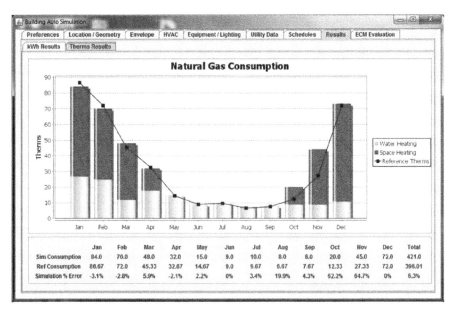

(b)

FIGURE 10.9   Results from automatic calibration tool for (a) electricity and (b) natural gas.

TABLE 10.4 Acceptance Criteria

Criteria	FEMP	ASHRAE 14
ERR_{Month}	$(\pm)15\%$	$(\pm)5\%$
ERR_{Year}	$(\pm)10\%$	—
$CV(RM\ SE_{Month})$	$(\pm)10\%$	$(\pm)5-15\%$

$$RSME = \sqrt{\frac{\sum_{month}(M-S)^2}{N_{month}}} \qquad (10.6)$$

The coefficient of variation can be found using Equation (10.7), where A_{month} is the average measured monthly consumption.

$$CV = \frac{RSME}{A_{month}} \qquad (10.7)$$

ASHRAE Guideline 14 and FEMP measurement and verification guidelines can be used to determine the acceptance criteria for calibrating the simulation models, as outlined in Table 10.4 (ASHRAE, 2002; FEMP, 2000).

10.4 Applications

The analysis methods presented in this chapter are used to measure and verify savings from energy retrofit projects. These methods are recommended as part of the International Performance for Monitoring and Verification Protocol (IPMVP, 2002, 2007), which identifies four different options for energy savings estimation procedures, depending on the type of measure, accuracy required, and cost involved:

- *Option A* uses simplified calibration methods to estimate stipulated savings from specific energy end uses (such as lighting and appliances). This option may use spot or short-term measurements to verify specific parameters, such as the electricity demand when lights are on. Common applications of this option involve lighting retrofits (such as replacement with more energy-efficient lighting systems and use of lighting controls).
- *Option B* typically involves long-term monitoring in an attempt to estimate the energy savings from measured data (rather than stipulated energy consumption). Simplified estimation methods can be used to determine the savings from the monitored data. The energy savings obtained from specific residential building retrofit measures (i.e., weather stripping, addition of thermal insulation, or replacement of appliances) is commonly estimated using option B.

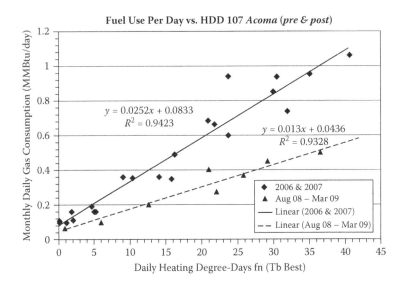

FIGURE 10.10   Regression models for pre- and postretrofit regression analysis of monthly natural gas usage.

- *Option C* is generally applied to estimate savings by monitoring whole building energy use and by using regression analysis to establish baseline models. The regression models can be developed from monthly or daily measured data. Examples of energy retrofit projects for which option C can be used include air conditioner, furnace, or boiler replacements. Figure 10.10 illustrates the pre- and postretrofit regression models used to estimate energy savings for a home.
- *Option D* is typically used when hourly savings need to be estimated. In addition to hourly monitoring, this option requires complex data analysis to establish baseline models. Dynamic models and calibrated simulation models are typically used to estimate hourly energy savings. The impact of energy management control systems is one of several applications for which option D can be considered.

Table 10.5 summarizes the requirements of each option, including typical accuracy levels of savings estimation and the average cost (expressed as a percent of the total cost required to implement the energy retrofit project) needed for metering equipment and for data analysis.

10.5 Uncertainty Analysis

ASHRAE Guideline 14 (2002) provides a method to estimate the uncertainty in determining the energy savings from retrofit projects. The uncertainty in estimating savings is a function of the coefficient of variance (CV) of the root mean square error (RMSE) and of the baseline model, corresponding *t*-statistic, fractional savings, and number of pre- and postretrofit months. The equation for the fractional uncertainty in savings measurements, either utility analysis or computer simulation model, is

TABLE 10.5 Basic Characteristics of the Various Options for Procedures to Estimate Energy Savings

	Models for			Metering	Applications		
Option	Data Analysis	Accuracy	Cost	Requirements	End Use	Load	Operation
A	Simplified methods	20–100%	1–5%	Spot/short term	Subsystem	Constant	Constant
B	Simplified methods	10–20%	3–10%	Long term	Subsystem	Constant/ variable	Constant/ variable
C	Regression analysis	5–10%	1–10%	Long term (daily or monthly)	Building	Variable	Variable
D	Dynamic analysis/ simulation	10–20%	3–10%	Long term (hourly)	Subsystem/ building	Variable	Variable

$$\frac{\Delta E_{save,m}}{E_{save,m}} = t \cdot \frac{1.26 \cdot CVRMSE\left[\left(1+\frac{2}{n}\right)\frac{1}{m}\right]^{1/2}}{F} \qquad (10.8)$$

where

$$RMSE = \left[\sum_{i=1}^{n}\left(E_i - \hat{E}_i\right)^2_{pre} / \left(n-p\right)\right]^{1/2} \qquad (10.9)$$

t = Student's t-statistic at selected confidence interval, n = number of preretrofit months used in model, m = number of postretrofit months used in savings analysis, F = percent savings, and p = number of model parameters (p = 3 for utility analysis, p = 1 for calibrated simulation).

This method of uncertainty analysis assumes zero measurement error, which holds true when using monthly utility data. The fractional uncertainty in the savings formula taken from Guideline 14 is shown graphically in Figure 10.11 for varying coefficients of variation, 12 months of preretrofit data, and a 68% confidence interval.

Figure 10.11 indicates that fractional uncertainty in predicted energy or cost savings decreases with the availability of more postretrofit data and better baseline models (i.e., lower CV values, as estimated using Equation (10.8)). The fractional uncertainty in predicted savings is also dependent on the estimated fraction of savings. For example, a model with a CV of 15%, an expected savings fraction of 15%, and 3 months of postretrofit data has a y-ordinate value of .02 from the above chart and yields a fractional uncertainty in savings of (.02/.15)*100 = 13%. The same model with an expected savings fraction of 40% would yield a fractional uncertainty in predicted savings of (.02/.4)*100 = 5%.

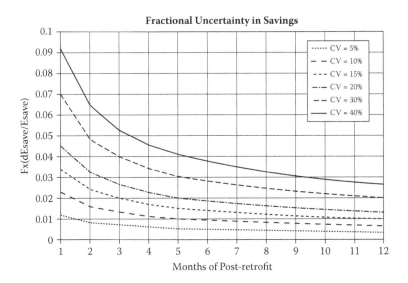

FIGURE 10.11   Fractional uncertainty in savings for varying CVs.

10.6 M&V Case Study

10.6.1 Background

In order to assess the effectiveness of a weatherization program for low-income houses, an evaluation was performed to assess the effectiveness of energy conservation measures and renewable energy technologies for 30 low-income housing units (Guiterman and Krarti, 2011). In particular, energy audits were performed, including walkthrough audits, blower door tests, and infrared scans. Historical utility data for all units were prescreened and analyzed to determine baseline energy use. Moreover, six representative units were modeled using a detailed whole building energy simulation tool and calibrated against the historical data.

The calibrated building energy models were used to simulate numerous energy conservation measures (ECMs). Recommended ECMs included reducing hot water heater set points, installing programmable thermostats, replacing light bulbs with CFLs, adding additional roof insulation and repairing existing insulation, insulating existing water heater tanks/pipes, installing faucet aerators, replacing furnaces with high-efficiency units, and installing tankless water heaters in 20 units. Reducing infiltration through leaky mechanical closets in the one- and two-bedroom units was given a high priority and was achieved by installing direct-vented furnace and water heater systems in these units.

Ground source heat pumps (GSHPs) and solar hot water panels were also included in the original analysis. Selected ECMs were implemented, as shown in Table 10.6, for various units, depending on their floor area.

TABLE 10.6 Implemented Energy Conservation Measures

No. Bedrooms	1	2	3	4
Area (ft²)	600	750	950	1,100
Number of units	10	8	10	2
New furnace installed	X			
Existing furnace tune-up		X	X	X
Programmable thermostat		X	X	X
Tankless water heater	X	X		
CFL bulbs	X	X	X	X
Roof insulation repaired	X	X	X	X
Roof insulation added	X	X	X	X
Reduce DHW temp. to 120°F	X	X	X	X

TABLE 10.7 Results of Pre- and Postretrofit Blower Door Tests (Air Changes per Hour)

Unit	Preretrofit (ACH)	Postretrofit (ACH)	% Change
Unit 1-A	0.89	0.35	61%
Unit 2-A	0.67	0.33	51%

To assess the effectiveness of the weatherization program for low-income homes, a verification analysis was conducted to measure the actual energy savings incurred for the 30 housing units. M&V of the energy savings follows ASHRAE Guideline 14-2002, *Measurement of Energy and Demand Savings* (ASHRAE, 2002). This protocol was chosen due to the clarity of the analysis procedure and broad acceptance of the guideline within the energy efficiency community.

The M&V of the energy savings due to the implemented ECMs began with on-site data collection and was completed with the collection of two full postretrofit years of utility data. Postretrofit blower door tests were completed, and data logging captured temperature settings in several units. On-site verification and occupant/staff interviews confirmed the installations of the specified measures.

10.6.2 M&V Methods

Blower door tests were performed on one- and two-bedroom units before and after the retrofits. Table 10.7 summarizes the results of the leakage tests and indicates that air infiltration was reduced by 61 and 51% for the one- and two-bedroom units, respectively.

Two M&V analysis procedures were utilized in this study and include the calibrated simulation approach and the whole building approach, both described in ASHRAE Guideline 14. Two methods are employed for the whole building approach, both using single-variable regression analysis of pre- and postretrofit utility data against actual weather data. The first method correlates energy use to average outdoor temperature,

TABLE 10.8   Calibration Errors for the Modeled Units

Unit ID	Electricity		Natural Gas	
	Annual Error	CV (RMSE)	Annual Error	CV (RMSE)
1-A	−1.9%	6.3%	N/A	N/A
1-B	3.5%	6.2%	N/A	N/A
1-C	−2.3%	6.7%	N/A	N/A
Triplex (1-A/1-B/1-C)	N/A	N/A	−4.6%	10.6%
2-A	−1.2%	6.3%	−2.0%	6.3%
2-B	−3.5%	5.9%	−1.9%	10.1%
3	0.2%	3.0%	−0.4%	7.2%
4	2.2%	3.8%	−3.8%	8.7%

referred to as the temperature-based method, and the second correlates energy use to heating degree-days, referred to as the degree-day-based method.

10.6.2.1 Calibrated Simulation Approach

DOE2.2-based software was utilized in the original preretrofit analysis to determine the most cost-effective ECMs to recommend for implementation. Seven units were modeled to reflect the range of unit types. Three one-bedroom units in a triplex building, with individual electric meters but one shared gas meter, were modeled. Two two-bedroom units and two three-bedroom units were also modeled. Table 10.8 shows the annual calibration errors for electricity and gas use for the six units modeled. All models are calibrated to 2 years of preretrofit utility data. Annual errors are within 5% of the utility data, and the coefficients of variation of the root mean squared error are less than 15%, the maximum value specified in ASHRAE Guideline 14. The energy model for the triplex unit was challenging to calibrate due to the fact that only one of the three occupants of the triplex was available for preretrofit interviews, so there is more uncertainty in the occupancy behavior and schedule patterns for two of the units. The whole building gas meter also adds additional complexity to the calibration process, as there are three separate models that need to be independently adjusted to match the utility data.

All available information from utility bills, occupant interviews, walkthrough audits, and blower door tests was utilized to accurately calibrate the energy models. Figure 10.12 shows the monthly natural gas consumption for the energy model compared to historical utility data.

Each ECM was simulated individually using the energy models, and all ECMs were simulated as one "bundled" package of ECMs to arrive at the total estimated savings and to capture any interactive effects, such as an increased heating load due to reduced electricity use with CFL bulbs.

10.6.2.2 Whole Building Approach

The temperature-based method provides information about the base water heating load, the fuel utilized for space heating, the building load coefficient (BLC), and the balance temperature of the house, as discussed in Chapters 3 and 5. This method was also used in the original preretrofit analysis to estimate the base water heating load and calibrate the

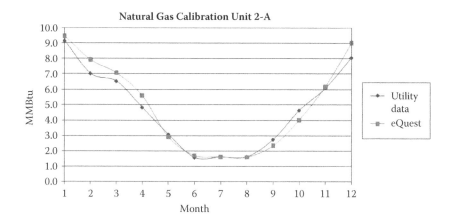

FIGURE 10.12 Results of the calibration procedure using natural gas use data for unit 2-A.

energy models. For the M&V analysis, it is used to estimate natural gas savings from ECM implementation, to provide a calculated balance temperature for the pre- and postretrofit periods, and to calculate the BLC for the pre- and postretrofit periods for comparison.

The degree-day-based method is based on the relationship between natural gas use and heating degree-days, resulting in a linear model that can be used to project energy use, assuming no retrofit took place and a model to project postretrofit gas use. This method is commonly referred to as the PRISM (Princeton Scorekeeping) method (Fels, 1986).

10.6.2.3 Temperature-Based Method

The temperature-based method utilizes monthly outdoor average temperature readings from the National Oceanic and Atmospheric Agency (NOAA) National Climactic Data Center (NCDC) online databases (NCDC, 2010). Monthly utility data are normalized over the days per month to derive a "monthly daily" fuel use per day in MMBtu/day (million Btu/day). This daily fuel use is then plotted against the average outdoor temperature for each month.

An example of the results of the regression analysis is shown in Figure 10.13 for both pre- and postretrofit periods. Base load gas use for water heating is reduced for this unit, and the more gradual slope of the heating line in the postretrofit case shows that demand for heating has decreased due to the retrofit measures.

The intersection of the regression line for the winter gas use with the base load represents the balance temperature of the house. The balance temperature represents the outdoor temperature below which space heating is required. The slope of the winter regression line allows for the BLC to be estimated using Equation (10.10), and the annual fuel used for heating is then determined from Equation (10.11), as noted in Chapter 3, since the units are conditioned 24 h per day.

$$BLC = Slope \times \frac{\eta_{heating}}{24} \qquad (10.10)$$

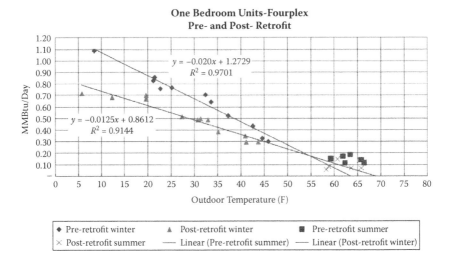

FIGURE 10.13   Pre- and postretrofit energy use vs. outdoor temperature.

$$E_{heating} = 24 \times BLC \times \frac{DD_{heating}}{\eta_{heating}} \tag{10.11}$$

Energy savings are calculated for the temperature-based method using the following equations:

$$\% \, Savings = 1 - \frac{E_{post}}{E_{pre}} \tag{10.12}$$

$$E_{pre} = \left[\left(24 \times BLC_{pre} \times \frac{DD}{\eta_{pre}} \right) + \left(DHW_{Summer_{pre}} \times 12 \right) \right] \tag{10.13}$$

$$E_{post} = \left[\left(24 \times BLC_{post} \times \frac{DD}{\eta_{post}} \right) + \left(DHW_{Summer_{post}} \times 12 \right) \right] \tag{10.14}$$

10.6.2.4 Degree-Day-Based Method

The degree-day-based method plots fuel use per day vs. daily heating degree-days (monthly total divided by number of days per month) for each month of the pre- and postretrofit periods. Unlike the temperature-based method, all months, including winter, summer, and swing months, are used in the analysis. The pre- and postretrofit

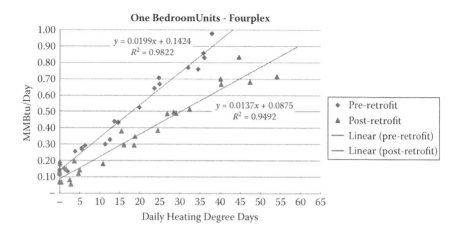

FIGURE 10.14 Degree-day-based natural gas use regression analysis.

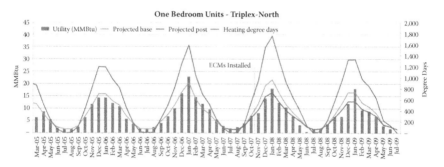

FIGURE 10.15 Degree-day-based projected energy use.

period degree-days are a function of the balance temperature selected. The calculated balance temperatures resulting from the temperature-based method described above provide a starting point for the regression analysis. The balance temperature can then be adjusted to ensure the highest R^2 value for each regression. This is also referred to in Kalinic (2009) as the best balance temperature. Figure 10.14 shows an example of the natural gas regression analysis for the degree-day-based method.

The slope and intercept from the linear equation describing the correlation between energy use and heating degree-days for each period provides the input parameters to the model that projects baseline energy use and postretrofit energy use, as shown in Figure 10.15.

The degree-day-based method allows for the calculation of energy savings following the ASHRAE Guideline 14 (ASHRAE, 2002) protocols, by subtracting the projected postretrofit energy use from the projected baseline energy use, as shown in Equation (10.15).

$$Energy\ Savings = E_{Projected\ baseline} - E_{Projected\ post} \qquad (10.15)$$

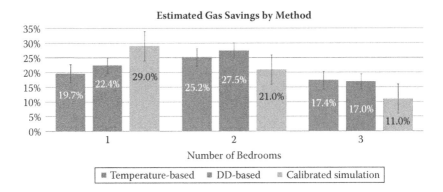

FIGURE 10.16   Estimated gas savings by analysis method and unit type.

10.6.3 Selected Results

Figure 10.16 shows average savings estimates for each unit type and M&V analysis method. For the one-bedroom units, the temperature-based method underpredicted savings relative to the degree-day-based method and the calibrated simulation approach by 3 and 10%, respectively. These units were retrofitted with new tankless water heaters and furnace replacements in addition to general weatherization. The one-bedroom units did not receive programmable thermostats. These units saved 22% of preretrofit natural gas use, but the total ECM costs were over $2,700, resulting in simple payback periods of nearly 18 years.

The temperature-based method and the degree-day-based method estimated savings at nearly identical levels, 27%. Both methods overpredicted savings relative to the calibrated simulation approach by approximately 6%. These units received new tankless water heaters in addition to general weatherization and programmable thermostats.

The temperature-based method and the degree-day-based method estimated savings at nearly identical levels, 17%. Both methods overpredicted savings relative to the calibrated simulation approach by approximately 6%. These units received general weatherization, including programmable thermostats. The three-bedroom units obtained the most cost-effective results, with savings of 17% of preretrofit natural gas use and ECM costs of only $361/unit, resulting in a simple payback period of just over 2 years.

Figure 10.17 shows the estimated savings by each method averaged over all the housing units.

The calibrated simulation approach and whole building approach methods estimated similar levels of total natural gas savings for all units, with all three approximating 18 to 21% savings, representing a total range of only 2.5%. The temperature-based method and degree-day-based method both predicted savings within approximately 1% of the other, at 20.0% and 21.2%, respectively, while the calibrated simulation approach estimated savings of 18.7%.

Based on these findings, verification of actual natural gas savings across all unit types is taken to be the average of all three methods, or 20.1% of preretrofit natural gas use. The results indicate that units receiving programmable thermostats achieved cost-effective

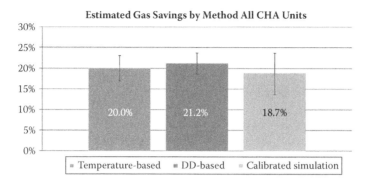

FIGURE 10.17 Estimated natural gas use savings for housing units and for each M&V analysis method.

savings, as the three-bedroom units received inexpensive measures yet reached average savings of 17%.

The energy savings from the new furnaces may not be fully realized, as these units were not retrofitted with programmable thermostats. It would be worthwhile to install programmable thermostats in a select number of one-bedroom units and monitor energy use to compare to the units that did not receive new thermostats.

Tankless water heaters are clearly expensive systems that can generate significant energy savings, yet payback periods for the units with tankless heaters are still greater than 10 years.

10.6.4 Summary and Conclusions

The measurement and verification analysis for the energy conservation measures installed for the 30 low-income housing units reveals that the units experienced significant natural gas savings and negligible electricity savings. The calibrated simulation approach and the two whole building approaches, the temperature-based and degree-day-based methods, all estimated natural gas savings within a range of 2.5%, from 18.7 to 21.2%.

The calibrated simulation approach underestimated gas savings for the units receiving general weatherization and new tankless water heaters, yet overestimated savings for the units receiving new furnaces. Calibrated simulation is a costly and time-intensive analysis method, and the energy savings estimates are within the error bounds of both of the whole building approach methods.

The temperature-based method requires the least amount of input weather data and processing to estimate energy savings, and yields results consistent with the degree-day-based method for units retrofitted with both HVAC and weatherization measures. It is therefore a viable and effective methodology of verifying the savings from energy conservation measures and would likely save time and reduce analysis efforts compared to both the degree-day-based method and calibrated simulation approach.

The two-bedroom units achieved the highest savings, more than the one-bedroom units that received new 90% efficient, condensing furnaces. The three-bedroom units achieved the most cost-effective savings, receiving only general weatherization and programmable thermostats at a cost of approximately $360/unit and saving 17% of natural gas use, or $160/year. Simple paybacks for these units are just over 2 years. The one- and two-bedroom units received over $2,000 per unit in retrofits and achieved savings ranging from 22 to 27% ($155 to $200/year) of preretrofit natural gas use, respectively. Payback periods ranged from 10.5 years for the two-bedroom units to 17.8 years for the one-bedroom units.

Programmable thermostats can help tenants achieve significant energy savings when their capabilities are utilized properly. The temperature monitoring shows that some units clearly do not take advantage of the automated scheduling for the thermostats. However, the energy savings achieved by the three- and two-bedroom units prove that several tenants do indeed utilize the setback capabilities for the programmable thermostats. Tankless water heaters are shown to be an expensive measure that yields rather long payback periods, typically over 10 years.

10.7 Summary

In this chapter, methods used to estimate energy savings from retrofit projects are briefly presented. The common applications of the presented methods are discussed with some indication of their typical expected accuracy and additional analysis efforts and costs. A case study has been presented to discuss application of common M&V approaches to estimate retrofit energy savings for 30 homes. As mentioned throughout the chapter, the presented measurement and verification protocols are applicable to estimate energy savings obtained from energy conservation measures applied to existing residential buildings. However, similar procedures can be applied for other applications, such as energy efficiency in new buildings, water efficiency, renewable technology, emissions trading, and indoor air quality.

11

Energy Audit Reporting Guidelines

Abstract

In this chapter, guidelines for reporting the energy audit process and findings are first provided. These guidelines are suitable to report results from a walkthrough energy audit as well as a standard and detailed energy audit. In particular, specific examples are provided to present findings from basic procedures to assess the energy performance of the residential buildings, including field tests to estimate building air leakage characteristics and temperature measurements. The reports also present analysis methods, described in previous chapters, to estimate energy use and cost savings from several energy conservation measures.

11.1 Reporting Guidelines

In order to summarize the findings of various tasks carried out in an energy audit, it is important to submit a well-documented report. The report should describe in detail the tasks completed and the analysis considered for the energy audit. The level of details in the report depends on the type of energy audit completed. Chapter 1 defines and discusses the various energy audit types. In the following sections, a set of guidelines is provided to help the auditor write the final report after completing an energy audit for an existing building.

11.1.1 Reporting for a Walkthrough Audit

The walkthrough audit can be a stand-alone task or part of a standard energy audit. Typically, this type of audit is sufficient for buildings with simple energy systems, including housing units. The basic tasks to be carried for a walkthrough audit include:

Task 1: Describe the basic energy systems of the building, including building envelope, mechanical systems, and electrical systems. Observations from the walkthrough as well as specifications from architectural, mechanical, and electrical drawings can be utilized to describe the building features.

Task 2: Perform basic testing and measurements to assess the basic performance of various energy systems. These measurements may depend on the type of building and its systems as well as time available for the auditor. For residential buildings, it is highly recommended to perform a pressurization or depressurization leakage testing using the blower door test kit, as outlined in Chapters 3 and 5. In all building types, spot measurement, and if possible, monitoring indoor air temperature and relative humidity within the space for at least 1 day, is helpful to estimate the indoor temperature settings and identify or check any comfort issues.

Task 3: Discuss with building occupants or building operators to identify any potential discomfort problems and sources of energy waste within the building. This task is often helpful to define potential operation and maintenance measures as well as energy conservation measures.

Task 4: Identify some potential operation and maintenance (O&M) measures and energy conservation measures (ECMs) as well as any measures required to improve comfort problems. In this task, the implementation details and associated costs should be estimated.

Task 5: Evaluate the energy savings (or requirements if measures are needed to improve comfort) using simplified analysis methods presented in Chapters 5 to 7.

Task 6: Perform cost analyses based on the simple payback period method to determine the cost-effectiveness of the identified O&Ms and ECMs. The cost data should be taken from actual estimations from local contractors.

The report of a walkthrough energy audit can be brief and includes at least the basic recommendations for cost-effective O&Ms and ECMs, that is, the results of task 6 described above. However, it is strongly recommended that a more detailed report be drafted and delivered to the client to document the findings and observations obtained from the completed tasks. In particular, the report should describe the basic features of the audited building as well as any potential problem areas identified during the walkthrough. Moreover, the calculations to estimate energy use and cost savings should be explained and presented in the report at least for the recommended energy conservation measures. In addition, references and specifications for implementing the recommended O&Ms and ECMs should be provided. A final report for a walkthrough energy audit can include the following sections:

1. Legible and complete drawings showing the floor plan and at least two elevation views. Examples of elevation views providing the dimensions of the walls and the windows are shown in Figure 11.1.

2. A brief description of the architectural features and energy systems of the building, including construction details, heating and cooling systems, and operation schedules. Table 11.1 provides a description of basic features of an audited house.

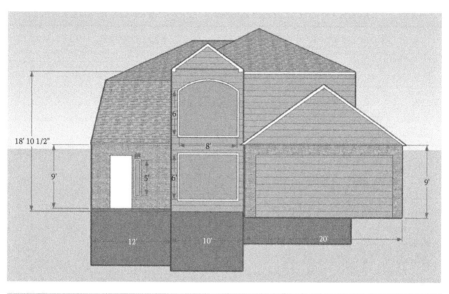

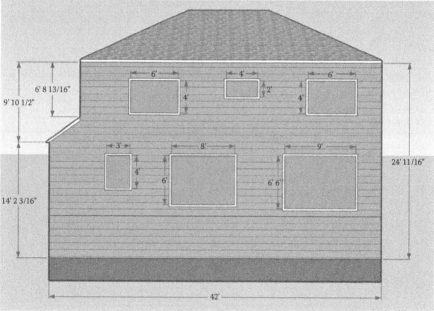

FIGURE 11.1 Elevation views for the audited house.

3. An analysis of the utility bills to estimate energy-use intensity, the building BLC, the balance temperature, and base loads. It is useful to perform this task before visiting the building. Figure 11.2 illustrates the utility analysis using the correlation between natural gas use and outdoor temperature to estimate the base load, the BLC, and the balance temperature of an audited house.

TABLE 11.1   Summary of Architectural Features and Energy Systems of the Audited House

	Specifications	Comments
Construction Details		
Roof	Shingles, 12–24 ft blown-in fiber glass insulation	
Walls	Siding, 2 × 6 frame construction, batt insulation, gypsum board	
Glazing	Clear double pane	
Doors	North façade: one 7 × 3 ft solid-wood door	Model rear door as window
Window/Wall Ratio		
North façade	15%	
East façade	16%	
South façade	23%	
West façade	4%	
Occupancy		
Schedule	2 adults, 24/7 2 children, 4 p.m.–8 a.m. M–F, 24/7 S/S	
LPD	0.55 W/ft^2	Sparing use during daylight
EPD	0.14 W/ft^2	
HVAC		
Type	Forced-air gas furnace	
Efficiency	80%	Assumed standard efficiency
Set point	65°, except 62° from 10 p.m. to 6 a.m.	Programmable thermostat in use
DHW		
Capacity	40 gallons	
Efficiency	73%	

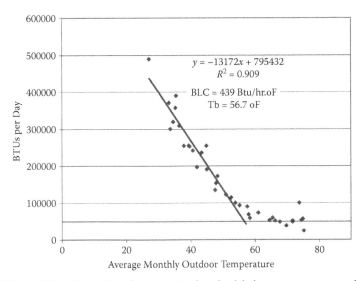

FIGURE 11.2   Utility data analysis for estimating base load, balance temperature, and BLC.

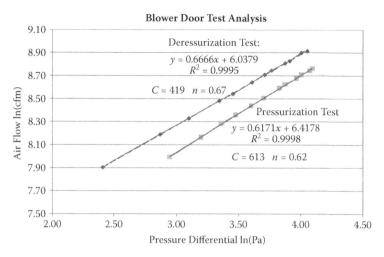

FIGURE 11.3 Analysis pressurization and depressurization tests.

4. A description of any testing procedures or measurements carried out during the walkthrough audit, including temperature and air leakage evaluations. For air leakage testing, all the relevant details of the testing and calculation analysis should be outlined. The analysis reporting should present the air leakage area as well as the infiltration rates (in ACH) under reference conditions (i.e., $\Delta P = 4$ Pa) and for average weather conditions (annual average and heating season average). Figure 11.3 illustrates the analysis of pressurization and depressurization to estimate the leakage characteristics of an audited house.

5. A discussion of the walkthrough audit tasks and its outcome. In particular, the report should highlight any occupant concerns and complaints and any identified potential O&Ms and ECMs. A list of O&Ms and ECMs identified during a walkthrough audit can include the following measures:

- Reducing the hot water temperature
- Replacing the domestic hot water heater with a high-efficiency model
- Weather stripping around exterior doors and windows
- Replacing all incandescent fixtures with CFLs
- Installing low-flow showerheads

6. A description of the calculation details to estimate energy use and cost savings for the considered O&Ms and ECMs. References for these calculations should be provided in the report, including any assumptions that were made to carry out the estimations. An example of simplified calculation of the energy savings associated with replacing the incandescent lamps with CFLs is shown in Table 11.2.

7. A summary of the energy and economic analysis results. In particular, the general procedure and the cost of implementing each ECM should be described. Table 11.3 summarizes the simple payback periods (SPPs) for selected O&Ms and ECMs considered for the walkthrough audit.

TABLE 11.2   Energy Savings Associated with Lighting Retrofit

Existing lighting kW	1,690 W	35 incandescent lamps
Retrofit lighting kW	768 W	35 CFLs
kW savings	922 W	
Based on occupant interviews and utility bills, the annual burn hours are estimated:		
Hours/day	8 h	
% of lamps in use	50%	
Effective annual full load hours:		
= 8 h/day * 25% * 365 days/year	1,460	
Total savings	1,346 kWh	
Cost of electricity	$0.10/kWh	
Total savings	$135	

TABLE 11.3   Summary of the Economic Analysis Results for Selected O&Ms and ECMs

Measure	kWh	Therms	EC	Savings	Initial Cost	SPP
Baseline case (existing house)	5,783	562	$1,174	n/a	n/a	
Lower DHW set point	5,783	512	$1,121	$53	$0	0
Replace DHW with high-efficiency model	5,783	496	$1,104	$70	$280	4.0
Replace incandescent with CFLs	4,246	592	$1,052	$122	$80	0.7
Reduce infiltration	5,731	429	$1,028	$146	$76	0.5
Replace with low-flow fixtures	5,783	537	$1,148	$27	$39	1.5

8. Specific recommendations to the client to reduce the utility bills or to improve the indoor environment within the building. For the O&Ms and ECMs evaluated in the audit and summarized in Table 11.3, the auditor may recommend that the homeowner implement the following measures by order of priority:

- Reducing the hot water temperature
- Weather stripping around exterior doors and windows
- Replacing all incandescent fixtures with CFL
- Installing low-flow showerheads
- Replacing the domestic hot water heater with a high-efficiency model

9. Some photos to highlight some of the features and the problem areas of the house. Some photos, including thermal images taken during the walkthrough of the energy audit, are shown in Figure 11.4.

11.1.2 Reporting for a Standard Audit

The report of a standard audit is more comprehensive than a report for a walkthrough audit outlined above. Indeed, a standard audit, as defined in Chapter 1, includes additional tasks and requires more effort and time to complete. This type of audit is typically suitable for large buildings, such as apartment buildings with complex energy systems.

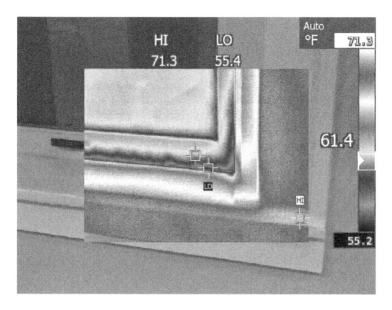

FIGURE 11.4   A thermal image of the ceiling fan indicating a source of high air leakage.

Moreover, the utility bills for large buildings are significantly high and can justify the level of details required by a standard audit. In addition to the tasks described for the walkthrough audit, the following tasks can be carried out as part of a standard audit:

Task 1: Carry out a detailed survey of lighting and electrical equipment. The main goal of this task is to estimate the lighting and equipment power densities within various spaces of the building.

Task 2: Identify heating, ventilating, and air conditioning (HVAC) systems and their operation schedules. This task is often crucial since energy used by HVAC systems is a significant portion of the total energy consumed in large buildings.

Task 3: Determine the main discomfort and complaints of occupants through a well-designed questionnaire. Surveying occupants very often provides valuable information about the performance of the building and its energy systems throughout the year.

Task 4: Collect and analyze utility data for at least 3 years. Utility data for only 1 year are often not sufficient to estimate the historical energy performance of a building. In some cases, special conditions such as special events or extreme weather may create biases in the energy use of the building.

Task 5: Perform any relevant measurements, such as lighting levels, thermal images, indoor temperatures, airflow rates supplied by air handling units, and electrical energy end uses, as well as indicators of electrical power quality.

Task 6: Model the existing building using a detailed energy simulation tool. Ensure that the simulation model is well calibrated using utility data. Typically, monthly calibration within 10% is required to increase the confidence level in the predictions of the building energy simulation model.

Task 7: Perform engineering calculations to estimate energy savings from potential energy conservation measures using both the calibrated energy simulation model and simplified calculation procedures outlined in this book.

Task 8: Perform an economic analysis using simple payback period, net present worth, or life cycle cost analysis methods for all the energy conservation measures. The implementation details and costs should be provided for each measure.

Task 9: Select the energy conservation measures to be recommended for implementation. In addition, specify the additional benefits of each measure (such as improving the thermal or visual comfort), the implementation costs, and any information that would help the client implement these measures.

The report of a standard energy audit should summarize the results of all the completed tasks. A recommended outline for the standard energy audit report is provided below. It should be noted that some tasks outlined above for the standard energy audit include the tasks discussed for the walkthrough energy audit.

1. Legible and complete drawings showing the floor plan and elevation views of the building.
2. A brief description of the features of the building and its systems, including building envelope, lighting systems, office equipment, appliances, HVAC systems, and heating and cooling plant.
3. A summary of the walkthrough audit findings and results from any testing and measurements. Typically for a standard audit, monitoring of indoor temperature and relative humidity is carried for at least 1 week.
4. A summary of the survey results for lighting, electrical, and HVAC systems as well as the results of the questionnaire of occupants.
5. A discussion of the results of the utility data analysis. The energy use intensity of the building can be compared to some available and established benchmarks.
6. Basic assumptions made to model the building using a detailed simulation tool (such as eQUEST and EnergyPlus). A three-dimensional rendering of an audited house modeled using eQUEST is shown in Figure 11.5.
7. Description of the calibration process, including discussion of the parameters that were utilized to match simulation tool predictions with utility data. Figure 11.6 illustrates the results of a calibration process of the eQUEST model of Figure 11.5.
8. A discussion of the implementation details for the various ECMs that are considered for the building. The assumptions made to estimate the energy use and cost savings should be provided.
9. A summary of the economic analysis using simple payback and LCC methods. Specific quotes from local contractors should be used in the economic analysis. However, established references such as RS Means (2008) can also be considered for cost estimates.
10. A list of the implementation priority for the recommended ECMs based on the economic analysis results and the budget constraints.

Case studies are presented in Chapter 12 to illustrate the analysis procedures and reporting guidelines discussed in this chapter for both walkthrough audits and standard energy audits performed for residential buildings.

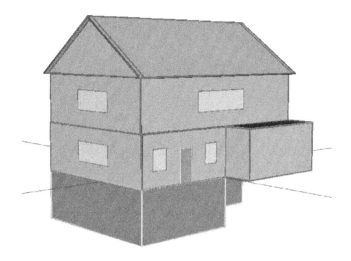

FIGURE 11.5 Simulation model of a three-dimensional rendering for an audited house.

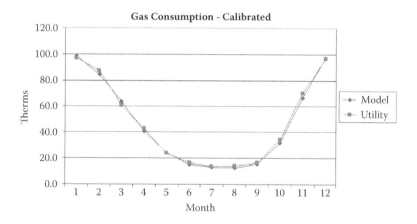

FIGURE 11.6 Predictions of calibrated eQUEST model against utility data for natural gas usage.

12

Case Studies

Abstract

In this chapter, two case studies are presented to illustrate the various analysis procedures presented throughout this book. The case studies include a walkthrough energy audit and a detailed energy audit of detached single-family houses. In the reports for the two audit types, typical findings are outlined as well as the basic procedures used to measure the energy performance of the buildings, including tests to estimate building air leakage characteristics. The reports also present analysis methods to estimate energy use and cost savings from several energy conservation measures. In particular, the results of a series of economic analyses are presented for the two case studies to assess the cost-effectiveness of the identified energy conservation measures.

12.1 Case Study 1: Walkthrough Audit

In this section, an example of a residential building walkthrough is presented. As outlined in Chapter 1, a walkthrough audit allows the collection of basic information about the building envelope (windows, walls, and doors), as well as the lighting fixtures, appliances, and heating and cooling systems. During a walkthrough audit, the auditor should meet and talk to the building owners and occupants to determine any problematic areas of the building related to thermal comfort and energy performance. The main purpose of a walkthrough audit is to provide recommendations for improving the energy efficiency of the residence by investigating selected operating and maintenance measures (O&Ms) and energy efficiency measures (EEMs) with short payback periods. A report of a walkthrough audit is provided in this section.

12.1.1 Building Description

Table 12.1 summarizes basic features of an audited house located in Colorado. The main concern communicated by the occupants is the draftiness of the house, especially the downstairs area. Another concern is the energy required to heat the hot tub, which is used from October through May.

TABLE 12.1 Basic Features of the Audited House

Age of building	70 years (upgrades made over the past 30 years)
Square footage	2,200 ft^2
Number of floors	2 floors mostly above grade
Basement or crawlspace	Some cellar/storage space
Method of construction	Original log construction downstairs; wood siding upstairs
Utilities	Propane and electricity
Condition of house	Fair-good
Additional notes	Mountain property

The following sections describe the basic findings of the walkthrough audit:

Building envelope: By inspecting the building envelope, the walls of the upper level of the house were found to have some insulation, including R-11 batt insulation. However, the lower floor exterior walls have no insulation (the original wall consists of logs plus drywall). The roof is built from a standard wood frame construction with an aluminum exterior finish. The roof appears to be in good condition with an adequate insulation level in the ceiling. The homeowner has indicated that R-38 insulation was added to the ceiling below the roof during a recent renovation of the house. Most of the windows are double panes. However, some window frames were found to be leaky and could be better sealed.

Building infiltration: A blower door test was performed using the pressurization technique. It was found that the rate of infiltration is 0.89 air changes per hour (i.e., 0.89 ACH). A typical house has a rate of infiltration of 0.4–0.75 ACH. An infiltration rate higher than 0.75 ACH indicates a leaky house. In the audited house, air infiltration is occurring through visible gaps in the window frames, gaps between the walls and the ceiling and the floor, through piping and plumbing fixtures, and through the fireplace doors.

HVAC system: A heating and cooling system consumes about 55% of a typical U.S. residence's energy use. The audited house has a unique HVAC system—a combination of passive solar heating, electric baseboard heaters, heat fans, a unit propane heater in the living area, and a furnace in the cellar area. There is no central A/C unit in the residence.

Water management: The domestic hot water tank was observed to be uninsulated, and the plumbing fixtures to be conventional with no low-flow devices.

Appliances: In the audited house, it was found that the appliances consuming the most energy are two refrigerators. Both refrigerators are fairly old models and could be replaced with new, energy-efficient models. Similarly, the clothes washer is rather old and could be replaced with a more efficient model.

Thermal comfort: The greatest concern outlined by the homeowner during the walkthrough audit regarding thermal comfort is that the lower floor is drafty, especially within the north television/guest room and bathroom. Due to the more recent renovations, the upper floor is less drafty and more comfortable.

12.1.2 Energy Efficiency Measures

This section describes the potential operation and maintenance and the energy efficiency measures to reduce the house energy use and cost. Eight measures have been identified involving the building envelope, appliances, water management, thermal comfort, and solar water heating for the hot tub. The following sections describe briefly the rationale for each EEM, the general plan for implementation, the estimated cost of installation (including labor and materials), and the payback period for its implementation.

12.1.2.1 Building Envelope

A typical house has a rate of air infiltration of 0.4 to 0.75 air changes per hour (ACH). An air infiltration rate higher than 0.75 ACH indicates a rather leaky house. This is the case for the audited house with a 0.89 ACH air infiltration rate.

12.1.2.1.1 EEM-1: Weather Stripping

For the audited house, basic weatherization measures such as applying weather stripping to unsealed windows and caulking gaps between the ceiling and floor can save energy and improve thermal comfort. Several studies for similar houses have indicated that a 20–50% reduction in air infiltration can be achieved after weatherization of residential buildings. For this house, a 33% (i.e., 1/3) reduction of air infiltration through the building envelope is assumed due to caulking and weather stripping of window frames and doors. The expected cost of basic weatherization for the house (including labor) is about $450. Using the degree-day method outlined in Chapter 5, the projected annual energy savings is $215 attributed to reduced heating energy cost, resulting in a simple payback period of 2.2 years. As a note, one easy recommendation is to apply magnetic tape around the fireplace to seal the fireplace doors. Simple black magnetic tape (available at most hardware stores) is an easy do-it-yourself fix to reduce air infiltration.

12.1.2.1.2 Water Management

Domestic hot water (DHW) accounts for 15–25% of the energy consumed in a typical U.S. residence. However, the actual DHW load can vary significantly, depending on the usage. Whether the water heater is gas or electric, significant energy savings can often be found through implementation of water heating efficiency improvements. In general, the four measures to reduce water heating costs are:

- Use less hot water
- Turn down thermostat on hot water tank
- Insulate hot water tank
- Replace existing water heater with a more efficient model
- Use low-flow plumbing fixtures

12.1.2.1.3 EEM-2: Hot Water Tank Insulation

For the audited house, the propane-fired domestic hot water tank is uninsulated. Even though the tank sits in a conditioned area of the house (the kitchen), it is still losing heat to the surroundings. The tank surface temperature is measured to be 74°F for a surrounding air temperature of 65°F. This 9° difference indicates that heat is being lost from the tank. Wrapping the hot water tank with R-12 insulation could save $111 annually. Tank insulation kits can be purchased from any hardware store and cost about $30–$60. Assuming the cost of implementing this measure is $50, the resulting simple payback period is 0.4 years.

12.1.2.1.4 EEM-3: Low-Flow Showerheads

Because propane is expensive relative to natural gas, water conservation measures such as low-flow fixtures can save energy and be cost-effective. Assuming that the occupants in the audited house take two showers per day and that a typical shower lasts 8 min, there would be a savings of about $175 in energy costs per year by replacing the conventional showerheads with low-flow fixtures. Moreover, the homeowner would also conserve about 12,000 gallons of water per year.

Low-flow showerheads cost between $25 and $50 and can be purchased at most hardware stores. For an installation cost of $100 for this measure (two showerheads), the simple payback period is less than 1 year.

12.1.2.1.5 EEM-4: Low-Flow Faucets

In the audited house, two faucets (4 gallons per minute (gpm)) can be replaced with low-flow fixtures (1.5 gpm) at a cost of $25 per faucet. Assuming each faucet is used for 3 min per day at an average temperature of 80°F, 5,370 gallons of water and $53 in energy cost per year can be saved. The initial investment of about $50 results in a simple payback period of 1 year.

12.1.2.2 Appliances

Appliances and electronics can be responsible for as much as 20% of a U.S. residence's energy bills. Refrigerators, clothes washers, and clothes dryers are often the greatest energy consumers. Substantial energy use savings can be achieved by altering the usage patterns and by investing in more energy-efficient Energy Star® appliances.

12.1.2.2.1 EEM-5: Cold Water Wash

Washing clothes in cold water is a straightforward strategy to save energy. Since hot water is not used, no energy is needed to heat the water. Based on the occupancy level of the audited house, the estimated laundry load is three washes per week. By washing all three loads in cold water, a savings in energy cost of $95 per year can be obtained. There is no cost required to implementing this ECM, resulting in an immediate simple payback period.

12.1.2.2.2 EEM-6: Phantom Loading

Several plugged appliances are utilized in the audited house. Appliances such as computers, televisions, and coffeemakers draw current even when turned off. These phantom loads can be eliminated by plugging these appliances into power strips. Using only five power strips around the house to reduce phantom loading can save about $70 per year. With an initial investment of $50 ($10/power strip), the simple payback period is about 0.7 year.

For this EEM to be effective, at least three appliances should be plugged into each strip, and the strip should be turned off for at least 12 h per day.

12.1.2.2.3 EEM-7: Energy Star Washing Machine

Energy Star washing machines use less water for each wash cycle, which saves both water and energy. The homeowner can save about $70 in annual energy use by replacing the conventional washing machine with an Energy Star model. With an initial investment of $600, the simple payback period is about 8.7 years.

12.1.2.2.4 EEM-8: Energy Star Refrigerator

In most homes, the refrigerator is the most energy-consuming appliance. Energy Star-qualified refrigerators use about half the energy consumed by models manufactured before 1993. Replacing the two refrigerators used in the audited house with new Energy Star models could reduce the refrigerator's energy consumption by 50%, saving about $131 annually. With an initial investment of $1,400 (for two new refrigerators), the simple payback period is about 10.7 years.

12.1.3 Economic Analysis

Table 12.2 summarizes the potential energy savings and the cost-effectiveness for the proposed EEMs. Specifically, Table 12.2 outlines the projected cost of installation and annual energy savings and then ranks each measure in terms of its simple payback period (initial investment divided by annual savings).

TABLE 12.2 Economic Analysis of Recommended EEMs

EEM Descriptions	Cost of Measure ($/total cost)	Annual Savings ($/year)	Simple Payback (years)
EEM-5: Cold water wash: Wash clothes in cold water instead of hot	$0	$95	Immediate
EEM-2: Hot water tank insulation: Wrap DHW tank with R-12 insulation	$50	$111	0.4
EEM-3: Low-flow showerheads: Replace the existing showerheads with low-flow fixtures	$100	$175	0.6
EEM-4: Low-flow faucets: Replace the existing faucets with low-flow fixtures	$50	$53	1.0
EEM-8: Energy Star* refrigerator: Save electricity by replacing conventional machine with energy-efficient model	$1,400	$131	10.7
Total	$2,920	$900	3.2 years

12.1.4 Recommendations

The total cost of implementing the eight recommended EEMs outlined in Table 12.2 is $2,920 and would yield an annual energy cost savings of $900. This estimate may vary depending on the actual behavioral patterns in the house (for example, the occupants may take less than two 8-min showers per day, in which case the energy savings realized by installing low-flow shower fixtures will be reduced).

EEMs with a simple payback period of less than 4 years, in addition to the weatherization EEM (to improve thermal comfort), would cost about $720 to implement. With an annual energy cost savings of $700, the simple payback period of these EEMs is slightly more than 1 year. The recommended EEMs would improve thermal comfort while saving energy in the home.

12.2 Case Study 2: Standard Audit of a Residence

In this section, a standard energy audit of a residential building is presented. As indicated in Chapter 1, the standard energy audit includes a walkthrough energy audit, a utility data analysis, a detailed energy modeling analysis, and an economic analysis to recommend cost-effective energy efficiency measures. The report outlining the findings of a standard audit for a house is provided in the following sections.

12.2.1 Architectural Characteristics

The house audited in this report is a 1,412 ft² single-family home located in Boulder, Colorado. It is occupied by two residents. The wood frame structure has three bedrooms, one full bath and one half bath, a living room, dining room, kitchen, and 667 ft² semiheated garage. It was originally constructed in 1963 and has undergone several minor renovations over the years, although no major additions or alterations to the floor plan have been undertaken. Figure 12.1 provides a picture of case study 2 house.

The house has a rectangular shape and is oriented with its long axis in the east-west direction. Figure 12.2 shows a floor plan layout of the house. Figure 12.3 shows an

FIGURE 12.1   Audited house for case study 2.

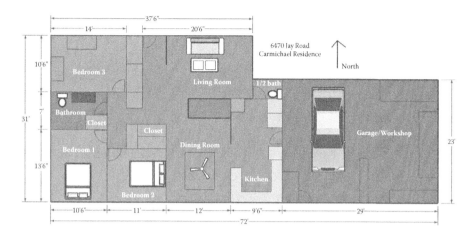

FIGURE 12.2 Floor plan and layout of the audited house.

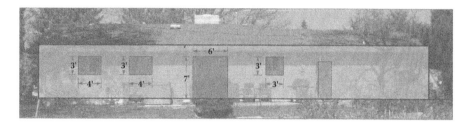

FIGURE 12.3 South-facing elevation and position of windows and main door.

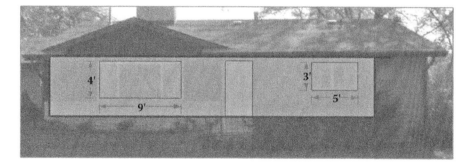

FIGURE 12.4 North-facing elevation and position of windows and back door.

elevation of the rear-facing south façade, while Figure 12.4 shows the north-facing front façade. The west façade is illustrated in Figure 12.5. The east façade of the house consists of the garage façade and has one 3 ft × 3 ft window. The roof slope is approximately 17°.

The audited house has a typical wood frame construction. The walls contain 4 in. of fiberglass batt insulation, while the ceiling has 12 in. of thermal insulation. An insulated 36 in. crawlspace separates the house from the ground. Figure 12.6 shows details of the

FIGURE 12.5 West-facing elevation and position of windows.

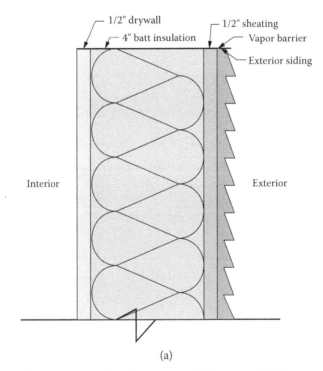

(a)

FIGURE 12.6 Construction details of (a) exterior wall, (b) roof, and (c) floor.

building envelope components, including the exterior wall, roof, and floor. No additional insulation has been added to the original walls, and there is minimal insulation in the crawlspace and attic. The windows are the weakest thermal component in the envelope and consist of two single panes in aluminum frame storm window units.

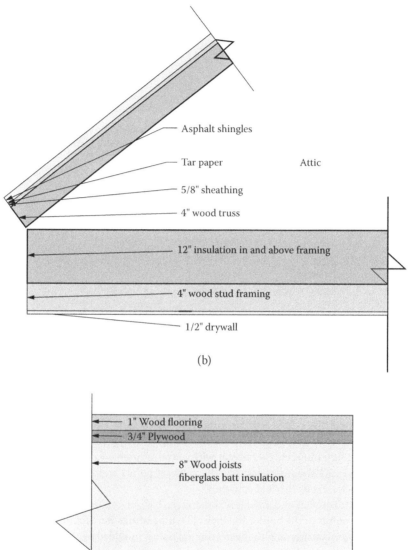

FIGURE 12.6 *(Continued)*

TABLE 12.3   Heating and Cooling Temperature Settings for House and Garage

	House	Garage
Heating Set Point		
Occupied	70°F	55°F
Unoccupied	64°F	55°F
Cooling Set Point		
Occupied	78°F	N/A (no cooling provided)
Unoccupied	85°F	N/A (no cooling provided)

TABLE 12.4   Heating System Characteristics for Both the House and the Garage

	Furnace for House	Furnace for Garage
Type	Lennox	Janitrol
Source	Natural Gas	Natural gas
Model no.	Elite Series	680307
Output rating	Not available	24,000 Btu/h
Efficiency	80%	80%

12.2.2 Heating and Cooling Systems

When originally built, the house had electric radiant heat in the ceilings, and later on, a forced-air system was added. The electric radiant system is still functional but is rarely used to heat the house. The furnace is estimated to be about 10 years old. The air conditioning unit was also added after the original construction. The heater in the garage is an older unit set to keep it above freezing temperature. The heating system is controlled with a programmable thermostat with lower set points for nonoccupied times. Tables 12.3 and 12.4 show, respectively, the zone set points and furnace performance parameters, including its thermal efficiency.

The temperature inside and outside was monitored over a 4-day period to confirm the thermostat set points in the house. The internal temperatures were found to range from 59 to 67°F and are consistent with the heating set points provided by the occupants.

12.2.3 Lighting and Appliances

There are several types of lighting lamps used throughout the audited house. The lighting schedule describing the various fixtures and their rated wattage is provided in Table 12.5. The total house square footage is 1,412 ft^2, and the total wattage of the luminaries (which are almost all compact fluorescent lamps or CFLs) listed in Table 11.3 is 256 W. Therefore, the total actual house lighting power density is 0.18 W/ft^2.

The appliances available in the audited house are generally fairly new, all Energy Star certified, and consist of a washer (0.5 kW) and dryer (5.0 kW), a new GE refrigerator, (112 W), a dishwasher (1.2 kW), a microwave (5.5 kW), and various other appliances.

TABLE 12.5 Type and Wattage of Lighting Fixtures Used in the Audited House

| Quantity | Lamp | | Description | Image |
	Type	Total Watts		
3	3-13WCFL Kichler K8046	39	Ceiling surface mounted directional	
1	3-13WCFL Westinghouse Vintage	39	Paddle fan with light	
2	2-13WCFL Dolan Design 492	26	Wall-mounted bathroom fixture	
3	3-16WCFL Kichler K8109	48	Flush-mount ceiling fixture	
4	1-40W GO 8201	40	Exterior classic wall sconce-med base	
3	2-32WT8	64	Linear fluorescent	

The water heater was also replaced last winter with a new gas unit that is Energy Star certified with an energy factor of 0.54. The last major energy user is the hot tub. It has a significant impact on electricity use when it is operated.

12.2.4 Utility Analysis

One of the first steps in evaluating a building's energy performance is to analyze its utility bills, as indicated in Chapter 5. In particular, the average daily fuel use (over a month) is plotted against the average ambient temperature in that month to assess the energy efficiency of the building envelope, by calculating the BLC as well as the balance temperature and the base load. Three years of utility data for the audited house were obtained. It should be noted that the building has two zones with separate heating systems and set-point temperatures, the house and the garage. However, there is only one meter, and the utility data are lumped for both zones. Figure 12.7 presents the correlation between the daily gas usage and the average monthly outdoor temperature.

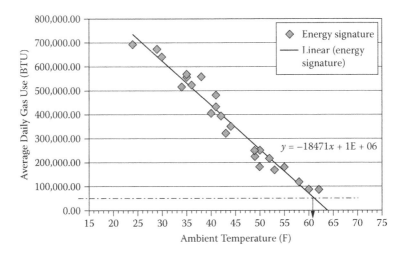

FIGURE 12.7   Natural gas consumption as a function of average monthly outdoor temperature.

The operating $BLC_{UTILITY}$ was determined by Equation (12.1):

$$BLC_{UTILITY} = \frac{-S \cdot \eta}{24 * f} \tag{12.1}$$

For this calculation, S is the slope (therms/average monthly temperature) of the linear regression shown in Figure 12.7, η is the efficiency of the heating system ($\eta = 0.80$), and $24*f$ is used since the house is heated less than 24 h per day during the winter due to the setback temperature settings during the heating season. The fraction f is determined using the simplified estimation approach, as discussed in Chapter 5:

$$f = \frac{N_{occ}}{24} + \frac{(24 - N_{occ})}{24} * \frac{(T_{b,\,setback} - T_{o,avg})}{(T_b - T_{o,avg})}$$

For the house, it is found that $f = 0.90$ and the BLC is calculated using Equation (12.1) to be 686 (Btu/hr.°F).

The base load, or fuel used to heat the domestic hot water (DHW), is determined by fitting a line to the bottom series of points and finding where it crosses the y-axis. The base load for this house is 48,954 Btu/day, or approximately 0.54 therms/day, and remains relatively constant throughout the year.

The point where the two linear regression lines meet in Figure 12.7 leads to the balance temperature. This temperature represents the outdoor temperature at which no additional heating is required to maintain the indoor temperature. For this house the balance temperature is approximately 61°F, which indicates the thermal performance of the envelope can be improved. The house shell is affected by both air infiltration and exterior insulation level.

There is also a method of calculating the BLC, BLC_{CALC}, based on the envelope thermal resistance levels and air infiltration rate. The BLC_{CALC} can be calculated using Equation (12.2):

$$BLC_{CALC} = \sum (U \cdot A) + (\dot{m} c_P)_{INF} \qquad (12.2)$$

where U is the thermal conductivity (Btu/h.°F.ft²) of each external surface, A is the corresponding surface area (ft²), $\dot{m}$ is the mass flow rate due to air infiltration (lbm/h), and c_p is the specific heat of air (Btu/lbm°F). Calculations based on the brick walls, double-paned glazing, R-19 insulation in the attic, and no floor insulation determined a (UA) value of 675. When combined with the air infiltration, $(\dot{m} c_p)_{INF}$, of 115.8 Btu/hr.°F, this residence has a calculated BLC of 791 Btu/h˚F. The modeled BLC_{CALC} was compared to the $BLC_{UTILITY}$ of 773 Btu/hr.°F to show good congruence. The BLC value obtained using the utility data is usually lower than the calculated BLC because of thermal coupling in the walls (i.e., infiltrating air recovers heat escaping the building envelope). This thermal coupling is described as the efficiency of thermal coupling between heat transmission and air infiltration/exfiltration within the wall, and is calculated using Equation (12.3):

$$\eta_{wall} = \frac{BLC_{CALC} - BLC_{UTILITY}}{BLC_{CALC}} \qquad (12.3)$$

The audited house had a thermal coupling wall efficiency of 2.2%, indicating that the air leakage sources are most likely made up of concentrated paths.

12.2.5 Air Leakage Testing

Air can flow in or out of the building envelope through leaks and infiltrate the conditioned house. Air infiltration can affect energy use, thermal comfort, and structural integrity (due to humidity transport) of the residential home. For this project, the air infiltration rate is measured using a blower door test, as outlined in Chapters 3 and 5.

The blower door test determines how the volumetric flow rate varies with the pressure difference between the interior of the house and outside. Before pressurizing the house, all window and exterior doors are closed, the furnace is turned off, and the water heater is set to the "pilot" setting. Moreover, all interior doors are left open and any exhaust fans and the vented dryer are turned off.

Two pressurization tests were carried out for both the audited house and the garage. Table 12.6 summarizes the measurements for the pressurization test of the house.

Using the data from Table 12.6, a correlation is found for $V_{inf} = C^* \Delta P^n$, where C and n are correlation coefficients determined by fitting the measured data of pressure differentials and flow rates as described in Chapter 5. The correlation regression is presented in Figure 12.8 for the pressurization test.

Using the correlation coefficients from Figure 12.8, the effective leakage area of the house can be estimated using Equation (12.4) as outlined in Chapter 5:

$$A_{leak} = V_{inf} * \sqrt{\frac{\rho}{2\Delta P}} \ (in^2) \qquad (12.4)$$

where $\Delta P = 4$ Pa.

TABLE 12.6   Data from
the House Pressurization Test

Measured Values (House)	
Pressure (Pa)	Airflow (cfm)
11.1	1,064
20.1	1,494
28.3	1,839
39.5	2,232
48.8	2,524
55.2	2,735
44.9	2,412
35.2	2,085
25.4	1,718
16.3	1,330

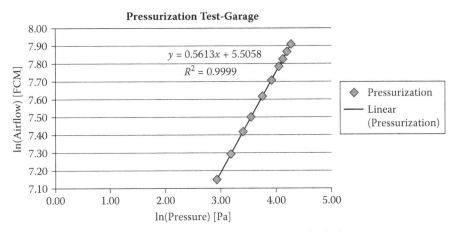

FIGURE 12.8   Regression correlation for the pressurization test for the house.

TABLE 12.7   Leakage Characteristics for Both the
House and Garage

	n	C	V_{ref} (cfm)	A_{leak} (in^2)
House	0.588	256	582	152
Garage	0.561	246	550	127

The correlation coefficients n and C as well as the leakage area and the reference air infiltration rate, V_{inf}, at $\Delta P = 4$ Pa are summarized in Table 12.7 from the pressurization tests for both the house and the garage.

The low values of n indicate that the air infiltration paths are made up of concentrated leaks, as was inferred by the low wall thermal efficiency of Equation (12.3). Potential

TABLE 12.8 Air Infiltration Rate Using LBL Model

Space	Stack Coefficient, f_s	Wind Coefficient, f_w	Average Winter ACH	Average Summer ACH
House	0.0156	0.0092	0.85	0.72
Garage	0.0156	0.0092	0.90	0.85

leakage areas were identified during a walkthrough inspection. The most prominent air leakages were the window frames in the house and the garage door.

As outlined in Chapter 3, the LBL model can be utilized to estimate the seasonal variation of the air infiltrating the house. The following assumptions are made for the LBL model to estimate the average air infiltration during the winter season:

- V_w = 8.5 mph (average winter wind speed for Denver)
- T_{indoor} = 65°F (for winter) and T_{indoor} = 75°F (for summer)
- T_{out} = 34.4°F (for winter) and T_{out} = 66.0°F (for summer)
- P_{air} = 0.063 lb/ft³
- One-story house with light local shielding
- $V_{inf} = A_{leak} * \sqrt{f_s \Delta T + f_w * v_w^2}$

Table 12.8 summarizes the calculations for the average air infiltration for the audited house and garage during both winter and summer seasons. The values of ACH indicate that the house and the garage are relatively leaky. Weather stripping measures to reduce air leakage should be considered as part of the energy retrofit analysis of the audited house.

12.2.6 BLC Calculation and Analysis

Once the building infiltration value was found, it was possible to calculate the BLC based on the building envelope's thermal characteristics, including heat transmission and air infiltration. In order to perform this analysis the *UA*-values for components of the shell were summed with the expected energy loss from infiltration to determine an overall building BLC. The garage is included in this analysis and is considered a conditioned space.

Table 12.9 summarizes the estimated *UA*-values for building components and estimates a UA for the building at 794.3 Btu/h.°F. Table 12.10 provides an estimate of the heat loss from air infiltration: 254.6 Btu/h.°F. The total house calculated BLC is 1,048.9 Btu/h.°F. It should be noted that though the *U*-value for the garage door seems high, it is typical for a single-layer uninsulated metal door.

The BLC calculations based on the calculated *UA*-values and the blower door infiltration tests (1,048.9 Btu/h.°F) are, as expected, higher than the BLC calculations generated through utility analysis (686.4 Btu/h.°F) by a difference of 34.5%. Thermal coupling between the infiltration and heat transmission through the walls can be significant in this house.

TABLE 12.9   Summary UA Calculations for the House Envelope Components

Component	Description	U-Value (Btu/ft²-h-F)	Area (ft²)	UA (Btu/h-F)	% of UA
House roof/ceiling	Flat wood frame ceiling (insulated with R-38 fiberglass) beneath attic with medium asphalt shingle roof	0.0	1,333.0	35.3	4.45%
Crawlspace	Wood frame floor (insulated with R-25 fiberglass), uninsulated walls	NA	NA	79.6	10.02%
House walls	2 × 4 wood stud framing (insulated with R-13 fiberglass)	0.1	1094.9	86.0	10.83%
Windows	Clear, double-pane glass in aluminum frames without thermal breaks	0.9	131.0	114.0	14.35%
Doors	Wood, solid core, 6% glazing	0.4	56.0	22.4	2.82%
Garage roof	2 × 4 frame roof (insulated with R-11 fiberglass)	0.1	690.5	79.8	10.05%
Slab	Concrete slab	NA	NA	112.5	14.16%
Garage walls	2 × 4 frame (insulated with R-11 fiberglass)	0.1	667.0	57.3	7.21%
Garage doors	Steel doors	1.2	162.0	194.4	24.47%
Total UA				794.3	100.00%

TABLE 12.10   Equivalent UA Calculation for Air Infiltration

	CFM	Flow (ft³/h)	Density (lb/ft³)	Cp (Btu/lb-F)	UA (Btu/h-F)	Weight	UA
House summer	135.3529	8121.173	0.063	0.24	122.79214	0.5	61.39607
House winter	159.3568	9561.41	0.063	0.24	144.56852	0.5	72.28426
Garage summer	130.6595	7839.572	0.063	0.24	118.53433	0.5	59.26716
Garage winter	136.007	8160.417	0.063	0.24	123.38551	0.5	61.69275
Total BLC							254.6402

As noted in Tables 12.9 and 12.10, air infiltration accounts for 24.3% of the total BLC of the house. Other key contributors include the garage doors (24.5%), the windows (14.3%), and the slab in the garage (14.1%). These should be the key areas to consider for retrofit measures.

12.2.7 Energy Modeling

The DOE-2 simulation tool, based on the eQUEST interface, was used to evaluate various energy conservation measures to reduce the energy use and cost for the audited house. The model is based on the characteristic features of the house summarized in Table 12.11. Air infiltration for the zones is based on the average annual infiltration determined with the blower door test. Three-dimensional renderings of the house model can be seen in Figure 12.9.

TABLE 12.11   Summary of the House Specifications

Component	Notes
	Envelope
House	
Roof	2 × 4 R-38 fiberglass
Walls	2 × 4 R-13 fiberglass
Windows	Storm windows
Doors	Wood core
Garage	
Roof	2 × 4 R-11 fiberglass
Walls	2 × 4 R-11 fiberglass
Doors	Steel
	HVAC
House	
Heating efficiency	0.8
Heating set point	70°F/60°F
Cooling COP	3.3
Cooling set point	78°F/85°F
Garage	
Heating efficiency	0.8
Heating set point	55°F
	Lighting
LDP	0.18 W/ft$_2$
	DHW
Energy factor	0.59
	Other Equipment
EPD	15.2 W/ft$_2$
	Occupancy
Weekday	Unoccupied 9-17
Weekend	Occupied

12.2.8 Model Calibration

After an initial model was created it was necessary to calibrate it against the actual building utility data. Before this calibration could be carried out, it was necessary to preprocess the utility data by removing electricity values from the time period before the hot tub was installed (14 months of data), since the model does not account for the energy use associated with this equipment. With the new utility data, it was possible to compare the model results to actual building performance and calibrate the model. Figure 12.10 shows the results of this initial analysis. The initial model had low natural gas usage but excessively high electrical energy usage.

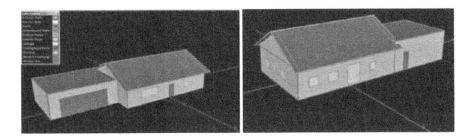

FIGURE 12.9   eQUEST house model renderings.

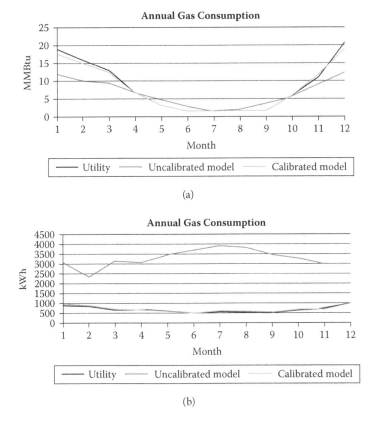

(a)

(b)

FIGURE 12.10   Comparative analysis of model prediction before and after calibration for (a) natural gas use and (b) electricity use.

The calibration process included several stages to improve the model predictions. The initial stage involved refining the building energy model and correcting errors in the initial simulation model development. Some input errors that were corrected included house envelope changes (adding thermal properties related to the doors of the garage), ensuring that cooling was not used during the winter and heat during the summer months, and that fans were only used with the HVAC systems.

TABLE 12.12 Annual Energy Use: Utility Data Compared to Predictions from the Calibrated Energy Model

Energy Type	Actual Use (kWh)	Calibrated Model	Percent Difference
Electricity (kWh)	8,451	8,522	1.0%
Natural gas (therms)	1,008	970	4.0%

After these steps were completed, various equipment and systems were scheduled to reflect more accurately the actual building usage and occupancy patterns. DHW was first calibrated to reflect occupancy pattern and utility data. Next, cooling usage and equipment power density were scheduled to match usage and utility data. Finally, heating loads were matched to utility data by adjusting the heating schedule to the occupant behavior. With these changes, a calibrated model is achieved within 1.0 and 4.0% for annual electrical and natural gas consumption, respectively, as summarized in Table 12.12.

12.2.9 Energy Conservation Measures

Using the calibrated simulation model for the audited house, a number of energy conservation measures (ECMs) have been analyzed. A brief description of each ECM and the results of both energy and economic analyses for the ECMs are summarized in this section.

12.2.9.1 ECM-1: Reduce Heating Set Point

The current heating set points for the house are 70°F during occupied hours and 64°F during unoccupied hours. This ECM calls for reducing the heating set point to 65°F during occupied hours and 60°F during unoccupied hours. These set points are lower than the acceptable ranges of operative temperatures recommended by ASHRAE 55-2004, but by wearing sweaters, this measure is suitable for the occupants, who are two active professionals. This measure can be implemented by simply changing the digital set points on the thermostat, and there are no costs associated with it.

12.2.9.2 ECM-2: Reduce Infiltration in the House

Given the high air infiltration rate revealed by the pressurization tests, the first retrofit measure that should be considered for the house is to seal the leaks in the building envelope. The leakage area for the house was found to be 0.11 in.^{2}/ft^2, which is considered leaky (ASHRAE, 2009). The impact of this ECM can be modeled by a reduction in air infiltration by 20%, which would bring the house closer to typical pre-1970 construction air leakage levels. This reduction in infiltration can be accomplished by a variety of weather-proofing measures. The first area to improve is the fireplace, by sealing the damper and the fireplace doors. Moreover, the windows with nonthermally broken aluminum frames contribute significantly to the heat loss through the glass and frames, as well as air infiltration. Replacing the windows with better units could provide multiple benefits.

Weather stripping of various air leakage sources throughout the house and the garage should also be undertaken. Rubber gaskets are available for switches and electrical outlets and are installed beneath the cover plate. Caulking can be used to seal cracks between stucco and wood siding. Caulking is also useful around exterior door and window casings and where pipes and wires pass through a wall. Expanding foam may be useful for filling larger areas, such as gaps around door and window jambs where they meet the surrounding wall framing.

Costs for various gaskets, caulking, and expanding foam were assumed to be $150. Installation time is anticipated to be roughly 4 h total and can be done by the homeowners.

12.2.9.3 ECM-3: Replace Garage Doors

As noted in the BLC analysis, the garage doors contribute significantly to the heat loss through the building envelope. It is therefore important to improve their energy efficiency by using insulated doors. The *U*-value of the insulated doors is 0.055. RS Means 2008 data were used to gather potential costs for the garage doors. Residential overhead garage doors, deluxe models, are roughly $1,020.50 per door. The total cost for this ECM is $2,041.

12.2.9.4 ECM-4: Upgrading Windows

Since most of the windows are single-pane storm windows with nonbroken aluminum frame, there is an opportunity to improve the energy performance of the house by upgrading to double low *e* windows. The windows have a *U*-value of 0.4, almost half the *U*-value of the existing windows of 0.82. As expected, this improvement reduces the UA of the windows by about 55%.

Based on a quote obtained from some local contractors, each 4 ft × 4 ft 6 in. casement unit with low *e* coating costs $796.81. Therefore, a 3 × 4 ft window (which is fairly typical for this residence) would cost approximately $530.61 per unit. The 9 × 4 ft front window on the north façade would cost approximately $1,558.98. Therefore, a total window replacement would cost approximately $5,274.25 (7 windows * $530.61 + 1 window * $1,559.98). Windows would need to be professionally installed. This cost could vary significantly, depending on vendor, but it is assumed that each window would take 3.5 h. At a laborer rate of $25/h, the installation costs would be approximately $700. Therefore, the total window upgrade costs would be approximately $5,974.25.

12.2.9.5 ECM-5: Upgrading Windows and Reducing Infiltration

In reality, replacing windows will not only provide a higher *U*-value, but it will also reduce infiltration. This ECM demonstrates a combination of ECM-2 and ECM-3. The capital cost for these measures is $6,124.25.

12.2.9.6 ECM-6: Reduced DHW Demand

Domestic hot water loads accounted for approximately 8.2% of the annual natural gas needs. By improving the fixture efficiency of those that use hot water, particularly the showerhead and bathroom and kitchen faucets, hot water demand can be reduced by up to 40%. The bathroom currently has a very old showerhead that was measured to use 3.2 gpm. Current regulations on fixture flow rates, primarily using EPACT 1992, require showerheads to be below 2.5 gpm, and high-efficiency heads can reduce flow rates to 1.25 gpm. These showerheads cost about $74.95 per unit. Moreover, aerators can be added to the faucets in the kitchen and bathroom for minimal costs of $3.24. Therefore, the total cost for this ECM is $81.43.

12.2.9.7 ECM-7: Improve Furnace Efficiency—House

The current furnace, with a thermal efficiency of 80%, can be upgraded to a more energy-efficient furnace of at least 90% energy efficiency. Note that this measure only addressed the furnace in the crawlspace that provides heat to the house, excluding the garage. It would require a full replacement of the unit in the crawlspace. It should be noted that by improving the envelope performance, the overall energy demands will be reduced and it may be possible to get a smaller unit. RS Means 2008 was used to gather costs for the furnace replacement. A comparable gas-fired furnace that could provide up to 75 MBH capacity would cost $895.01.

12.2.9.8 ECM-8: Improve Furnace Efficiency—Garage

Similar to the main house furnace, the garage furnace can be replaced with a more energy-efficient unit. Since it is a smaller unit, the cost is slightly less than the whole house furnace. RS Means 2008 was used to estimate the cost for the furnace replacement. A comparable gas-fired furnace that could provide up to 45 MBH rating would cost $798.79.

12.2.9.9 ECM-9: Heat Pump

This ECM involved replacing the two existing furnaces to a single, electric air-to-air heat pump. It would involve significant reconfiguration of the existing mechanical system, including providing ductwork into the garage from the centrally located heat pump. This measure should also consider other factors, such as the current and future cost of electricity against natural gas. Also, switching to mostly electricity use gives the resident the ability to offset more of the energy use (through photovoltaics, for instance) with the goal of a carbon-neutral home. According to RS Means 2008, an air source, split-heat pump of up to 10 tons would cost $933.75.

12.2.9.10 ECM-10: PV System

A 4 kW system will provide enough energy to balance out on an annual basis and will more than meet the peak demand of the home. There are several older trees to the south that could potentially provide some shade, so the system was slightly oversized to compensate. The cost of the PV system was estimated to be $10 per watt, a generally accepted industry cost estimate.

12.2.10 Energy and Economical Analyses of ECMs

The annual energy savings for each ECM was calculated by changing parameters in the calibrated energy simulation model of the house. Whenever it is more convenient, hand calculations are used to estimate the energy savings associated to some ECMs. Then, the cost-effectiveness of each ECM is determined using both simple payback period and life cycle cost analysis methods. Findings of the energy analysis are summarized in Table 12.13.

TABLE 12.13 Evaluation of the Cost-Effectiveness of the ECMs Using Detailed Simulation Modeling

	Natural Gas Use (Therm)	Electrical Energy Use (kWh)	Annual Energy Cost ($)	Initial Cost ($)	Simple Payback Period (years)	Life Cycle Cost ($)
Base	970	8,523	1,590	0	—	22,407
ECM-1	935	8,523	1,557	0	0	21,939
ECM-2	948	8,529	1,569	150	7.3	22,268
ECM-3	826	8,523	1,455	2,041	15.1	22,545
ECM-4	909	8,510	1,532	5,974	103.1	27,565
ECM-5	887	8,515	1,511	6,124	78.0	27,424
ECM-6	924	8,523	1,547	81	1.9	21,878
ECM-7	946	8,523	1,568	895	40.2	22,988
ECM-8	906	8,523	1,530	799	13.2	22,356
ECM-9	758	11,440	1,625	933	49.6	23,833
ECM-10	908	3,123	1,158	40,000	92.6	56,318

For the economic analysis, a discount rate $d = 5.0\%$ is used with a life time period of 25 years. The cost for implementing each measure was estimated using RS Means cost data (RS Means, 2008). Based on the utility rate, the natural gas price was found to be $0.94/therm and the electricity cost was $0.08/kWh. As noted in Chapter 2, the following equation can be used to calculate the $USPW(d,N)$:

$$USPW(d,N) = \frac{1-(1+d)^{-N}}{d} \qquad (12.5)$$

Based on the energy savings and the economic analysis, the best ECMs for energy use and cost savings include the use of low-flow fixtures, the temperature setback, and the weather stripping measures, as indicated by the results of Table 11.8. These measures have the shortest payback period and the lowest life cycle cost (LCC) values. Other economically feasible measures are replacement of the garage doors and the upgrade of the garage furnace with a more efficient model. Due to very high installation costs, upgrading the windows, the upgrade of the house furnace, and the installation of a PV system (without any rebates) did not prove to be cost-effective.

12.2.11 Conclusions and Recommendations

A detailed energy audit is carried out for a family residence in Boulder, Colorado. The building load coefficient (BLC) was calculated using two different methods. The first is based on the analysis of monthly utility data, and the other used the actual *UA*-values and measured infiltration rates to directly estimate the house BLC. The first method provided a BLC of 686.4 Btu/h-F, whereas the second method calculated it at 1,048.9 Btu/h-F, a difference of 34.5%. This is a fairly large difference, but it is typical to find some variations due to thermal coupling between air infiltration and heat transmission as well as uncertainties on the actual thermal properties of building envelope components. Both methods indicate that the energy efficiency of the house can be improved through weatherization measures. Using a blower door test, building infiltration was found to be 0.737 ACH, while the conditioned garage had 0.847 ACH. The house values may be influenced by the unsealed window. These values are fairly typical numbers, but again, indicate room for improvement.

Energy and cost analyses of several energy conservation measures have been performed. Based on these analyses, the following ECMs are recommended for the house in order:

1. ECM-6: Reducing DHW demand
2. ECM-1: Reduce heating set point
3. ECM-2: Lower infiltration in house
4. ECM-8: Improve garage furnace efficiency
5. ECM-3: Replacing garage doors

Other measures to consider in future analysis include:

- Investigating a solar hot water system
- Improving wall and ceiling insulation values (since these were relatively low contributors to the UA, they were not considered in this analysis).

Appendix A: Housing Archetypes

TABLE A.1   U.S. Housing Archetypes by Census Division: New England (Boston)

	Prototype A	Prototype A1	Prototype B	Prototype B1	Prototype C
Year built	Pre-1950s	Pre-1950s	1950–1979	1950–1979	1980–2000
No. bedrooms	3	3	3	3	3
No. bathrooms	1	1	1	1	2
No. stories	2	2	2	2	2
Floor conditioned area (ft^2)	2,090	2,090	1,626	1,626	2,092
Window area (ft^2)	233	233	212	212	242
Window-to-wall ratio	0.12	0.12	0.12	0.12	0.12
Glazing layers	1	2	1	2	2
Windows	Wood	Wood	Wood	Wood	Wood
Wall siding type	Wood	Wood	Wood	Wood	Wood
Wall type	Plaster	Plaster	Drywall	Drywall	Drywall
R-Values					
Wall	0	7	0	7	13
Ceiling	0	22	22	22	27
Floor	0	0	0	0	0
Foundation type	Basement	Basement	Basement	Basement	Basement
Foundation insulation	None	None	None	None	None
Garage	Detached	Detached	Attached	Attached	Attached
Heating equipment	Natural gas	Natural gas	Natural gas	Natural gas	Natural gas
Furnace efficiency AFUE	80%	80%	80%	80%	80%
Cooling equipment	Window	Window	Window	Window	Window
A/C efficiency SEER	7	7	8	8	10
Cooling set points	76	76	76	76	76
Heating set points	67	67	67	67	67
Infiltration ACH 50 Pa	0.7	0.7	0.7	0.7	0.4
Appliances	Non-ES[a]	Non-ES	Non-ES	Non-ES	Non-ES
Lighting (fluorescent)	14%	14%	14%	14%	14%

[a] Non-Energy Star.

TABLE A.2   U.S. Housing Archetypes by Census Division: Middle Atlantic (New York)

	Prototype A	Prototype A1	Prototype B	Prototype B1	Prototype C
Year built	Pre-1950s	Pre-1950s	1950–1979	1950–1979	1980–2000
No. bedrooms	3	3	3	3	3
No. bathrooms	1	1	1	1	2
No. stories	2	2	2	2	2
Floor conditioned area (ft^2)	1,789	1,789	1,804	1,804	2,111
Window area (ft^2)	239	239	196	196	226
Window-to-wall ratio	0.12	0.12	0.12	0.12	0.12
Glazing layers	1	2	1	2	2
Windows	Wood	Wood	Wood	Wood	Wood
Wall siding type	Wood	Wood	Wood	Wood	Wood
Wall type	Plaster	Plaster	Drywall	Drywall	Drywall
R-Values					
Wall	0	7	0	7	13
Ceiling	0	7	7	11	27
Floor	0	0	0	0	19
Foundation type	Basement	Basement	Basement	Basement	Basement
Foundation insulation	None	None	None	None	None
Garage	Detached	Detached	Attached	Attached	Attached
Heating equipment	Natural gas	Natural gas	Natural gas	Natural gas	Natural gas
Furnace efficiency AFUE	80%	80%	80%	80%	80%
Cooling equipment	Window	Window	Window	Window	Window
A/C efficiency SEER	7	7	8	8	10
Cooling set points	78	78	78	78	78
Heating set points	67	67	67	67	67
Infiltration ACH 50 Pa	0.7	0.7	0.7	0.7	0.4
Appliances	Non-ES[a]	Non-ES	Non-ES	Non-ES	Non-ES
Lighting (fluorescent)	14%	14%	14%	14%	14%

[a] Non-Energy Star.

TABLE A.3   U.S. Housing Archetypes by Census Division: East North (Chicago)

	Prototype A	Prototype A1	Prototype B	Prototype B1	Prototype C
Year built	Pre-1950s	Pre-1950s	1950–1979	1950–1979	1980–2000
No. bedrooms	3	3	3	3	3
No. bathrooms	1	1	1	1	2
No. stories	2	2	1	1	2
Floor conditioned area (ft²)	1,646	1,646	1,557	1,557	2,223
Window area (ft²)	199	199	167	167	190
Window-to-wall ratio	0.12	0.12	0.12	0.12	0.12
Glazing layers	1	2	1	2	2
Windows	Wood	Wood	Wood	Wood	Wood
Wall siding type	Wood	Wood	Brick	Brick	Aluminum
Wall type	Plaster	Plaster	Drywall	Drywall	Drywall
R-Values					
Wall	0	7	0	7	13
Ceiling	0	11	11	19	25
Floor	0	0	0	0	0
Foundation type	Basement	Basement	Basement	Basement	Basement
Foundation insulation	None	None	None	None	None
Garage	Detached	Detached	Attached	Attached	Attached
Heating equipment	Natural gas	Natural gas	Natural gas	Natural gas	Natural gas
Furnace efficiency AFUE	80%	80%	80%	80%	80%
Cooling equipment	Window	Window	Central	Central	Central
A/C efficiency SEER	7	7	8	8	10
Cooling set points	74	74	74	74	74
Heating set points	61	67	61	67	68
Infiltration ACH 50 Pa	0.7	0.7	0.7	0.7	0.4
Appliances	Non-ES[a]	Non-ES	Non-ES	Non-ES	Non-ES
Lighting (fluorescent)	14%	14%	14%	14%	14%

[a] Non-Energy Star.

TABLE A.4   U.S. Housing Archetypes by Census Division: West North Central (Minneapolis)

	Prototype A	Prototype A1	Prototype B	Prototype B1	Prototype C
Year built	Pre-1950s	Pre-1950s	1950–1979	1950–1979	1980–2000
No. bedrooms	3	3	3	3	3
No. bathrooms	1	1	1	1	2
No. stories	2	2	1	1	2
Floor conditioned area (ft²)	1,621	1,621	1,487	1,487	2,048
Window area (ft²)	209	209	174	174	231
Window-to-wall ratio	0.12	0.12	0.12	0.12	0.12
Glazing layers	1	1	1	2	2
Windows	Wood	Wood	Wood	Wood	Wood
Wall siding type	Wood	Wood	Wood	Wood	Wood
Wall type	Plaster	Plaster	Drywall	Drywall	Drywall
R-Values					
Wall	0	7	0	7	19
Ceiling	0	11	11	22	27
Floor	0	0	0	0	0
Foundation type	Basement	Basement	Basement	Basement	Basement
Foundation insulation	None	None	None	None	R-5, 4 ft basement wall
Garage	Detached	Detached	Attached	Attached	Attached
Heating equipment	Natural gas	Natural gas	Natural gas	Natural gas	Natural gas
Furnace efficiency AFUE	80%	80%	80%	80%	80%
Cooling equipment	Window	Window	Central	Central	Central
A/C efficiency SEER	7	7	8	8	10
Cooling set points	74	74	72	72	72
Heating set points	59	66	59	66	68
Infiltration ACH 50 Pa	0.3	0.3	0.3	0.3	0.1
Appliances	Non-ES[a]	Non-ES	Non-ES	Non-ES	Non-ES
Lighting (fluorescent)	14%	14%	14%	14%	14%

[a] Non-Energy Star.

TABLE A.5 U.S. Housing Archetypes by Census Division: West North Central (Kansas City)

	Prototype A	Prototype A1	Prototype B	Prototype B1	Prototype C
Year built	Pre-1950s	Pre-1950s	1950–1979	1950–1979	1980–2000
No. bedrooms	3	3	3	3	3
No. bathrooms	1	1	1	1	2
No. stories	2	2	1	1	2
Floor conditioned area (ft²)	1,621	1,621	1,487	1,487	2,048
Window area (ft²)	209	209	174	174	231
Window-to-wall ratio	0.12	0.12	0.12	0.12	0.12
Glazing layers	1	1	2	2	2
Windows	Wood	Wood	Aluminum	Aluminum	Wood
Wall siding type	Wood	Wood	Wood	Wood	Wood
Wall type	Plaster	Plaster	Drywall	Drywall	Drywall
R-Values					
Wall	0	7	0	7	19
Ceiling	0	11	11	22	27
Floor	0	0	0	0	0
Foundation type	Basement	Basement	Basement	Basement	Basement
Foundation insulation	None	None	None	None	None
Garage	Detached	Detached	Attached	Attached	Attached
Heating equipment	Natural gas	Natural gas	Natural gas	Natural gas	Natural gas
Furnace efficiency AFUE	80%	80%	80%	80%	80%
Cooling equipment	Window	Window	Central	Central	Central
A/C efficiency SEER	7	7	8	8	10
Cooling set points	74	74	72	72	72
Heating set points	62	67	62	67	69
Infiltration ACH 50 Pa	0.7	0.7	0.7	0.7	0.4
Appliances	Non-ES[a]	Non-ES	Non-ES	Non-ES	Non-ES
Lighting (fluorescent)	14%	14%	14%	14%	14%

[a] Non-Energy Star.

TABLE A.6 U.S. Housing Archetypes by Census Division: South Atlantic (Washington, D.C.)

	Prototype A	Prototype A1	Prototype B	Prototype B1	Prototype C
Year built	Pre-1950s	Pre-1950s	1950–1979	1950–1979	1980–2000
No. bedrooms	3	3	3	3	3
No. bathrooms	1	1	1	1	2
No. stories	1	1	1	1	2
Floor conditioned area (ft²)	1,831	1,831	1,643	1,643	2,211
Window area (ft²)	163	163	165	165	160
Window-to-wall ratio	0.12	0.12	0.12	0.12	0.12
Glazing layers	1	1	1	1	2
Windows	Wood	Wood	Aluminum	Aluminum	Wood
Wall siding type	Wood	Wood	Brick	Brick	Aluminum
Wall type	Plaster	Plaster	Drywall	Drywall	Drywall
R-Values					
Wall	0	7	0	7	11
Ceiling	0	7	7	22	23
Floor	0	0	0	0	19
Foundation type	Crawlspace	Crawlspace	Crawlspace	Crawlspace	Basement
Foundation insulation	None	None	None	None	None
Garage	Detached	Detached	Attached	Attached	Attached
Heating equipment	Natural gas	Natural gas	Electric Furnace	Electric Furnace	Electric Furnace
Furnace efficiency AFUE	80%	80%	99%	99%	99%
Cooling equipment	Window	Window	Central	Central	Central
A/C efficiency SEER	7	7	8	8	10
Cooling set points	77	77	77	77	77
Heating set points	75	75	75	75	75
Infiltration ACH 50 Pa	0.3	0.3	0.3	0.3	0.1
Appliances	Non-ES[a]	Non-ES	Non-ES	Non-ES	Non-ES
Lighting (fluorescent)	14%	14%	14%	14%	14%

[a] Non-Energy Star.

TABLE A.7 U.S. Housing Archetypes by Census Division: South Atlantic (Atlanta)

	Prototype A	Prototype A1	Prototype B	Prototype B1	Prototype C
Year built	Pre-1950s	Pre-1950s	1950–1979	1950–1979	1980–2000
No. bedrooms	3	3	3	3	3
No. bathrooms	1	1	1	1	2
No. stories	1	1	1	1	2
Floor conditioned area (ft²)	1831	1831	1643	1643	2211
Window area (ft²)	170	170	161	161	158
Window-to-wall ratio	0.12	0.12	0.12	0.12	0.12
Glazing layers	1	1	1	1	2
Windows	Wood	Wood	Aluminum	Aluminum	Aluminum
Wall siding type	Wood	Wood	Brick	Brick	Wood
Wall type	Plaster	Plaster	Drywall	Drywall	Drywall
R-Values					
Wall	0	7	0	7	11
Ceiling	0	7	7	22	24
Floor	0	0	0	0	19
Foundation type	Crawlspace	Crawlspace	Crawlspace	Crawlspace	Basement
Foundation insulation	None	None	None	None	None
Garage	Detached	Detached	Attached	Attached	Attached
Heating equipment	Natural gas	Natural gas	Electric furnace	Electric furnace	Electric furnace
Furnace efficiency AFUE	80%	80%	99%	99%	99%
Cooling equipment	Window	Window	Central	Central	Central
A/C efficiency SEER	7	7	8	8	10
Cooling set points	77	77	77	77	77
Heating set points	75	75	75	75	75
Infiltration ACH 50 Pa	0.3	0.3	0.3	0.3	0.1
Appliances	Non-ES[a]	Non-ES	Non-ES	Non-ES	Non-ES
Lighting (fluorescent)	14%	14%	14%	14%	14%

[a] Non-Energy Star.

TABLE A.8 U.S. Housing Archetypes by Census Division: South Atlantic (Miami)

	Prototype A	Prototype A1	Prototype B	Prototype B1	Prototype C
Year built	Pre-1950s	Pre-1950s	1950–1979	1950–1979	1980–2000
No. bedrooms	3	3	3	3	3
No. bathrooms	1	1	1	1	2
No. stories	1	1	1	1	1
Floor conditioned area (ft²)	1,831	1,831	1,643	1,643	2,211
Window area (ft²)	163	163	165	165	160
Window-to-wall ratio	0.12	0.12	0.12	0.12	0.12
Glazing layers	1	1	1	1	1
Windows	Wood	Wood	Aluminum	Aluminum	Aluminum
Wall siding type	Wood	Wood	Brick	Brick	Stucco
Wall type	Plaster	Plaster	Drywall	Drywall	Drywall
R-Values					
Wall	0	7	0	7	11
Ceiling	0	7	7	22	23
Floor	0	0	0	0	0
Foundation type	Crawlspace	Crawlspace	Crawlspace	Crawlspace	Slab
Foundation insulation	None	None	None	None	None
Garage	Detached	Detached	Attached	Attached	Attached
Heating equipment	Natural gas	Natural gas	Electric furnace	Electric furnace	Electric furnace
Furnace efficiency AFUE	80%	80%	99%	99%	99%
Cooling equipment	Window	Window	Central	Central	Central
A/C efficiency SEER	7	7	8	8	10
Cooling set points	77	77	77	77	77
Heating set points	75	75	75	75	75
Infiltration ACH 50 Pa	1.0	1.1	1.1	1.1	0.7
Appliances	Non-ES[a]	Non-ES	Non-ES	Non-ES	Non-ES
Lighting (fluorescent)	14%	14%	14%	14%	14%

[a] Non-Energy Star.

TABLE A.9 U.S. Housing Archetypes by Census Division: West South Central (Fort Worth)

	Prototype A	Prototype A1	Prototype B	Prototype B1	Prototype C
Year built	Pre-1950s	Pre-1950s	1950–1979	1950–1979	1980–2000
No. bedrooms	3	3	3	3	3
No. bathrooms	1	1	1	1	2
No. stories	1	1	1	1	1
Floor conditioned area (ft²)	1,247	1,247	1,493	1,493	2,055
Window area (ft²)	180	180	150	150	168
Window-to-wall ratio	0.12	0.12	0.12	0.12	0.12
Glazing layers	1	1	1	1	2
Windows	Wood	Wood	Aluminum	Aluminum	Aluminum
Wall siding type	Wood	Wood	Brick	Brick	Wood
Wall type	Plaster	Plaster	Drywall	Drywall	Drywall
R-Values					
Wall	0	7	0	7	11
Ceiling	0	7	7	19	24
Floor	0	0	0	0	0
Foundation type	Slab	Slab	Slab	Slab	Slab
Foundation insulation	None	None	None	None	R-5, 2 ft slab edge
Garage	Detached	Detached	Attached	Attached	Attached
Heating equipment	Natural gas	Natural gas	Natural gas	Natural gas	Natural gas
Furnace efficiency AFUE	80%	80%	80%	80%	80%
Cooling equipment	Window	Window	Central	Central	Central
A/C efficiency SEER	7	7	8	8	10
Cooling set points	78	78	78	78	78
Heating set points	72	72	72	72	72
Infiltration ACH 50 Pa	1.1	1.1	1.1	1.1	0.7
Appliances	Non-ES[a]	Non-ES	Non-ES	Non-ES	Non-ES
Lighting (fluorescent)	14%	14%	14%	14%	14%

[a] Non-Energy Star.

TABLE A.10 U.S. Housing Archetypes by Census Division: West South Central (New Orleans)

	Prototype A	Prototype A1	Prototype B	Prototype B1	Prototype C
Year built	Pre-1950s	Pre-1950s	1950–1979	1950–1979	1980–2000
No. bedrooms	3	3	3	3	3
No. bathrooms	1	1	1	1	2
No. stories	1	1	1	1	1
Floor conditioned area (ft²)	1,247	1,247	1,493	1,493	2,055
Window area (ft²)	180	180	150	150	168
Window-to-wall ratio	0.12	0.12	0.12	0.12	0.12
Glazing layers	1	1	1	1	2
Windows	Wood	Wood	Aluminum	Aluminum	Aluminum
Wall siding type	Wood	Wood	Brick	Brick	Brick
Wall type	Plaster	Plaster	Drywall	Drywall	Drywall
R-Values					
Wall	0	7	0	7	11
Ceiling	0	7	7	19	24
Floor	0	0	0	0	0
Foundation type	Slab	Slab	Slab	Slab	Slab
Foundation insulation	None	None	None	None	None
Garage	Detached	Detached	Attached	Attached	Attached
Heating equipment	Natural gas	Natural gas	Electric furnace	Electric furnace	Electric furnace
Furnace efficiency AFUE	80%	80%	99%	99%	99%
Cooling equipment	Window	Window	Central	Central	Central
A/C efficiency SEER	7	7	8	8	10
Cooling set points	78	78	78	78	78
Heating set points	72	72	72	72	72
Infiltration ACH 50 Pa	1.1	1.1	1.1	1.1	0.7
Appliances	Non-ES[a]	Non-ES	Non-ES	Non-ES	Non-ES
Lighting (fluorescent)	14%	14%	14%	14%	14%

[a] Non-Energy Star.

TABLE A.11 U.S. Housing Archetypes by Census Division: Mountain (Denver)

	Prototype A	Prototype A1	Prototype B	Prototype B1	Prototype C
Year built	Pre-1950s	Pre-1950s	1950–1979	1950–1979	1980–2000
No. bedrooms	3	3	3	3	3
No. bathrooms	1	1	1	1	2
No. stories	1	1	1	1	2
Floor conditioned area (ft²)	1,823	1,823	1,734	1,734	2,017
Window area (ft²)	159	159	141	141	238
Window-to-wall ratio	0.12	0.12	0.12	0.12	0.12
Glazing layers	1	1	1	1	2
Windows	Wood	Wood	Wood	Wood	Wood
Wall siding type	Wood	Wood	Brick	Brick	Wood
Wall type	Plaster	Plaster	Drywall	Drywall	Drywall
R-Values					
Wall	0	7	0	7	13
Ceiling	0	11	11	11	30
Floor	0	0	0	0	11
Foundation type	Basement	Basement	Slab	Slab	Basement
Foundation insulation	None	None	None	None	None
Garage	Detached	Detached	Attached	Attached	Attached
Heating equipment	Natural gas	Natural gas	Natural gas	Natural gas	Natural gas
Furnace efficiency AFUE	80%	80%	80%	80%	80%
Cooling equipment	Window	Window	Central	Central	Central
A/C efficiency SEER	7	7	8	8	10
Cooling set points	77	77	77	77	77
Heating set points	70	70	70	70	70
Infiltration ACH 50 Pa	1.1	1.1	1.1	1.1	0.7
Appliances	Non-ES[a]	Non-ES	Non-ES	Non-ES	Non-ES
Lighting (fluorescent)	14%	14%	14%	14%	14%

[a] Non-Energy Star.

TABLE A.12   U.S. Housing Archetypes by Census Division: Mountain (Albuquerque)

	Prototype A	Prototype A1	Prototype B	Prototype B1	Prototype C
Year built	Pre-1950s	Pre-1950s	1950–1979	1950–1979	1980–2000
No. bedrooms	3	3	3	3	3
No. bathrooms	1	1	1	1	2
No. stories	1	1	1	1	1
Floor conditioned area (ft²)	1,823	1,823	1,734	1,734	2,017
Window area (ft²)	159	159	141	141	238
Window-to-wall ratio	0.12	0.12	0.12	0.12	0.12
Glazing layers	1	2	1	2	2
Windows	Wood	Wood	Aluminum	Aluminum	Wood
Wall siding type	Wood	Wood	Brick	Brick	Stucco
Wall type	Plaster	Plaster	Drywall	Drywall	Drywall
R-Values					
Wall	0	7	0	7	13
Ceiling	0	11	11	11	30
Floor	0	0	0	0	0
Foundation type	Basement	Basement	Slab	Slab	Slab
Foundation insulation	None	None	None	None	R-5, 2 ft slab edge
Garage	Detached	Detached	Attached	Attached	Attached
Heating equipment	Natural gas	Natural gas	Natural gas	Natural gas	Natural gas
Furnace efficiency AFUE	80%	80%	80%	80%	80%
Cooling equipment	Window	Window	Central	Central	Central
A/C efficiency SEER	7	7	8	8	10
Cooling set points	77	77	77	77	77
Heating set points	70	70	70	70	70
Infiltration ACH 50 Pa	1.1	1.1	1.1	1.1	0.7
Appliances	Non-ES[a]	Non-ES	Non-ES	Non-ES	Non-ES
Lighting (fluorescent)	14%	14%	14%	14%	14%

[a] Non-Energy Star.

TABLE A.13 U.S. Housing Archetypes by Census Division: Mountain (Phoenix)

	Prototype A	Prototype A1	Prototype B	Prototype B1	Prototype C
Year built	Pre-1950s	Pre-1950s	1950–1979	1950–1979	1980–2000
No. bedrooms	3	3	3	3	3
No. bathrooms	1	1	1	1	2
No. stories	1	1	1	1	1
Floor conditioned area (ft²)	1823	1823	1734	1734	2017
Window area (ft²)	159	159	141	141	238
Window-to-wall ratio	0.12	0.12	0.12	0.12	0.12
Glazing layers	1	1	1	1	1
Windows	Wood	Wood	Aluminum	Aluminum	Aluminum
Wall siding type	Wood	Wood	Brick	Brick	Stucco
Wall type	Plaster	Plaster	Drywall	Drywall	Drywall
R-Values					
Wall	0	7	0	7	13
Ceiling	0	11	11	11	30
Floor	0	0	0	0	0
Foundation type	Basement	Basement	Slab	Slab	Slab
Foundation insulation	None	None	None	None	None
Garage	Detached	Detached	Attached	Attached	Attached
Heating equipment	Natural gas	Natural gas	Natural gas	Natural gas	Natural gas
Furnace efficiency AFUE	80%	80%	80%	80%	80%
Cooling equipment	Window	Window	Central	Central	Central
A/C efficiency SEER	7	7	8	8	10
Cooling set points	80	80	78	78	78
Heating set points	70	70	70	70	70
Infiltration ACH 50 Pa	0.4	0.3	0.4	0.3	0.1
Appliances	Non-ES[a]	Non-ES	Non-ES	Non-ES	Non-ES
Lighting (fluorescent)	14%	14%	14%	14%	14%

[a] Non-Energy Star.

TABLE A.14 U.S. Housing Archetypes by Census Division: Pacific North (Seattle)

	Prototype A	Prototype A1	Prototype B	Prototype B1	Prototype C
Year built	Pre-1950s	Pre-1950s	1950–1979	1950–1979	1980–2000
No. bedrooms	3	3	3	3	3
No. bathrooms	1	1	1	1	2
No. stories	1	1	1	1	2
Floor conditioned area (ft²)	1,786	1,786	1,722	1,722	1,978
Window area (ft²)	180	180	142	142	194
Window-to-wall ratio	0.12	0.12	0.12	0.12	0.12
Glazing layers	1	1	1	1	1
Windows	Wood	Wood	Wood	Wood	Wood
Wall siding type	Wood	Wood	Wood	Wood	Wood
Wall type	Plaster	Plaster	Drywall	Drywall	Drywall
R-Values					
Wall	0	7	0	7	11
Ceiling	0	11	11	19	23
Floor	0	0	0	0	19
Foundation type	Crawlspace	Crawlspace	Crawlspace	Crawlspace	Crawlspace
Foundation insulation	None	None	None	None	None
Garage	Detached	Detached	Attached	Attached	Attached
Heating equipment	Natural gas	Natural gas	Natural gas	Natural gas	Natural gas
Furnace efficiency AFUE	80%	80%	80%	80%	80%
Cooling equipment	Window	Window	Central	Central	Central
A/C efficiency SEER	7	7	8	8	10
Cooling set points	75	75	73	73	73
Heating set points	70	70	70	70	70
Infiltration ACH 50 Pa	0.4	0.4	0.4	0.4	0.3
Appliances	Non-ES[a]	Non-ES	Non-ES	Non-ES	Non-ES
Lighting (fluorescent)	14%	14%	14%	14%	14%

[a] Non-Energy Star.

TABLE A.15   U.S. Housing Archetypes by Census Division: Pacific South (San Francisco)

	Prototype A	Prototype A1	Prototype B	Prototype B1	Prototype C
Year built	Pre-1950s	Pre-1950s	1950–1979	1950–1979	1980–2000
No. bedrooms	3	3	3	3	3
No. bathrooms	1	1	1	1	2
No. stories	1	1	1	1	2
Floor conditioned area (ft²)	1,786	1,786	1,722	1,722	1,978
Window area (ft²)	180	180	142	142	194
Window-to-wall ratio	0.12	0.12	0.12	0.12	0.12
Glazing layers	1	1	1	1	2
Windows	Wood	Wood	Aluminum	Aluminum	Aluminum
Wall siding type	Wood	Wood	Stucco	Stucco	Stucco
Wall type	Plaster	Plaster	Drywall	Drywall	Drywall
R-Values					
Wall	0	7	0	7	11
Ceiling	0	11	11	19	23
Floor	0	0	0	0	19
Foundation type	Crawlspace	Crawlspace	Crawlspace	Crawlspace	Crawlspace
Foundation insulation	None	None	None	None	None
Garage	Detached	Detached	Attached	Attached	Attached
Heating equipment	Natural gas	Natural gas	Natural gas	Natural gas	Natural gas
Furnace efficiency AFUE	80%	80%	80%	80%	80%
Cooling equipment	Window	Window	Central	Central	Central
A/C efficiency SEER	7	7	8	8	10
Cooling set points	75	75	73	73	73
Heating set points	70	70	70	70	70
Infiltration ACH 50 Pa	1.1	1.1	1.1	1.1	0.4
Appliances	Non-ES[a]	Non-ES	Non-ES	Non-ES	Non-ES
Lighting (fluorescent)	14%	14%	14%	14%	14%

[a] Non-Energy Star.

TABLE A.16 U.S. Housing Archetypes by Census Division: Pacific South (Los Angeles)

	Prototype A	Prototype A1	Prototype B	Prototype B1	Prototype C
Year built	Pre-1950s	Pre-1950s	1950–1979	1950–1979	1980–2000
No. bedrooms	3	3	3	3	3
No. bathrooms	1	1	1	1	2
No. stories	1	1	1	1	2
Floor conditioned area (ft²)	1,786	1,786	1,722	1,722	1,978
Window area (ft²)	180	180	142	142	194
Window-to-wall ratio	0.12	0.12	0.12	0.12	0.12
Glazing layers	1	1	1	1	2
Windows	Wood	Wood	Aluminum	Aluminum	Aluminum
Wall siding type	Wood	Wood	Stucco	Stucco	Stucco
Wall type	Plaster	Plaster	Drywall	Drywall	Drywall
R-Values					
Wall	0	7	0	7	11
Ceiling	0	11	11	19	23
Floor	0	0	0	0	19
Foundation type	Crawlspace	Crawlspace	Crawlspace	Crawlspace	Slab
Foundation insulation	None	None	None	None	None
Garage	Detached	Detached	Attached	Attached	Attached
Heating equipment	Natural gas	Natural gas	Natural gas	Natural gas	Natural gas
Furnace efficiency AFUE	80%	80%	80%	80%	80%
Cooling equipment	Window	Window	Central	Central	Central
A/C efficiency SEER	7	7	8	8	10
Cooling set points	75	75	73	73	73
Heating set points	70	70	70	70	70
Infiltration ACH 50 Pa	1.1	1.1	1.1	1.1	0.4
Appliances	Non-ES[a]	Non-ES	Non-ES	Non-ES	Non-ES
Lighting (fluorescent)	14%	14%	14%	14%	14%

[a] Non-Energy Star.

Appendix B:
Thermal Properties
of Building Materials

TABLE B.1A　Properties of Building Materials (IP)

Material	Thickness, L in.	Density, ρ lbm/ft³	Thermal Conductivity, k Btu/ft·°F	Specific Heat, C_p Btu/lbm·°F	R-Value (for listed thickness, L/k), °F·h·ft²/Btu
Building Boards					
Asbestos-cement board	—	120	4.00	0.24	—
Gypsum of plaster board	—	40	1.10	0.27	—
Plywood (Douglas fir)	1/2 in.	29	—	0.45	0.79
	5/8 in.	34	—	0.45	0.85
Insulated board and sheeting (regular density)	1/2 in.	18	—	0.31	1.32
Hardboard (high density, standard tempered)	—	63	1.00	0.32	—
Particle board					
Medium density	—	50	0.94	0.31	—
Underlayment	5/8 in.	40	—	0.29	0.82
Building Membranes					
Vapor-permeable felt	—	—	—	—	0.06
Vapor seal (2 layers of mopped 17.3 lbm/ft² felt)	—	—	—	—	0.12
Flooring Materials					
Carpet and rebounded urethane pad	3/4 in.	7	—	—	2.38
Carpet and rubber pad (one piece)	3/8 in.	20	—	—	0.68
Linoleum/cork tile	1/4 in.	29	—	—	0.51
Masonry Materials					
Masonry units:					
Brick, fire clay		150	8.4–10.2	—	—
		120	5.6–6.8	0.19	—
		70	2.5–3.1	—	—

Material	Size				
Concrete blocks					
Limestone aggregate					
36 lb, 138 lb/ft³ concrete, 2 cores with perlite-filled cores	8 in.	—	—	2.1	—
55 lb, 138 lb/ft³ concrete, 2 cores with perlite-filled cores	12 in.	—	—	3.7	—
Concrete:					
Lightweight aggregates (including expanded shale, clay, or slate; expanded slags; cinders; pumice; and scoria)		120	6.4–9.1	—	—
		100	4.7–6.2	0.2	—
		80	3.3–4.1	0.2	—
		60	2.1–2.5	—	—
		40	1.3	—	—
Cement/lime, mortar, and stucco		120	9.7	—	—
		100	6.7	—	—
		80	4.5	—	—
Roofing					
Asbestos–cement shingles		120	—	0.24	0.21
Asphalt roll roofing		70	—	0.36	0.15
Asphalt shingles		70	—	0.30	0.44
Built-in roofing	3/8 in.	70	—	0.35	0.33
Slate	1/2 in.	—	—	0.30	0.05
Wood shingles (plain and plastic film faced)		—	—	0.31	0.94
Plastering Materials					
Cement plaster, sand aggregate		116	5.0	0.20	—
Gypsum plaster		70	2.6	—	—
		80	3.2	—	—
Lightweight aggregate	1/2 in.	45	—	—	0.32
Sand aggregate		105	5.6	0.20	—
Perlite aggregate		45	1.5	0.32	—

TABLE B.1A (*Continued*)   Properties of Building Materials (IP)

Material	Thickness, L in.	Density, ρ lbm/ft^3	Thermal Conductivity, k Btu/ft·°F	Specific Heat, C_p Btu/lbm·°F	R-Value (for listed thickness, L/k), °F·h·ft^2/Btu
Siding Material (on flat surfaces)					
Asbestos-cement shingles	—	120	—	—	0.21
Hardboard siding	$^7/_{16}$ in.	—	—	0.28	0.67
Wood (drop) siding	1 in.	—	—	0.28	0.79
Wood (plywood) siding, lapped	$^3/_8$ in.	—	—	0.29	0.59
Aluminum or steel siding (over sheeting)					
Hollow backed		—	—	0.29	0.62
Insulating-board backed	$^3/_8$ in.	—	—	0.32	1.82
Architectural glass	—	158	6.9	0.20	—
Woods					
Hardwoods (maple, oak, etc.)	—	39.8–46.8	1.09–1.25	0.39	—
Softwoods (fir, pine, etc.)	—	24.5–41.2	0.74–1.12	0.39	—
Metals					
Aluminum (alloy 1100)	—	171	128	0.214	—
Steel, mild	—	489	26.2	0.120	—
Blanket and Batt					
Mineral fiber (fibrous form processed from rock, slag, or glass)	2 to $2^3/_4$ in.	0.3–2.0	—	0.17–0.23	7
	3 to $3^1/_2$ in.	0.3–2.0	—	0.17–0.23	11
	$5^1/_4$ to $6^1/_2$ in.	0.3–2.0	—	0.17–0.23	19

Board and Slab

Material				
Cellular glass	8.5	0.38	0.24	—
Glass fiber (organic bonded)	4–9	0.25	0.23	—
Expanded polystyrene (molded beads)	1.0	0.28	0.29	—
Expanded polyurethane (R-11 expanded)	1.5	0.16	0.38	—
Expanded perlite (organic bonded)	1.0	0.36	0.30	—
Expanded rubber (rigid)	4.5	0.22	0.40	—
Mineral fiber with resin binder	15	0.29	0.17	—
Cork	7.5	0.27	0.43	—

Sprayed or Formed in Place

Material				
Polyurethane foam	1.5–2.5	0.16–0.18	—	—
Glass fiber	3.5–4.5	0.26–0.27	—	—
Urethane, two-part mixture (rigid foam)	4.4	0.18	0.25	—
Mineral wool granules with asbestos/inorganic binders (sprayed)	12	0.32	—	—

Loose Fill

Material					
Mineral fiber (rock, slag, or glass)	3.75–5 in.	0.6–0.20	—	0.17	11
	6.5–8.75 in.	0.6–0.20	—	0.17	19
	7.5–10 in.	—	—	0.17	22
	7.25 in.	—	—	0.17	30
Silica aerogel		7.6			
Vermiculite (expanded)		7–8			
Perlite, expanded		2–4.1			
Sawdust or shavings		8–15			

TABLE B.1A (Continued)   Properties of Building Materials (IP)

Material	Thickness, L in.	Density, ρ lbm/ft^3	Thermal Conductivity, k Btu/ft·°F	Specific Heat, C_p Btu/lbm·°F	R-Value (for listed thickness, L/k), °F·h·ft^2/Btu
Cellulosic insulation (milled paper or wood pump)		0.3–3.2			
Cork, granulated		10			
Roof Insulation					
Cellular glass	—	9	0.4	0.24	—
Preformed, for use above deck	½ in.	—	—	0.24	1.39
	1 in.	—	—	0.50	2.78
	2 in.	—	—	0.94	5.56
Reflective Insulation					
Silica powder (evacuated)		10	0.0118	—	—
Aluminum foil separating fluffy glass mats; 10–12 layers(evacuated); for cryogenic applications (270 R)		2.5	0.0011	—	—
Aluminum foil and glass paper laminate; 75–150 layers (evacuated); for cryogenic applications (270 R)		7.5	0.00012	—	—

Source: The properties provided in this table are adapted from ASHRAE, *Handbook of Fundamentals*, American Society of Heating, Refrigerating, and Air-Conditioning Engineers, Atlanta, GA, 2009.

Note: All thermal properties are provided at a mean temperature of 75°F.

TABLE B.1B Properties of Building Materials (SI)

Material	Thickness, mm	Density, ρ kg/m^3	Thermal Conductivity, k W/m.°C	Specific Heat, C_p kJ/kg.°C	R-Value (for listed thickness, L/k), °C·h·m^2/W
Building Boards					
Asbestos-cement board	—	1,900	0.57	1.00	—
Gypsum of plaster board	—	640	0.16	1.15	—
Plywood (Douglas fir)	12.7 mm	460	—	1.88	0.14
	15.9 mm	540	—	1.88	0.15
Insulated board and sheeting (regular density)	12.7 mm	290	—	1.30	0.23
Hardboard (high density, standard tempered)	—	1010	0.144	1.34	—
Particle board:					
Medium density	—	800	0.135	1.30	—
Underlayment	15.9 mm	640	—	1.21	1.22
Building Membranes					
Vapor-permeable felt	—	—	—	—	0.011
Vapor seal (2 layers of mopped 17.3 lbm/ft^2 felt)	—	—	—	—	0.21
Flooring Materials					
Carpet and rebounded urethane pad	19 mm	110	—	—	0.42
Carpet and rubber pad (one piece)	9.5 mm	320	—	—	0.12
Linoleum/cork tile	6.4 mm	465	—	—	0.09
Masonry Materials					
Masonry units:					
Brick, fire clay		2,400	1.21–1.47	—	—
		1,920	0.81–0.98	0.80	—
		1,120	0.36–0.45	—	—

TABLE B.1B (Continued)   Properties of Building Materials (SI)

Material	Thickness, mm	Density, ρ kg/m³	Thermal Conductivity, k W/m·°C	Specific Heat, C_p kJ/kg·°C	R-Value (for listed thickness, L/k), °C·h·m²/W
Concrete blocks					
Limestone aggregate					
16.3 kg, 2,200 kg/m³ concrete, 2 cores with perlite-filled cores	200 mm	—	—	0.37	—
25 kg, 2,200kg/m³ concrete, 2 cores with perlite-filled cores	300 mm	—	—	0.65	—
Concretes:					
Lightweight aggregates (including expanded shale, clay, or slate; expanded slags; cinders; pumice; and scoria)		1,920	0.9–1.3	—	—
		1,600	0.68–0.89	0.84	—
		1,280	0.48–0.59	0.84	—
		960	0.30–0.36	—	—
		640	0.18	—	—
Cement/lime, mortar, and stucco		1,920	1.40	—	—
		1,600	0.97	—	—
		1,280	0.65	—	—
Roofing					
Asbestos-cement shingles		1,120	—	1.00	0.037
Asphalt roll roofing		920	—	1.51	0.027
Asphalt shingles		920	—	1.26	0.078
Built-in roofing	10 mm	920	—	1.47	0.059
Slate	13 mm	—	—	1.26	0.009
Wood shingles (plain and plastic film faced)		—	—	1.30	0.166
Plastering Materials					
Cement plaster, sand aggregate	—	1,860	0.72	0.84	—
Gypsum plaster		1,120	0.38	—	—
		1,280	0.46	—	—

	Thickness	Density	Conductivity	Specific heat	Resistance
Lightweight aggregate	13 mm	720	—	—	0.056
Sand aggregate	—	1,680	0.81	0.84	—
Perlite aggregate	—	720	0.22	1.34	—
Siding Material (on flat surfaces)					
Asbestos-cement shingles	—	1,900	—	—	0.037
Hardboard siding	11 mm	—	—	1.17	0.12
Wood (drop) siding	25 mm	—	—	1.17	0.14
Wood (plywood) siding, lapped	9.5 mm	—	—	1.22	0.10
Aluminum or steel siding (over sheeting):					
Hollow backed	—	—	—	1.22	0.11
Insulating-board backed	9.5 mm	—	—	1.34	0.32
Architectural glass	—	2,500	1.0	0.84	—
Woods					
Hardwoods (maple, oak, etc.)	—	635–750	0.15–0.18	1.63	—
Softwoods (fir, pine, etc.)	—	390–660	0.10–0.16	1.63	—
Metals					
Aluminum (alloy 1100)	—	2,740	221	896	—
Steel, mild	—	7,830	45.3	500	—
Blanket and Batt					
Mineral fiber (fibrous form processed from rock, slag, or glass)	50–70 mm	4.8–32	—	0.71–0.96	1.23
	75–90 mm	4.8–32	—	0.71–0.96	1.94
	135–165 mm	4.8–32	—	0.71–0.96	3.32
Board and Slab					
Cellular glass	—	136	0.055	1.0	—
Glass fiber (organic bonded)	—	64–144	0.036	0.096	—

TABLE B.1B (Continued)   Properties of Building Materials (SI)

Material	Thickness, mm	Density, ρ kg/m^3	Thermal Conductivity, k W/m·°C	Specific Heat, C_p kJ/kg·°C	R-Value (for listed thickness, L/k), °C·h·m^2/W
Expanded polystyrene (molded beads)		16	0.040	1.2	—
Expanded polyurethane (R-11 expanded)		24	0.023	1.6	—
Expanded perlite (organic bonded)		16	0.052	1.26	—
Expanded rubber (rigid)		72	0.032	1.68	—
Mineral fiber with resin binder		240	0.042	0.71	—
Cork		120	0.039	1.80	—
Sprayed or Formed in Place					
Polyurethane foam		24–40	0.023–0.026	—	—
Glass fiber		56–72	0.038–0.039	—	—
Urethane, two-part mixture (rigid foam)		70	0.026	1.045	—
Mineral wool granules with asbestos/inorganic binders (sprayed)		190	0.046	—	—
Loose Fill					
Mineral fiber (rock, slag, or glass)	75–125 mm	9.6–32	—	0.71	1.94
	165–222 mm	9.6–32	—	0.71	3.35
	191–254 mm	—	—	0.71	3.87
	185 mm	—	—	0.71	5.28
Silica aerogel		122	0.025	—	—
Vermiculite (expanded)		122	0.068	—	—
Perlite, expanded		32–66	0.039–0.045	1.09	—
Sawdust or shavings		128–240	0.065	1.38	—
Cellulosic insulation (milled paper or wood pump)		37–51	0.039–0.046	—	—

Roof Insulation					
Cellular glass	—	144	0.058	1.0	—
Preformed, for use above deck	13 mm	—	—	1.0	0.24
	25 mm	—	—	2.1	0.49
	50 mm	—	—	3.9	0.93
Reflective Insulation					
Silica powder (evacuated)		160	0.0017	—	—
Aluminum foil separating fluffy glass mats; 10–12 layers (evacuated); for cryogenic applications (150 K)		40	0.00016	—	—
Aluminum foil and glass paper laminate; 75–150 layers (evacuated); for cryogenic applications (150 K)		120	0.000017	—	—

Source: The properties provided in this table are adapted from ASHRAE, *Handbook of Fundamentals*, American Society of Heating, Refrigerating, and Air-Conditioning Engineers, Atlanta, GA, 2009.

Note: All thermal properties are provided at a mean temperature of 24°C.

TABLE B.2   Surface Conductance and Resistance for Air [IP]

Position of Surface	Direction of Heat Flow	Surface Emittance, ε			
		Nonreflective ε = 0.90		Reflective ε = 0.05	
		h_i	R	h_i	R
Inside air					
Horizontal	Upward	1.63	0.61	0.76	1.32
Vertical	Horizontal	1.46	0.68	0.59	1.7
Horizontal	Downward	1.08	0.92	0.22	4.55
Outside air		ho	R		
15 mph wind (for winter)	Any	6	0.17	—	—
7.5 mph wind (for summer)	Any	4	0.25	—	—

TABLE B.3   Surface Conductance and Resistance for Air [SI]

Position of Surface	Direction of Heat Flow	Surface Emittance, ε			
		Nonreflective ε = 0.90		Reflective ε = 0.05	
		h_i	R	h_i	R
Inside air					
Horizontal	Upward	9.26	0.11	4.32	0.23
Vertical	Horizontal	8.29	0.12	3.35	0.3
Horizontal	Downward	6.13	0.16	1.25	0.8
Outside air		ho	R		
Wind at 6.7 m/s (for winter)	Any	34	0.03	~	~
Wind at 3.4 m/s (for summer)	Any	22.7	0.044	~	~

TABLE B.4A IP Thermal Resistances of Plane Air Spaces [h·ft²·°F/Btu]

Position of Air Space	Direction of Heat Flow	Air Space		Effective Emittance, ε_{eff}			
		Mean Temp., °F	Temp. Diff., °F	0.5 in. Air Space		1.5 in. Air Space	
				0.05	0.82	0.05	0.82
Horizontal	Up	0	20	1.70	0.91	1.95	0.97
		0	10	2.04	1.00	2.35	1.06
		50	30	1.57	0.75	1.81	0.80
		50	10	2.05	0.84	2.40	0.89
Vertical	Horizontal	0	20	2.72	1.13	2.66	1.12
		0	10	2.82	1.15	3.35	1.23
		50	30	2.46	0.90	2.46	0.90
		50	10	2.54	0.91	3.55	1.02
Horizontal	Down	0	20	2.83	1.15	6.43	1.49
		0	10	2.85	1.16	6.66	1.51
		50	30	2.54	0.91	5.63	1.14
		50	10	2.55	0.92	5.90	1.15

TABLE B.4B SI Thermal Resistances of Plane Air Spaces [m²·K/W]

Position of Air Space	Direction of Heat Flow	Air Space		Effective Emittance, ε_{eff}			
		Mean Temp., °F	Temp. Diff., °F	13 mm Air Space		40 mm Air Space	
				0.05	0.82	0.05	0.82
Horizontal	Up	10.0	16.7	0.28	0.13	0.32	0.14
		10.0	5.6	0.36	0.15	0.42	0.16
		−17.8	11.1	0.30	0.16	0.34	0.17
		−17.8	5.6	0.36	0.18	0.41	0.19
Vertical	Horizontal	10.0	16.7	0.43	0.16	0.43	0.16
		10.0	5.6	0.45	0.16	0.62	0.18
		−17.8	11.1	0.48	0.20	0.47	0.20
		−17.8	5.6	0.50	0.20	0.59	0.22
Horizontal	Down	10.0	16.7	0.45	0.16	0.99	0.20
		10.0	5.6	0.45	0.16	1.04	0.20
		−17.8	11.1	0.50	0.20	1.13	0.26
		−17.8	5.6	0.50	0.20	1.17	0.27

Appendix C: Heating and Cooling Degree Days (SI Units)

Annual Heating Degree Days (HDD), Cooling Degree Days (CDD), and Cooling Degree Hours (CDH) International Sites (SI Units)

Country	City	HDD Tb=10°C	HDD Tb=18.3°C	CDD Tb=10°C	CDD Tb=18.3°C	CDH Tb=10°C	CDH Tb=18.3°C
Albania	TIRANA	292	1644	2321	633	5927	2318
Algeria	ANNABA	25	924	2999	856	7302	2459
American Samoa	PAGO PAGO WSO AP	0	0	6442	3400	37333	11474
Antarctica	DAVIS	7436	10478	0	0	0	0
Antigua and Barbuda	VC BIRD INTL AIRPOR	0	0	6249	3208	32044	8623
Argentina	EZEIZA AERO	129	1211	2597	637	5565	1923
Armenia	YEREVAN/YEREVAN-ARA	1211	2804	2191	744	8569	3923
Aruba	QUEEN BEATRIX AIRPO	0	0	6790	3748	41777	14922
Australia	CANBERRA AIRPORT	390	2113	1564	246	2823	1076
Austria	SALZBURG-FLUGHAFEN	1343	3372	1166	153	1382	371
Azerbaijan	LANKARAN	598	2107	2183	651	5013	1405
Bahamas	NASSAU AIRPORT NEW	0	9	5643	2609	23534	7304
Bahrain	BAHRAIN (INT. AIRPORT)	0	103	6153	3214	43452	26351
Barbados	GRANTLEY ADAMS	0	0	6308	3267	32611	8441
Belarus	MINSK	2154	4405	876	84	742	147
Belgium	BRUXELLES NATIONAL	864	2933	1069	96	823	200
Belize	BELIZE/PHILLIP GOLD	0	0	6145	3103	30986	10421
Benin	BOHICON	0	0	6416	3374	32351	12746
Bermuda	BERMUDA INTL	0	88	4596	1643	11676	2542
Bolivia	LA PAZ/ALTO	925	3941	27	0	0	0
Bosnia and Herzegovina	SARAJEVO-BJELAVE	1288	3186	1359	217	1938	596
Botswana	SERETSE KHAMA INTER	5	435	3976	1364	14449	6227
Brazil	BRASILIA (AEROPORTO)	0	23	4364	1346	9734	2571
Bulgaria	SOFIA (OBSERV.)	1318	3169	1431	239	2431	784
Burkina Faso	OUAGADOUGOU	0	0	6844	3802	49567	27676

Country	Location						
Cape Verde	SAL	0	0	5125	2083	11379	1826
Chad	NDJAMENA	0	1	6870	3829	50915	30309
Chile	PUERTO MONTT	462	2916	592	2	47	6
China	BEIJING	1310	2830	2370	848	7954	3031
Christmas Island	CHRISTMAS ISLAND AE	0	0	5414	2373	12402	643
Cocos (Keeling) Islands	COCOS ISLAND AERO	0	0	6203	3161	30969	5908
Colombia	BOGOTA/ELDORADO	1	1752	1292	0	3	0
Congo	BRAZZAVILLE/MAYA-M	0	0	5841	2799	21976	7507
Cook Islands	AMURI/AITUTAKI ISL.	0	0	5858	2817	23341	3743
Costa Rica	JUAN SANTAMARIA INT	0	0	4867	1826	8564	1674
Cote d'Ivoire	ABIDJAN	0	0	6239	3197	30476	8942
Croatia	ZAGREB/MAKSIMIR	1107	2873	1576	301	2636	791
Cuba	AEROPUERTO JOSE MAR	0	21	5411	2391	19541	6592
Cyprus	PAPHOS AIRPORT	16	765	3198	906	7760	1927
Czech Rep.	PRAHA/RUZYNE	1565	3754	941	89	918	231
Dem. People's Rep. of Korea	PYONGYANG	1629	3298	1947	574	3808	917
Denmark	KOEBENHAVN/KASTRUP	1357	3653	790	45	199	17
Diego Garcia	DIEGO GARCIA NAF	0	0	6427	3386	35509	9748
Dominican Rep.	SANTO DOMINGO	0	0	6023	2982	25939	8221
Ecuador	QUITO AEROPUERTO	0	1402	1640	1	4	1
Egypt	CAIRO AIRPORT	1	393	4416	1767	19113	9392
Estonia	TALLINN	2193	4649	617	31	244	32
Falkland Islands (Malvinas)	MOUNT PLEASANT AIRP	1477	4364	153	0	2	0
Faroe Islands	TORSHAVN	1291	4239	94	0	0	0
Fiji	NADI AIRPORT	0	0	5675	2633	22351	6795
Finland	HELSINKI-VANTAA	2411	4856	637	39	337	39
France	PARIS-AEROPORT CHAR	743	2649	1301	164	1404	413
French Polynesia	TAHITI-FAAA	0	0	6026	2984	27652	7815

Annual Heating Degree Days (HDD), Cooling Degree Days (CDD), and Cooling Degree Hours (CDH) International Sites (SI Units) (Continued)

Country	City	HDD Tb=10°C	HDD Tb=18.3°C	CDD Tb=10°C	CDD Tb=18.3°C	CDH Tb=10°C	CDH Tb=18.3°C
Gabon	LIBREVILLE	0	0	6016	2974	27177	6108
Gambia	BANJUL/YUNDUM	0	1	6154	3113	28227	12066
Georgia	TBILISI	806	2371	2134	659	5657	2244
Germany	BERLIN/DAHLEM	1270	3390	1039	118	979	234
Gibraltar	GIBRALTAR	3	620	3122	697	2989	654
Greece	ATHINAI (AIRPORT)	119	1165	3076	1079	10117	3957
Greenland	NUUK (GODTHAAB)	4173	7203	12	0	0	0
Grenada	POINT SALINES AIRPO	0	0	6378	3336	34507	9339
Guam	ANDERSEN AFB	0	0	6229	3187	30847	6544
Guatemala	GUATEMALA (AEROPUERTO)	0	71	3616	645	1874	125
Guernsey	GUERNSEY AIRPORT	426	2567	924	23	77	11
Guiana	ROCHAMBEAU	0	0	6052	3010	25655	7857
Guyana	TIMEHRI/CHEDDI JAG	0	0	6136	3094	25396	8681
Honduras	TEGUCIGALPA	0	12	4525	1495	9513	2407
Hungary	BUDAPEST/FERIHEGY I	1322	3188	1433	258	2803	928
Iceland	REYKJAVIK	2049	4990	103	0	0	0
India	NEW DELHI/PALAM	1	286	5767	3011	42516	25343
Indonesia	JAKARTA/SOEKARNO-HA	0	0	6439	3398	34051	12496
Iran (Islamic Rep. of)	TEHRAN-MEHRABAD	428	1588	3421	1540	20017	10807
Ireland	DUBLIN AIRPORT	717	3135	630	6	12	0
Isle of Man	ISLE OF MAN/RONALDS	620	3079	586	4	11	0
Israel	BEN-GURION INT. AIR	10	619	3721	1289	12346	4565
Italy	ROMA FIUMICINO	201	1525	2273	555	4304	1022
Jamaica	KINGSTON/NORMAN MAN	0	0	6608	3567	39379	14343
Japan	TOKYO	341	1611	2671	902	7421	2427

Country	Location						
Jersey	JERSEY AIRPORT	484	2581	982	37	170	30
Jordan	AMMAN AIRPORT	184	1291	2974	1037	10217	4263
Kazakhstan	ASTANA	3571	5717	1101	206	2291	736
Kenya	NAIROBI/KENYATTA AI	0	104	3459	523	3023	333
Kerguelen	PORT-AUX-FRANCAIS	1918	4925	36	0	0	0
Kiribati	TARAWA	0	0	6698	3656	42259	14305
Korea (Rep. of)	SEOUL	1183	2721	2202	699	4928	1318
Kuwait	KUWAIT INTERNATIONA	11	426	5987	3360	54703	38978
Kyrgyzstan	BISHKEK	1531	3218	1950	596	6961	3046
Latvia	RIGA	1910	4193	827	69	451	81
Lebanon	BEYROUTH (AEROPORT)	4	464	3966	1383	11850	3063
Libya	TRIPOLI INTERNATION	8	668	4016	1633	18871	10227
Liechtenstein	VADUZ (LIECHTENSTEIN)	1147	3119	1239	169	1087	232
Lithuania	VILNIUS	2077	4361	830	72	624	124
Luxembourg	LUXEMBOURG/LUXEMBOU	1217	3383	994	119	893	217
Macao	TAIPA GRANDE	8	309	4728	1987	18408	5667
Macedonia	SKOPJE- AIRPORT	1031	2653	1920	500	6006	2633
Madagascar	ANTANANARIVO/IVATO	0	330	3356	643	3006	406
Malaysia	KUALA LUMPUR SUBANG	0	0	6601	3559	35356	13689
Maldives	MALE	0	0	6779	3737	45547	16941
Mali	BAMAKO/SENOU	0	0	6572	3530	42710	23263
Malta	LUQA	11	782	3293	1024	7332	2244
Marshall Islands	MAJURO WSO AP	0	0	6547	3505	39609	11397
Martinique	LE LAMENTIN	0	0	6087	3045	27854	7906
Mauritania	NOUAKCHOTT	0	2	5998	2959	29233	13642
Mauritius	VACOAS (MAURITIUS)	0	25	4245	1229	4048	302
Mayotte	DZAOUDZI/PAMANZI	0	0	6102	3060	29903	7871
Mexico	AEROP. INTERNACIONA	2	563	2670	190	1905	241

Annual Heating Degree Days (HDD), Cooling Degree Days (CDD), and Cooling Degree Hours (CDH) International Sites (SI Units) (Continued)

Country	City	HDD Tb=10°C	HDD Tb=18.3°C	CDD Tb=10°C	CDD Tb=18.3°C	CDH Tb=10°C	CDH Tb=18.3°C
Micronesia	YAP ISLAND WSO AP	0	0	6458	3416	35953	10617
Moldova	KISINEV	1511	3337	1541	325	2332	631
Mongolia	ULAANBAATAR	4546	6964	701	79	1016	303
Morocco	RABAT-SALE	6	800	2780	533	2858	722
Mozambique	MAPUTO/MAVALANE	0	22	4958	1938	14446	4288
Namibia	WINDHOEK	8	379	3719	1049	9511	3222
Netherlands	AMSTERDAM AP SCHIPH	883	3038	951	65	486	103
New Caledonia	NOUMEA (NLLE-CALEDONIE)	0	2	4849	1810	10103	1880
New Zealand	WELLINGTON AIRPORT	135	1877	1346	47	42	1
Nicaragua	MANAGUA A.C.SANDINO	0	0	6457	3416	35517	15527
Niger	NIAMEY-AERO	0	0	7192	4151	57896	34902
Niue Is.	ALOFI	0	0	5438	2397	16208	3099
Norfolk Island	NORFOLK ISLAND AERO	0	308	3232	499	408	2
Northern Mariana Islands	SAIPAN	0	0	6541	3499	36653	10852
Norway	OSLO/GARDERMOEN	2447	4982	528	21	232	26
Oman	SEEB INTL AIRPORT	0	1	6692	3652	48896	28848
Pakistan	ISLAMABAD AIRPORT	16	652	4388	1982	25718	13954
Palau	KOROR WSO	0	0	6523	3481	36552	10433
Panama	MARCOS A GELABERT I	0	0	6621	3579	36052	12472
Paraguay	ASUNCION/AEROPUERTO	6	254	4841	2049	20329	8939
Peru	LIMA-CALLAO/AEROP	0	165	3698	822	2619	344
Philippines	MANILA	0	0	6737	3695	43723	16608
Poland	WARSZAWA-OKECIE	1637	3771	1021	112	1079	266
Portugal	LISBOA/PORTELA	35	1012	2664	599	4084	1513
Puerto Rico	SAN JUAN INTL ARPT	0	0	6159	3118	29006	8330

Country	Station						
Qatar	DOHA INTERNATIONAL	0	73	6496	3527	49126	31403
Reunion Is.	SAINT-DENIS/GILLOT	0	0	5154	2113	13305	2522
Romania	BUCURESTI AFUMATI	1332	3069	1686	382	4134	1507
Russia	MOSKVA	2502	4747	903	107	759	145
Saint Helena Is.	ST. HELENA IS.	0	309	3057	326	48	1
Saint Lucia	HEWANORRA INTL AIRP	0	0	6429	3388	36468	10501
Samoa	APIA	0	0	6160	3118	30271	8176
Saudi Arabia	RIYADH OBS. (O.A.P.)	6	301	6009	3264	51690	35063
Senegal	DAKAR/YOFF	0	1	5363	2322	17989	4546
Serbia	BEOGRAD	961	2558	1944	498	4316	1583
Seychelles	SEYCHELLES INTERNAT	0	0	6329	3288	33339	8399
Singapore	SINGAPORE/CHANGI AI	0	0	6579	3537	37142	12225
Slovakia	BRATISLAVA-LETISKO	1251	3099	1458	265	2533	813
Slovenia	LJUBLJANA/BEZIGRAD	1123	2953	1482	269	2178	636
Solomon Islands	HONIARA/HENDERSON	0	0	6048	3007	26449	8396
South Africa	PRETORIA (IRENE)	23	811	2728	473	3217	666
Spain	MADRID/BARAJAS RS	451	2023	2084	612	8047	3864
Sri Lanka	KATUNAYAKE	0	0	6454	3412	36566	12590
Suriname	ZANDERIJ	0	0	6264	3222	26783	10101
Svalbard and Jan Mayen	HOPEN	5460	8502	0	0	0	0
Sweden	STOCKHOLM/ARLANDA	2015	4409	692	46	414	69
Switzerland	ZURICH-KLOTEN	1240	3303	1112	132	1334	358
Syria	DAMASCUS INT. AIRPO	298	1527	2874	1060	14780	8331
Taiwan	TAIBEI	1	237	4850	2044	19554	7584
Tajikistan	DUSHANBE	603	1941	2655	952	12143	6284
Tanzania	DAR ES SALAAM AIRPO	0	0	5851	2809	24708	8192
Thailand	BANGKOK METROPOLIS	0	0	6915	3873	46537	20777
Togo	LOME	0	0	6356	3314	33283	10981

Annual Heating Degree Days (HDD), Cooling Degree Days (CDD), and Cooling Degree Hours (CDH) International Sites (SI Units) *(Continued)*

Country	City	HDD Tb=10°C	HDD Tb=18.3°C	CDD Tb=10°C	CDD Tb=18.3°C	CDH Tb=10°C	CDH Tb=18.3°C
Tonga	FUAAMOTU	0	3	5001	1963	11098	1666
Tunisia	TUNIS-CARTHAGE	19	814	3432	1186	11128	4799
Turkey	ESENBOGA	1373	3299	1344	227	3548	1271
Turkmenistan	ASHGABAT KESHI	633	1909	3221	1454	19659	11118
Tuvalu	FUNAFUTI NF	0	0	6694	3653	42887	14536
Ukraine	KYIV	1878	3907	1194	180	1249	272
United Arab Emirates	ABU DHABI INTER. AI	0	30	6577	3565	49047	31114
United Arab Emirates	DUBAI INTERNATIONAL	0	24	6461	3442	47226	29061
United Kingdom	LONDON WEATHER CENT	464	2344	1291	129	694	155
United States Minor Outlying Islands	MIDWAY ISLAND NAS	0	28	4867	1854	12637	2543
Uruguay	CARRASCO	98	1221	2379	461	2736	719
Uzbekistan	TASHKENT	780	2162	2679	1019	13323	7046
Vanuatu	ANEITYUM	0	0	5002	1961	12066	1685
Venezuela	CARACAS/MAIQUETIA A	0	0	6284	3242	30706	9906
Vietnam	HA NOI	0	168	5223	2348	23638	9309
Wallis and Futuna	HIHIFO (ILE WALLIS)	0	0	6291	3249	33102	8432
Zimbabwe	HARARE (KUTSAGA)	0	348	3424	731	4586	889

Annual Heating Degree Days (HDD), Cooling Degree Days (CDD), and Cooling Degree Hours (CDH) US Sites (SI Units)

State	City	HDD Tb=10°C	HDD Tb=8.3°C	CDD Tb=10°C	CDD Tb=18.3°C	CDH Tb=10°C	CDH Tb=18.3°C
AK	JUNEAU INT'L ARPT	1937	4629	352	2	49	3
AL	MONTGOMERY DANNELLY FIELD	252	1191	3372	1268	12048	5233
AR	LITTLE ROCK ADAMS FIELD	489	1653	3058	1180	12054	5413
AZ	PHOENIX SKY HARBOR INTL AP	17	523	5067	2532	38386	24726
CA	SACRAMENTO METROPOLITAN AP	179	1327	2676	782	9969	5181
CO	DENVER INTL AP	1434	3301	1606	432	5288	2342
CT	HARTFORD BRADLEY INTL AP	1499	3329	1621	411	3772	1299
DE	DOVER AFB	930	2509	2116	653	5314	1804
FL	TALLAHASSEE REGIONAL AP	128	852	3741	1424	13428	5569
GA	ATLANTA HARTSFIELD INTL AP	381	1497	2949	1023	8753	3266
HI	HONOLULU INTL ARPT	0	0	5624	2583	19731	5259
IA	DES MOINES INTL AP	1734	3467	1887	578	5294	1929
ID	BOISE AIR TERMINAL	1295	3143	1687	494	6908	3416
IL	SPRINGFIELD CAPITAL AP	1382	3016	2038	631	6112	2255
IN	INDIANAPOLIS INTL AP	1315	2957	1986	586	5003	1647
KS	TOPEKA FORBES FIELD	1198	2773	2202	736	7287	3069
KY	LOUISVILLE STANDIFORD FIELD	888	2316	2445	831	7694	2883
LA	BATON ROUGE RYAN ARPT	152	894	3772	1474	13556	5429
MA	BOSTON LOGAN INT'L ARPT	1297	3123	1632	417	3116	1045
MD	BALTIMORE BLT-WASHNGTN INT'L	959	2537	2145	682	6287	2397
ME	AUGUSTA AIRPORT	2059	4097	1232	228	1705	449
MI	LANSING CAPITAL CITY ARPT	1884	3827	1416	317	3017	895
MN	ST PAUL DOWNTOWN AP	2271	4183	1500	372	3103	971
MO	JEFFERSON CITY MEM	1008	2508	2309	767	7604	3049
MS	JACKSON INTERNATIONAL AP	309	1284	3326	1258	11998	5184
MT	HELENA REGIONAL AIRPORT	2137	4266	1119	208	3138	1301

Annual Heating Degree Days (HDD), Cooling Degree Days (CDD), and Cooling Degree Hours (CDH) US Sites (SI Units) (Continued)

State	City	HDD Tb=10°C	HDD Tb=8.3°C	CDD Tb=10°C	CDD Tb=18.3°C	CDH Tb=10°C	CDH Tb=18.3°C
NC	RALEIGH DURHAM INTERNATIONAL	551	1846	2627	877	7764	3036
ND	BISMARCK MUNICIPAL ARPT	2658	4706	1293	299	3857	1598
NE	LINCOLN MUNICIPAL ARPT	1616	3329	1987	658	7293	3206
NH	CONCORD MUNICIPAL ARPT	1973	3989	1281	256	2848	931
NJ	TRENTON MERCER COUNTY AP	1151	2858	1883	548	4887	1763
NM	ALBUQUERQUE INTL ARPT	762	2261	2293	749	7964	3439
NV	LAS VEGAS MCCARRAN INTL AP	176	1169	3908	1860	27352	16813
NY	ALBANY COUNTY AP	1773	3671	1472	329	2798	814
NY	NEW YORK J F KENNEDY INT'L AR	1007	2682	1911	543	3369	909
OH	COLUMBUS PORT COLUMBUS INTL A	1290	2957	1914	539	4783	1606
OK	OKLAHOMA CITY WILL ROGERS WOR	663	1953	2822	1070	11496	5467
OR	SALEM MCNARY FIELD	626	2542	1289	162	2603	1021
PA	HARRISBURG CAPITAL CITY ARPT	1211	2904	1899	550	4863	1724
RI	PROVIDENCE T F GREEN STATE AR	1293	3106	1633	405	3004	919
SC	COLUMBIA METRO ARPT	332	1405	3139	1171	11588	5059
SD	PIERRE MUNICIPAL AP	2071	3948	1684	520	6404	3117
TN	NASHVILLE INTERNATIONAL AP	666	1968	2676	935	8907	3522
TX	AUSTIN/BERGSTROM	167	919	3949	1661	17897	8648
UT	SALT LAKE CITY INT'L ARPT	1298	3067	1935	663	7963	3837
VA	VIRGINIA TECH ARPT	1036	2689	1798	409	3250	881
VA	WASHINGTON DC REAGAN AP	772	2223	2439	847	7497	2825
VT	MONTPELIER AP	2413	4565	1036	146	1355	294
WA	OLYMPIA AIRPORT	822	2984	937	56	1345	422
WI	MADISON DANE CO REGIONAL ARPT	2064	3998	1447	338	3160	954
WV	CHARLESTON YEAGER ARPT	947	2468	2113	592	5132	1726
WY	CHEYENNE MUNICIPAL ARPT	1866	3971	1117	180	2523	818

Annual Heating Degree Days (HDD), Cooling Degree Days (CDD), and Cooling Degree Hours (CDH) Canadian Sites (SI Units)

State	City	HDD Tb=10°C	HDD Tb=18.3°C	CDD Tb=10°C	CDD Tb=18.3°C	CDH Tb=10°C	CDH Tb=18.3°C
AB	CALGARY INT'L A	2655	5086	648	37	727	166
AB	EDMONTON CITY CENTRE A	2947	5275	778	63	639	125
BC	VANCOUVER INT'L A	801	2932	951	41	146	11
BC	VICTORIA INT'L A	779	3022	821	22	316	55
MB	WINNIPEG RICHARDSON INT'L A	3552	5750	1011	168	1723	490
NB	FREDERICTON A	2497	4692	979	132	1219	304
NF	ST JOHN'S A	2381	4907	543	28	109	6
NS	HALIFAX STANFIELD INT'L A	2142	4356	927	98	555	92
NT	YELLOWKNIFE A	5714	8306	482	33	170	18
NU	IQALUIT A	7056	10076	23	0	1	0
ON	OTTAWA MACDONALD-CARTIER INT'	2517	4563	1233	236	1819	476
ON	TORONTO LESTER B. PEARSON INT	1954	3956	1316	276	2190	612
PE	CHARLOTTETOWN A	2438	4703	873	94	406	43
QC	MONTREAL/MIRABEL INT'L A	2710	4849	1064	162	1291	273
QC	QUEBEC/JEAN LESAGE INTL A	2891	5094	970	132	988	201
QC	SHERBROOKE A	2803	5058	880	93	956	173
SK	REGINA A	3436	5701	902	126	1774	581
SK	SASKATOON DIEFENBAKER INT'L A	3558	5861	844	105	1488	462
YT	WHITEHORSE A	4070	6803	313	6	197	36

Appendix D: Heating and Cooling Degree Days (IP Units)

Annual Heating Degree Days (HDD), Cooling Degree Days (CDD), and Cooling Degree Hours (CDH) International Sites (IP Units)

Country	City	HDD Tb=50°F	HDD Tb=65°F	CDD Tb=50°F	CDD Tb=65°F	CDH Tb=74°F	CDH Tb=80°F
Albania	TIRANA	525	2959	4178	1140	10669	4173
Algeria	ANNABA	45	1663	5399	1541	13143	4427
American Samoa	PAGO PAGO WSO AP	0	0	11595	6120	67200	20654
Antarctica	DAVIS	13385	18860	0	0	0	0
Antigua and Barbuda	VC BIRD INTL AIRPOR	0	0	11249	5774	57679	15522
Argentina	EZEIZA AERO	232	2180	4674	1147	10017	3462
Armenia	YEREVAN/YEREVAN-ARA	2179	5047	3944	1340	15424	7062
Aruba	QUEEN BEATRIX AIRPO	0	0	12222	6747	75199	26859
Australia	CANBERRA AIRPORT	702	3804	2816	443	5082	1936
Austria	SALZBURG-FLUGHAFEN	2417	6069	2098	275	2488	667
Azerbaijan	LANKARAN	1076	3792	3930	1172	9024	2529
Bahamas	NASSAU AIRPORT NEW	0	16	10157	4697	42361	13148
Bahrain	BAHRAIN (INT. AIRPORT)	0	186	11075	5785	78213	47431
Barbados	GRANTLEY ADAMS	0	0	11355	5880	58700	15194
Belarus	MINSK	3877	7929	1577	151	1335	265
Belgium	BRUXELLES NATIONAL	1555	5279	1924	173	1482	360
Belize	BELIZE/PHILLIP GOLD	0	0	11061	5586	55774	18757
Benin	BOHICON	0	0	11549	6074	58232	22942
Bermuda	BERMUDA INTL	0	159	8273	2957	21016	4576
Bolivia	LA PAZ/ALTO	1665	7093	48	0	0	0
Bosnia and Herzegovina	SARAJEVO-BJELAVE	2319	5735	2447	390	3488	1072
Botswana	SERETSE KHAMA INTER	9	783	7156	2455	26009	11209
Brazil	BRASILIA (AEROPORTO)	0	42	7855	2422	17521	4627
Bulgaria	SOFIA (OBSERV.)	2372	5704	2575	430	4376	1412
Burkina Faso	OUAGADOUGOU	0	0	12319	6844	89220	49816

Country	Station						
Cape Verde	SAL	0	0	9225	3750	20482	3287
Chad	NDJAMENA	0	2	12366	6893	91647	54556
Chile	PUERTO MONTT	832	5249	1066	4	84	11
China	BEIJING	2358	5094	4266	1527	14318	5456
Christmas Island	CHRISTMAS ISLAND AE	0	0	9746	4271	22323	1158
Cocos (Keeling) Islands	COCOS ISLAND AERO	0	0	11165	5690	55745	10635
Colombia	BOGOTA/ELDORADO	2	3153	2326	0	5	0
Congo	BRAZZAVILLE/MAYA-M	0	0	10514	5039	39556	13512
Cook Islands	AMURI/AITUTAKI ISL	0	0	10545	5070	42013	6738
Costa Rica	JUAN SANTAMARIA INT	0	0	8761	3286	15415	3014
Cote d'Ivoire	ABIDJAN	0	0	11230	5755	54857	16096
Croatia	ZAGREB/MAKSIMIR	1993	5171	2837	542	4745	1423
Cuba	AEROPUERTO JOSE MAR	0	37	9740	4304	35173	11865
Cyprus	PAPHOS AIRPORT	28	1377	5756	1630	13968	3469
Czech Rep.	PRAHA/RUZYNE	2817	6757	1694	160	1652	415
Dem. People's Rep. of Korea	PYONGYANG	2933	5937	3504	1034	6855	1650
Denmark	KOEBENHAVN/KASTRUP	2442	6575	1422	81	359	31
Diego Garcia	DIEGO GARCIA NAF	0	0	11569	6094	63917	17547
Dominican Rep.	SANTO DOMINGO	0	0	10842	5367	46690	14798
Ecuador	QUITO AEROPUERTO	0	2523	2952	1	8	1
Egypt	CAIRO AIRPORT	2	708	7949	3180	34403	16906
Estonia	TALLINN	3948	8369	1110	55	439	58
Falkland Islands (Malvinas)	MOUNT PLEASANT AIRP	2659	7855	276	0	3	0
Faroe Islands	TORSHAVN	2324	7630	169	0	0	0
Fiji	NADI AIRPORT	0	0	10215	4740	40232	12231
Finland	HELSINKI-VANTAA	4340	8740	1146	71	607	71
France	PARIS-AEROPORT CHAR	1338	4768	2342	295	2528	744
French Polynesia	TAHITI-FAAA	0	0	10846	5371	49774	14067

Annual Heating Degree Days (HDD), Cooling Degree Days (CDD), and Cooling Degree Hours (CDH) International Sites (IP Units) *(Continued)*

Country	City	HDD Tb=50°F	HDD Tb=65°F	CDD Tb=50°F	CDD Tb=65°F	CDH Tb=74°F	CDH Tb=80°F
Gabon	LIBREVILLE	0	0	10829	5354	48918	10995
Gambia	BANJUL/YUNDUM	0	1	11078	5604	50808	21719
Georgia	TBILISI	1450	4267	3841	1186	10182	4039
Germany	BERLIN/DAHLEM	2286	6102	1871	213	1763	422
Gibraltar	GIBRALTAR	5	1116	5619	1255	5380	1177
Greece	ATHINAI (AIRPORT)	215	2097	5536	1942	18210	7123
Greenland	NUUK (GODTHAAB)	7512	12965	21	0	0	0
Grenada	POINT SALINES AIRPO	0	0	11480	6005	62113	16810
Guam	ANDERSEN AFB	0	0	11212	5737	55524	11780
Guatemala	GUATEMALA (AEROPUERTO)	0	128	6509	1161	3374	225
Guernsey	GUERNSEY AIRPORT	766	4621	1663	42	139	19
Guiana	ROCHAMBEAU	0	0	10893	5418	46179	14143
Guyana	TIMEHRI/CHEDDI JAG	0	0	11044	5569	45713	15625
Honduras	TEGUCIGALPA	0	21	8145	2691	17123	4333
Hungary	BUDAPEST/FERIHEGY I	2379	5739	2579	464	5045	1670
Iceland	REYKJAVIK	3689	8982	185	0	0	0
India	NEW DELHI/PALAM	2	515	10381	5419	76529	45618
Indonesia	JAKARTA/SOEKARNO-HA	0	0	11591	6116	61291	22492
Iran (Islamic Rep. of)	TEHRAN-MEHRABAD	770	2858	6158	2772	36030	19452
Ireland	DUBLIN AIRPORT	1290	5643	1134	11	22	0
Isle of Man	ISLE OF MAN/RONALDS	1116	5543	1055	8	19	0
Israel	BEN-GURION INT. AIR	18	1115	6697	2320	22223	8217
Italy	ROMA FIUMICINO	362	2745	4092	999	7748	1840
Jamaica	KINGSTON/NORMAN MAN	0	0	11895	6420	70882	25818
Japan	TOKYO	614	2900	4808	1623	13358	4368

Jersey	JERSEY AIRPORT	872	4645	1768	66	306	54
Jordan	AMMAN AIRPORT	332	2324	5353	1866	18390	7673
Kazakhstan	ASTANA	6427	10291	1981	371	4123	1324
Kenya	NAIROBI/KENYATTA AI	0	187	6227	942	5441	600
Kerguelen	PORT-AUX-FRANCAIS	3453	8865	64	0	0	0
Kiribati	TARAWA	0	0	12056	6581	76066	25749
Korea (Rep. of)	SEOUL	2130	4897	3964	1258	8870	2372
Kuwait	KUWAIT INTERNATIONA	19	766	10777	6048	98465	70161
Kyrgyzstan	BISHKEK	2756	5793	3510	1072	12529	5483
Latvia	RIGA	3438	7548	1489	124	812	146
Lebanon	BEYROUTH (AEROPORT)	8	835	7139	2490	21330	5513
Libya	TRIPOLI INTERNATION	15	1202	7228	2940	33967	18408
Liechtenstein	VADUZ (LIECHTENSTEIN)	2065	5615	2230	305	1956	417
Lithuania	VILNIUS	3739	7849	1494	130	1124	223
Luxembourg	LUXEMBOURG/LUXEMBOU	2190	6089	1790	215	1608	390
Macao	TAIPA GRANDE	14	556	8511	3577	33134	10201
Macedonia	SKOPJE- AIRPORT	1856	4776	3456	900	10811	4740
Madagascar	ANTANANARIVO/IVATO	0	594	6041	1158	5411	730
Malaysia	KUALA LUMPUR SUBANG	0	0	11882	6407	63640	24641
Maldives	MALE	0	0	12202	6727	81985	30494
Mali	BAMAKO/SENOU	0	0	11829	6354	76878	41874
Malta	LUQA	19	1407	5928	1843	13197	4040
Marshall Islands	MAJURO WSO AP	0	0	11784	6309	71297	20514
Martinique	LE LAMENTIN	0	0	10956	5481	50138	14230
Mauritania	NOUAKCHOTT	0	4	10797	5327	52619	24555
Mauritius	VACOAS (MAURITIUS)	0	45	7641	2212	7286	543
Mayotte	DZAOUDZI/PAMANZI	0	0	10983	5508	53825	14168
Mexico	AEROP. INTERNACIONA	4	1014	4806	342	3429	434

Annual Heating Degree Days (HDD), Cooling Degree Days (CDD), and Cooling Degree Hours (CDH) International Sites (IP Units) (Continued)

Country	City	HDD Tb=50°F	HDD Tb=65°F	CDD Tb=50°F	CDD Tb=65°F	CDH Tb=74°F	CDH Tb=80°F
Micronesia	YAP ISLAND WSO AP	0	0	11624	6149	64716	19110
Moldova	KISINEV	2719	6006	2773	585	4198	1136
Mongolia	ULAANBAATAR	8182	12536	1261	142	1828	545
Morocco	RABAT-SALE	11	1440	5004	960	5144	1299
Mozambique	MAPUTO/MAVALANE	0	39	8924	3488	26003	7719
Namibia	WINDHOEK	14	683	6694	1889	17120	5799
Netherlands	AMSTERDAM AP SCHIPH	1589	5468	1712	117	874	186
New Caledonia	NOUMEA (NLLE-CALEDONIE)	0	3	8729	3258	18185	3384
New Zealand	WELLINGTON AIRPORT	243	3378	2423	84	76	1
Nicaragua	MANAGUA A.C.SANDINO	0	0	11623	6148	63931	27948
Niger	NIAMEY-AERO	0	0	12946	7471	104212	62823
Niue Is.	ALOFI	0	0	9789	4315	29174	5579
Norfolk Island	NORFOLK ISLAND AERO	0	554	5817	898	735	3
Northern Mariana Islands	SAIPAN	0	0	11774	6299	65975	19534
Norway	OSLO/GARDERMOEN	4405	8967	951	37	418	47
Oman	SEEB INTL AIRPORT	0	2	12045	6573	88013	51927
Pakistan	ISLAMABAD AIRPORT	29	1174	7899	3568	46293	25118
Palau	KOROR WSO	0	0	11741	6266	65793	18779
Panama	MARCOS A GELABERT I	0	0	11918	6443	64893	22450
Paraguay	ASUNCION/AEROPUERTO	11	457	8714	3688	36593	16091
Peru	LIMA-CALLAO/AEROP.	0	297	6657	1480	4715	620
Philippines	MANILA	0	0	12126	6651	78701	29895
Poland	WARSZAWA-OKECIE	2947	6787	1838	201	1943	478
Portugal	LISBOA/PORTELA	63	1822	4796	1079	7352	2724
Puerto Rico	SAN JUAN INTL ARPT	0	0	11087	5612	52210	14994

Country	Station						
Qatar	DOHA INTERNATIONAL	0	132	11693	6349	88426	56526
Reunion Is.	SAINT-DENIS/GILLOT	0	0	9278	3803	23949	4539
Romania	BUCURESTI AFUMATI	2398	5525	3035	687	7442	2713
Russia	MOSKVA	4504	8545	1625	193	1366	261
Saint Helena Is.	ST. HELENA IS.	0	556	5503	586	86	2
Saint Lucia	HEWANORRA INTL AIRP	0	0	11573	6098	65643	18902
Samoa	APIA	0	0	11088	5613	54488	14717
Saudi Arabia	RIYADH OBS. (O.A.P.)	10	541	10817	5875	93042	63113
Senegal	DAKAR/YOFF	0	1	9653	4180	32381	8182
Serbia	BEOGRAD	1730	4605	3500	897	7768	2849
Seychelles	SEYCHELLES INTERNAT	0	0	11393	5918	60010	15119
Singapore	SINGAPORE/CHANGI AI	0	0	11842	6367	66856	22005
Slovakia	BRATISLAVA-LETISKO	2252	5579	2625	477	4559	1463
Slovenia	LJUBLJANA/BEZIGRAD	2022	5315	2668	485	3920	1145
Solomon Islands	HONIARA/HENDERSON	0	0	10887	5412	47608	15112
South Africa	PRETORIA (IRENE)	42	1459	4910	852	5791	1199
Spain	MADRID/BARAJAS RS	812	3641	3752	1102	14484	6955
Sri Lanka	KATUNAYAKE	0	0	11617	6142	65819	22662
Suriname	ZANDERIJ	0	0	11275	5800	48210	18181
Svalbard and Jan Mayen	HOPEN	9828	15303	0	0	0	0
Sweden	STOCKHOLM/ARLANDA	3627	7937	1246	82	745	124
Switzerland	ZURICH-KLOTEN	2232	5945	2002	238	2402	644
Syria	DAMASCUS INT. AIRPO	537	2748	5173	1908	26604	14995
Taiwan	TAIBEI	1	426	8730	3680	35198	13651
Tajikistan	DUSHANBE	1085	3493	4779	1713	21857	11312
Tanzania	DAR ES SALAAM AIRPO	0	0	10531	5056	44474	14746
Thailand	BANGKOK METROPOLIS	0	0	12447	6972	83767	37399
Togo	LOME	0	0	11441	5966	59910	19766

Annual Heating Degree Days (HDD), Cooling Degree Days (CDD), and Cooling Degree Hours (CDH) International Sites (IP Units) *(Continued)*

Country	City	HDD Tb=50°F	HDD Tb=65°F	CDD Tb=50°F	CDD Tb=65°F	CDH Tb=74°F	CDH Tb=80°F
Tonga	FUAAMOTU	0	5	9002	3533	19977	2999
Tunisia	TUNIS-CARTHAGE	34	1466	6177	2135	20030	8638
Turkey	ESENBOGA	2471	5939	2420	409	6387	2287
Turkmenistan	ASHGABAT KESHI	1139	3436	5798	2617	35387	20013
Tuvalu	FUNAFUTI NF	0	0	12050	6575	77197	26165
Ukraine	KYIV	3380	7033	2149	324	2248	489
United Arab Emirates	ABU DHABI INTER. AI	0	54	11838	6417	88284	56005
United Arab Emirates	DUBAI INTERNATIONAL	0	43	11629	6196	85007	52310
United Kingdom	LONDON WEATHER CENT	835	4219	2324	232	1250	279
United States Minor Outlying Islands	MIDWAY ISLAND NAS	0	51	8760	3337	22747	4577
Uruguay	CARRASCO	176	2198	4282	829	4925	1294
Uzbekistan	TASHKENT	1404	3892	4822	1835	23981	12682
Vanuatu	ANEITYUM	0	0	9003	3529	21719	3033
Venezuela	CARACAS/MAIQUETIA A	0	0	11311	5836	55271	17831
Vietnam	HA NOI	0	302	9401	4227	42548	16756
Wallis and Futuna	HIHIFO (ILE WALLIS)	0	0	11324	5849	59584	15177
Zimbabwe	HARARE (KUTSAGA)	0	627	6164	1316	8254	1601

Annual Heating Degree Days (HDD), Cooling Degree Days (CDD), and Cooling Degree Hours (CDH) US Sites (IP Units)

State	City	HDD Tb=50°F	HDD Tb=65°F	CDD Tb=50°F	CDD Tb=65°F	CDH Tb=74°F	CDH Tb=80°F
AK	JUNEAU INT'L ARPT	3487	8333	634	3	89	6
AL	MONTGOMERY DANNELLY FIELD	454	2143	6069	2282	21686	9420
AR	LITTLE ROCK ADAMS FIELD	881	2976	5505	2124	21697	9744
AZ	PHOENIX SKY HARBOR INTL AP	30	941	9120	4557	69095	44506
CA	SACRAMENTO METROPOLITAN AP	323	2389	4817	1408	17945	9325
CO	DENVER INTL AP	2581	5942	2890	777	9519	4216
CT	HARTFORD BRADLEY INTL AP	2699	5992	2917	739	6789	2338
DE	DOVER AFB	1674	4517	3808	1176	9566	3247
FL	TALLAHASSEE REGIONAL AP	230	1534	6733	2563	24170	10024
GA	ATLANTA HARTSFIELD INTL AP	686	2694	5309	1841	15755	5878
HI	HONOLULU INTL ARPT	0	0	10124	4649	35515	9466
IA	DES MOINES INTL AP	3121	6240	3397	1041	9530	3472
ID	BOISE AIR TERMINAL	2331	5658	3036	890	12435	6149
IL	SPRINGFIELD CAPITAL AP	2487	5429	3668	1135	11001	4059
IN	INDIANAPOLIS INTL AP	2367	5322	3575	1055	9005	2965
KS	TOPEKA FORBES FIELD	2157	4992	3964	1325	13116	5524
KY	LOUISVILLE STANDIFORD FIELD	1599	4168	4401	1496	13849	5190
LA	BATON ROUGE RYAN ARPT	274	1610	6790	2653	24400	9772
MA	BOSTON LOGAN INT'L ARPT	2334	5621	2938	750	5609	1881
MD	BALTIMORE BLT-WASHNGTN INT'L	1726	4567	3861	1228	11317	4315
ME	AUGUSTA AIRPORT	3707	7375	2217	410	3069	809
MI	LANSING CAPITAL CITY ARPT	3392	6889	2548	570	5431	1611
MN	ST PAUL DOWNTOWN AP	4087	7529	2700	669	5586	1747
MO	JEFFERSON CITY MEM	1815	4514	4156	1380	13687	5489
MS	JACKSON INTERNATIONAL AP	557	2311	5986	2265	21596	9332
MT	HELENA REGIONAL AIRPORT	3847	7679	2014	374	5649	2342

Annual Heating Degree Days (HDD), Cooling Degree Days (CDD), and Cooling Degree Hours (CDH) US Sites (IP Units) (Continued)

State	City	HDD Tb=50°F	HDD Tb=65°F	CDD Tb=50°F	CDD Tb=65°F	CDH Tb=74°F	CDH Tb=80°F
NC	RALEIGH DURHAM INTERNATIONAL	991	3322	4728	1579	13975	5465
ND	BISMARCK MUNICIPAL ARPT	4785	8471	2328	539	6942	2877
NE	LINCOLN MUNICIPAL ARPT	2908	5993	3576	1184	13127	5770
NH	CONCORD MUNICIPAL ARPT	3551	7180	2305	461	5126	1676
NJ	TRENTON MERCER COUNTY AP	2072	5144	3389	987	8796	3173
NM	ALBUQUERQUE INTL ARPT	1372	4069	4128	1348	14336	6190
NV	LAS VEGAS MCCARRAN INTL AP	316	2105	7034	3348	49234	30264
NY	ALBANY COUNTY AP	3192	6608	2649	592	5036	1465
NY	NEW YORK J F KENNEDY INT'L AR	1813	4828	3439	978	6064	1636
OH	COLUMBUS PORT COLUMBUS INTL A	2322	5322	3446	971	8610	2890
OK	OKLAHOMA CITY WILL ROGERS WOR	1193	3516	5079	1926	20693	9840
OR	SALEM MCNARY FIELD	1127	4576	2320	292	4685	1837
PA	HARRISBURG CAPITAL CITY ARPT	2180	5228	3419	990	8754	3104
RI	PROVIDENCE T F GREEN STATE AR	2327	5591	2940	729	5408	1654
SC	COLUMBIA METRO ARPT	598	2529	5651	2108	20858	9106
SD	PIERRE MUNICIPAL AP	3727	7107	3031	936	11527	5610
TN	NASHVILLE INTERNATIONAL AP	1199	3542	4817	1683	16032	6340
TX	AUSTIN/BERGSTROM	300	1654	7108	2989	32215	15566
UT	SALT LAKE CITY INT'L ARPT	2337	5521	3483	1193	14334	6907
VA	VIRGINIA TECH ARPT	1865	4841	3236	737	5850	1586
VA	WASHINGTON DC REAGAN AP	1389	4001	4390	1524	13494	5085
VT	MONTPELIER AP	4343	8217	1865	262	2439	530
WA	OLYMPIA AIRPORT	1479	5372	1686	101	2421	759
WI	MADISON DANE CO REGIONAL ARPT	3715	7197	2604	608	5688	1718
WV	CHARLESTON YEAGER ARPT	1704	4443	3803	1066	9238	3107
WY	CHEYENNE MUNICIPAL ARPT	3358	7148	2011	324	4542	1472

Annual Heating Degree Days (HDD), Cooling Degree Days (CDD), and Cooling Degree Hours (CDH) Canadian Sites (IP Units)

State	City	HDD Tb=50°F	HDD Tb=65°F	CDD Tb=50°F	CDD Tb=65°F	CDH Tb=74°F	CDH Tb=80°F
AB	CALGARY INT'L A	4779	9154	1167	67	1308	298
AB	EDMONTON CITY CENTRE A	5305	9495	1401	114	1150	225
BC	VANCOUVER INT'L A	1442	5278	1712	73	262	19
BC	VICTORIA INT'L A	1402	5439	1478	40	568	99
MB	WINNIPEG RICHARDSON INT'L A	6393	10350	1819	303	3102	882
NB	FREDERICTON A	4494	8445	1763	238	2194	547
NF	ST JOHN'S A	4285	8832	977	51	196	11
NS	HALIFAX STANFIELD INT'L A	3856	7840	1669	177	999	166
NT	YELLOWKNIFE A	10286	14951	867	59	306	33
NU	IQALUIT A	12701	18136	41	0	1	0
ON	OTTAWA MACDONALD-CARTIER INT'	4531	8213	2219	425	3274	857
ON	TORONTO LESTER B. PEARSON INT	3518	7120	2368	496	3942	1102
PE	CHARLOTTETOWN A	4389	8465	1571	170	730	77
QC	MONTREAL/MIRABEL INT'L A	4878	8729	1915	292	2324	492
QC	QUEBEC/JEAN LESAGE INTL A	5203	9169	1746	238	1779	361
QC	SHERBROOKE A	5045	9105	1584	167	1721	311
SK	REGINA A	6185	10262	1623	227	3193	1045
SK	SASKATOON DIEFENBAKER INT'L A	6405	10550	1520	189	2679	832
YT	WHITEHORSE A	7326	12246	564	10	354	64

References

Akbari, H. 1993. *Monitoring Peak Power and Cooling Energy Savings of Shade Trees and White Surfaces in the Sacramento Municipal Utility District (SMUD) Service Area Data Analysis, Simulations and Results.* LBNL Report 34411. Berkeley, CA.

Akbari, H., Huang, J., and Davis, S. 1997. Peak Power and Cooling Energy Savings of Shade Trees. *Energy and Buildings*, 25, 139–147.

Albertsen, S., Krarti, M., and Bianchi, M. 2011. Estimating Optimal Energy Savings from Retrofitting US Existing House Stock, Submitted to ASME Energy Sustainability Conference.

Alspector, D., *Simulation Environment for automated Calibration of Building Energy Models,* MS Thesis, University of Colorado, Boulder, CO, 2008.

Alspector, D., and Krarti, M. 2009. Simulation Environment for Automated Calibration of Building Energy Models. Energy Sustainability Conference.

Anderlind, G. 1985. Energy Consumption Due to Air Infiltration. *Proceedings of the 3rd ASHRAE/DOE/BTECC Conference on Thermal Performance of the Exterior Envelope of Buildings* (pp. 201–208). Clearwater Beach, FL.

Andersson, B., Carroll, W.L., and Martin, M.R. 1985. Aggregation of US Population Centers Using Climate Parameters Related to Building Energy Use. *Journal of Climate and Applied Meteorology*, 25, 596–614.

Andreas, J. 1992. *Energy-Efficient Motors: Selection and Application.* Marcel Dekker, New York.

ASHRAE, Standard 55: Thermal Environmental Conditions for Human Occupancy, American Society of Heating, refrigerating and Air-Conditioning Engineers, Inc., Atlanta, GA, 2004.

ASHRAE. 1997. *Proposed ASHRAE Guideline 14P, Measurement of Energy and Demand Savings.* American Society of Heating, Refrigerating and Air-Conditioning Engineers, Inc., Atlanta, GA.

ASHRAE. 2002. *ASHRAE Guideline 14, Measurement of Energy and Demand Savings.* American Society of Heating, Refrigerating and Air-Conditioning Engineers, Inc., Atlanta, GA.

ASHRAE. 2004. *Ventilation for Acceptable Indoor Air Quality.* Standard 62-2004. American Society of Heating, Refrigerating and Air-Conditioning Engineers, Inc., Atlanta, GA.

ASHRAE. 2007. *Handbook of HVAC Applications*. American Society of Heating, Refrigerating and Air-Conditioning Engineers, Inc., Atlanta, GA.

ASHRAE. 2008. *Handbook of HVAC Systems and Equipment*. American Society of Heating, Refrigerating and Air-Conditioning Engineers, Inc., Atlanta, GA.

ASHRAE. 2009. *Handbook of Fundamentals*. American Society of Heating, Refrigerating and Air-Conditioning Engineers, Inc., Atlanta, GA.

AWWA. 1999. *Water Use Inside the Home*, Report of American Water Works Research Foundation.

Azebergi, R., Hunsberger, R., and Zhou, N. 2000. *A Residential Building Energy Audit*. Report for Class Project CVEN5020. University of Colorado, Boulder.

Balcomb, J.D., Barley, D., McFarland, R., Perry, J., Wray, W., and Noll, S. 1980. *Passive Solar Design Handbook: Passive Solar Design Analysis*. Vol. 2, DOE/CS-0127/2. U.S. Department of Energy.

Bichiou, Y., and Krarti, M. 2011. Optimization of Envelope and HVAC Systems Selection for Residential Buildings. *Energy and Buildings,* Vol. 43(12), 3373–3382.

Brandl, H. 2006. Energy Foundations and Other Thermo-Active Ground Structures. *Géotechnique*, 56(2), 81–122.

CADDET. 1999. *Radiant Heating Panels Save Energy in Homes*. US-1999-502, Result 337. Centre for the Analysis and Dissemination of Demonstrated Energy Technologies.

Caldas, L.G., and Norford, L.K. 2003. Genetic Algorithms for Optimization of Building Envelopes and the Design and Control of HVAC Systems. *Journal of Solar Energy Engineering*, 125, 343–351.

Casey, S., Krarti, M., Bianchi, M., and Roberts, D. 2010. Identifying Inefficient Single-Family Homes with Utility Bill Analysis. Proceedings of the ASME 2010 Conference on Energy Sustainability.

Celenec, EN 814: Air Conditioners and Heat Pumps with Electrically Driven Compressors, European Committee for Standardization, Brussels, 1997.

Christensen, C., Barker, G., and Horowitz, S. 2004. A Sequential Search Technique for Identifying Optimal Building Designs on the Path to Zero Net Energy. Proceedings of the Solar 2004, Portland, OR. American Solar Energy Society.

Christian, J.E., and Kosny, J. 1996. Thermal Performance and Wall Ratings. *ASHRAE Journal*, 56–65.

Claridge, D. 1988. *Design Methods for Earth-Contact Heat Transfer*. Progress in Solar Energy, American Solar Energy Society, Boulder, CO.

Claridge, D.E., Haberl, J., Turner, W., O'Neal, D., Heffington, W., Tombari, C., Roberts, M., and Jaeger, S. 1991. Improving Energy Conservation Retrofits with Measured Savings. *ASHRAE Journal*, 33(10), 14.

Claridge, D.E., Krarti, M., and Bida, M. 1987. A Validation Study of Variable-Base Degree-Day Cooling Calculations. *ASHRAE Transactions*, 93(2), 90–104.

Conchilla, M. 1999. Interactions of Water and Energy Resources in Residential Buildings—A Modeling Study. MS thesis, University of Colorado, Boulder.

Conchilla, M., and Krarti, M. 2002. Interactions of Water and Energy Use in Residential Buildings: Part II Results. *ASHRAE Transactions*, 108, Part 2.

Davis Energy Group. 1994. Coachella Valley Project: La Paloma Site Final Design Report. Davis Energy Group, Davis, CA.

DeMonsabert, S. 1996. WATERGY: A Water and Energy Conservation Model for Federal Facilities. CONSERV 96 Conference Proceedings.

Desmedt, J., and Hoes, H. 2007. Case Study of a BTES and Energypiles Application for a Belgian Hospital. EcoStock 2007, Stockton, NJ.

Dickson, A., Akmal, M., and Thorpe, S. 2003. *Australian Energy: National and State Projections to 2019-20.* ABARE eReport 03.10 for the Ministerial Council on Energy, Canberra, June.

Diekerhoff, D.J., Grimsrud, D.T., and Lipschutz, R.D. 1982. Component Leakage Testing in Residential Buildings. Proceedings of the American Council for an Energy Efficient Economy, 1982 Summer Study, Santa Cruz, CA.

DOE. 2005. *A Look at Residential Energy Consumption in 1997.* DOE/EIA-0632 (97). U.S. Department of Energy.

DOE. 2009. *Building Technologies Program: Building America.* U.S. Department of Energy.

DOE. 2011. Energy Efficiency for Buildings, Directory of Building Energy Software. U.S. Department of Energy, Washington, DC. An updated list: http://apps1.eren.doe.gov/buildings/tools_directory/.

EIA. 2007. *Annual Energy Review: Energy Consumption by Sector.* Energy Information Administration. http://www.eia.doe.gov/emeu/aer/consump.html.

EIA. 2008. *2005 Residential Energy Consumption Survey.* U.S. Energy Information Administration, Washington, DC.

EIA. 2009. Annual Energy Review. Department of Energy, Energy Information Administration. http://www.doe.eia.gov.

EIA. 2010. *Annual Energy Outlook 2010 Early Release with Projections to 2035.* DOE/EIA-0383. Energy Information Administration.

Energy Information Administration. *Annual Energy Outlook 2010 Early Release with Projections to 2035,* DOE/EIA-0383, 2009.

EnergyPlus, U.S. Department of Energy. 2009. Energy Efficiency for Buildings, Building Technologies Program. Washington, DC. An updated list: http://apps1.eren.doe.gov/buildings/energyplus/.

Energy Star, U.S. Environmental Protection Agency. 2009. Information provided on the Energy Star website: http://www.energystar.gov.

Energy Trust of Oregon. 2009. *Home Energy Solutions—Existing Homes.* http://www.energytrust.org/residential/existinghomes/index.html.

EPA, U.S. Environmental Protection Agency. 1995. Community Water System Survey. Office of Ground Water and Drinking Water. Washington, DC.

Erbs, D.G., Klein, S.A., and Beckman, W.A. 1983. Estimation of Degree Days and Ambient Temperature Bin Data from Monthly-Average Temperatures, *ASHRAE Journal,* 25(6), 60–66.

Eto, J. 1988. On Using Degree-Days to Account for the Effects of Weather on Annual Energy Use in Office Buildings. *Energy and Buildings,* 12(2), 113.

Fanger, P.O. 1970. *Thermal Comfort.* Danish Technical Press, Copenhagen.

Fels, M. 1986. Special Issue Devoted to Measuring Energy Savings: The Scorekeeping Approach. *Energy and Buildings,* 9(2).

Fels, J. 1988. Special Issue Devoted to Measuring Energy Savings: The Scorekeeping Approach. *Energy and Buildings,* 12(2), 113.

Fels, M., and Keating, K. 1993. Measurement of Energy Savings from Demand-Side Management Programs in US Electric Utilities. *Annual Review of Energy and Environment*, 18, 57.

FEMP. 2000. *M&V Guidelines: Measurement and Verification for Federal Energy Projects.* Version 2.2. Federal Energy Management Program.

FEMP. 2002. *Energy Policy Act of 1992 Becomes Law.* FEMP Focus Special Edition 2. Federal Energy Management Program.

FEMP. 2008. *M&V Guidelines: Measurement and Verification for Federal Energy Projects.* Version 3.0. Federal Energy Management Program.

Fong, K., Hanby, V., and Chowa, T. 2006. HVAC System Optimization for Energy Management by Evolutionary Programming. *Energy and Buildings*, 38, 220–231.

FSEC. 2007. EnergyGauge Pro Software. Florida Solar Energy Center, Cocoa, FL. http://energygauge.com/FlaRes/features/pro.htm.

Gagge, A.P., Stolwijk, J.J., and Nishi, Y. 1970. An Effective Temperature Scale Based on a Simple Model of Human Physiological Regulatory Response. *ASHRAE Transactions*, 70(1).

Grant, P., Burch J., and Krarti, M. 2011. Modeling Gas Tankless Water Heaters and Impact of Draw Profile on System Efficiency. Proceedings of ASME Energy Sustainability Conference 2011, Washington, DC.

Greely, K., Harris, J., and Hatcher, A. 1990. *Measured Savings and Cost-Effectiveness of Conservation Retrofits in Commercial Buildings.* Lawrence Berkeley National Laboratory Report-27586. Berkeley, CA.

Green Seal, *Environmental Standards for Major Household Appliances,* Green Seal, Washington DC, 1993.

Griego, D., Krarti, M., and Hernandez-Guerrero, A. 2011. Optimization of Energy Efficiency and Thermal Comfort for Residential Buildings in Salamanca, Mexico. Proceedings of ECOS 2011, International Conference on Efficiency, Cost, Optimization, and Simulation, Serbia.

Guiterman, T., and Krarti, M. 2011. Analysis of Measurement and Verification Methods for Energy Retrofits Applied to Residential Buildings. *ASHRAE Transactions*, 117(2).

Haberl, J.S., and Abbas, M. 1998a. Development of Graphical Indices for Viewing Building Energy Data: Part I. *ASME Solar Energy Engineering Journal*, 120(3), 156.

Haberl, J.S., and Abbas, M. 1998b. Development of Graphical Indices for Viewing Building Energy Data: Part II. *ASME Solar Energy Engineering Journal*, 120(3), 162.

Haberl, J.S., and Bou-Saada, T.E. 1998. Procedures for Calibrating Hourly Simulation Models to Measured Energy and Environmental Data. *ASME Solar Energy Engineering Journal*, 120(3), 193.

Hamada, Y., Nakamura, M., Kubota, H., and Ochifuji, K. 2007. Field Performance of an Energy Pile System for Space Heating. *Energy and Buildings*, 39(5).

Han, J., Mol, A.P.J., and Lu, Y. 2010. Solar Water Heaters in China: A New Day Dawning. *Energy Policy*, 38(1), 383–391.

Harrje, D.T., and Born, G.J. 1982. Cataloguing Air Leakage Components in Houses. Proceedings of the American Council for an Energy Efficient Economy, 1982 Summer Study, Santa Cruz, CA.

Hashimoto, K. 2006. Technology and Market Development of Heat Pump Water Heaters in Japan. Central Research Institute of Electric Power Industry.

Hewett, M., Dunsworth, T., and Miller, T. 1986. Measured versus Predicted Savings from Single Retrofits: Sample Study. *Energy and Buildings*, 9, 65–73.

Hoshide, R.K. 1994. Electric Motor Do's and Don'ts. *Energy Engineering*, 1(1), 6–24.

Huang, J. 1987. The Potential of Vegetation in Reducing Summer Cooling Load in Residential Buildings. *Journal of Climate and Applied Meteorology*, 26(9), 1103.

Huang, J., and Gu, L. 2002. *Prototype Residential Buildings to Represent the U.S. Housing Stock*. Draft LBNL Report. Lawrence Berkeley National Laboratory, Berkeley, CA.

Huang, W., and Lam, H.N. 1997. Using Genetic Algorithms to Optimize Controller Parameters for HVAC Systems. *Energy and Buildings*, 26(3), 277–282.

IECC. "International Energy Conservation Code – 2009 Edition." Washington, DC. International Code Council, 2009.

IEEE. 1992. *Guide for Harmonic Control and Reactive Compensation of Static Power Converters*. IEEE 519-1992.

Ihm, P., and Krarti, M. 2005. Controlling Hydronic Radiant Heating and Cooling Systems—Optimal Control Strategies for Heated Radiant Floor Systems. *ASHRAE Transactions*, 111(1), 535–546.

International Code Council. 2009. International Energy Conservation Code. 2009 edition.

IPCC. 1996. IPCC Technical Paper on Technologies, Policies and Measures for Mitigating Climate Change. Intergovernmental Panel on Climate Change, Geneva.

IPMVP. 1997. *International Performance Monitoring and Verification Protocol*. U.S. Department of Energy DOE/EE-0157. U.S. Government Printing Office, Washington, DC.

IPMVP. 2002. *International Performance Monitoring and Verification Protocol, Concepts and Options for Determining Energy and Water Savings*. Vol. 1, U.S. Department of Energy DOE/GO-102002-1554. U.S. Government Printing Office, Washington, DC.

IPMVP. 2007. *International Performance Monitoring and Verification Protocol, Concepts and Options for Determining Energy and Water Savings*. http://www.evo-world.org.

Johnson, B., and Zoi, C. 1992. EPA Energy Star Computers: The Next Generation of Office Equipment. Proceedings of the Conference on ACEEE 1992 Summer Study on Energy Efficiency in Buildings, Panel 6, Pacific Grove, CA.

Kalinic, N. 2009. *Measurement and Verification of Savings from Implemented ECMs*. MS report, University of Colorado, Boulder.

Katipamula, S., Reddy, T.A., and Claridge, D.E. 1994. Development and Application of Regression Models to Predict Cooling Energy Use in Large Commercial Buildings. In *Proceedings of the ASME/JSES/JSES International Solar Energy Conference*, San Francisco, CA, p. 307.

Kearns, P.A., and Krarti, M. 2011. Residential Energy Analysis: Regression Analysis of Heating Degree Days with Temperature Setback for Selected ASHRAE Climate Zones. Proceedings of ASME Energy Sustainability Conference 2011, Washington, DC.

Kissock, K., Claridge, D., Haberl, J., and Reddy, A. 1992. Measuring Retrofit Savings for the Texas LoanSTAR Program: Preliminary Methodology and Results. In *Proceedings of the ASME/JSES/JSES International Solar Energy Conference*, Maui, HI, p. 299.

Kissock, K., and Fels, M. 1995. An Assessment of PRISM's Reliability for Commercial Buildings. National Energy Program Evaluation Conference, Chicago.

Kissock, K. Haberl J. and Claridge, D, "Inverse Modeling Toolkit (1050RP): Numerical Algorithms", *ASHRAE Transactions*, Vol. 109(2), 2003.

Kissock, K., Reddy, T.A., and Claridge, D.E. 1998. Ambient-Temperature Regression Analysis for Estimating Retrofit Savings in Commercial Buildings, *ASME Journal of Solar Energy Engineering*, 120(3), 168–176.

Koomey, J.G., Dunham, C., and Lutz, J.D. 1994. *The Effect of Efficiency Standards on Water Use and Water Heating Energy Use in the US: A Detailed End-Use Treatment*. LBL-35475. Lawrence Berkeley National Laboratory, Berkeley, CA.

Krarti, M. 1994. Effect of Air Flow on Heat Transfer in Walls. *Journal of Solar Energy Engineering*, 116, 35–42.

Krarti, M. 2000. Ground-Coupled Heat Transfer. In *Advances in Solar Energy*, ed. Y. Goswami. ASES, Boulder, CO.

Krarti, M. 2010. *Energy Audit of Building Systems: An Engineering Approach*. 2nd ed. Taylor and Francis Publishing, CRC Press, Boca Raton, FL.

Krarti, M., and Chuangchid, P. 1999. *Cooler Floor Heat Gain in Refrigerated Structures*. Final Report, ASHRAE Project RP-953. American Society of Heating, Refrigerating, and Air Conditioning Engineering, Atlanta, GA.

Krarti, M., Erickson, P., and Hillman, T. 2005. A Simplified Method to Estimate Energy Savings of Artificial Lighting Use from Daylighting. *Building and Environment*, 40, 747–754.

Krarti, M., Lopez-Alonzo, C., Claridge, D.E., and Kreider, J.F. 1994. Analytical Model to Predict Annual Soil Surface Temperature Variation. *ASME Journal of Solar Energy Engineering*, 117(2), 91.

Kreider, J.F., Blanc, S.L., Kammerud, R.C., and Curtiss, P.S. 1997. Operational Data as the Basis for Neural Network Prediction of Hourly Electrical Demand. *ASHRAE Transactions*, 103(2).

Kusuda, T., and Achenbach, P.R. 1965. Earth Temperatures and Thermal Diffusivity at Selected Stations in the United States. *ASHRAE Transactions*, 71(1), 61.

LBL. 1980. *DOE-2 User Guide*. Version 2.1, LBL Report LBL-8689, Rev. 2. Lawrence Berkeley Laboratory, Berkeley, CA.

LBL. 1982. *DOE-2 Engineers Manual*. Lawrence Berkeley Laboratory Report LBL-11353. National Technical Information Services, Springfield, VA.

LBNL. 2004. Measurements of Whole-House Standby Power Consumption in California Homes, by Ross, J.P., and Meier, A.K., and published in the *International Energy Journal*, 27(9), 861–868.

Maguire, J., Fang, X., and Krarti, M. 2011. An Analysis Model for Domestic Hot Water Distribution Systems. Proceedings for ASME Energy Sustainability Conference, Washington, DC.

Mayer, P.W. 1995. Residential Water Use and Conservation Effectiveness: A Process Approach. MS thesis, University of Colorado, Boulder.

McCartney, J.S., LaHaise, D., and LaHaise, T. 2009. Feasibility of Incorporating Geothermal Heat Sinks/Sources into Deep Foundations Based on Experience from North Dakota. GSP. The Art of Foundation Engineering Practice. ASCE Conference Proceedings.

McKinsey Global Energy and Materials. 2009. *Unlocking Energy Efficiency in the U.S. Economy.*

Meier, A.K. 1982. *Supply Curves of Conserved Energy*. LBL-14686. Lawrence Berkeley National Laboratory, Berkeley, CA.

Nadel, S., Shepard, M., Greenberg, S., Katz, G., and de Almeida, A. 1991. *Energy-Efficient Motor Systems: A Handbook of Technology, Program, and Policy Opportunities.* American Council for Energy-Efficient Economy, Washington, DC.

NAECA. 2004. National Appliance Energy Conservation Act Standard for Water Heaters. Department of Energy, Washington, DC.

NAESCO. 1993. NAESCO Standard for Measurement of Energy Savings for Electric Utility Demand Side Management (DSM) Projects. National Association of Energy Services Companies, Washington, DC.

NCDC. 2010. National Climatic Data Center. U.S. Department of Commerce. http://www.ncdc.noaa.gov.

NEC. 1996. National Electrical Code. National Fire Protection Association, Quincy, MA.

NEC. 2008. National Electrical Code. National Fire Protection Association, Quincy, MA.

NEMA. 2006. *Energy Management Guide for Selection and Use of Polyphase Motors.* Standard MG-10-1994. National Electrical Manufacturers Association, Rosslyn, VA.

NLPIP. 1995. Power Quality. Lighting Answers. *Newsletter by the National Lighting Product Information Program*, 2(2), 5 (updated report on 2005).

OECD. 2011. Economic Statistics. http://www.ocde.org.

Omer, A. 2006. Ground-Source Heat Pump Systems and Applications. *Renewable and Sustainable Energy Reviews*, 12 (2).

Ooka, R., Sekine, K., Mutsumi, Y., Yoshiro, S., and SuckHo, H. 2007. Development of a Ground Source Heat Pump System with Ground Heat Exchanger Utilizing the Cast in Place Concrete Pile Foundations of a Building. EcoStock 2007, Stockton, NJ.

ORNL. 2005. http://www.ornl.gov/sci/roofs+walls/whole_wall/.

Periera, L.S. 1996. Evapotranspiration: Review of Concepts and Future Trends. Procedures: Evapotranspiration and Irrigation Schedules. In *American Society of Agricultural Engineers Conference Proceedings*, San Antonio, TX, pp. 109–115.

Pescatore, P., Roth, K.W., and Brodick, J. 2004. Condensing Natural Gas Water Heaters. *ASHRAE Journal*, 46(2), 51–52.

Phelan, J., Brandemuehl, M.J., and Krarti, M. 1996. *Methodology Development to Measure In-Situ Chiller, Fan, and Pump Performance.* Final Report, ASHRAE Project RP-827. American Society of Heating, Refrigerating, and Air Conditioning Engineering, Atlanta, GA.

Raffio, R., Isambert, O., Mertz, G., Schreier, C., and Kissock, K. 2007. Targeting Residential Energy Assistance. Proceedings of the ASME 2007 Conference on Energy Sustainability.

Reddy, T.A., Deng, S., and Claridge, D.E. 1999. Development of an Inverse Method to Estimate Overall Building and Ventilation Parameters of Large Commercial Buildings. *ASME Journal of Solar Energy Engineering*, 121(1), 40–46.

Ritschard, R.L., Handford, J.W., and Sezgen, A.O. 1992. *Single-Family Heating and Cooling Requirements: Assumptions, Methods, and Summary Results.* Topical Report LBL-30377. Gas Research Institute, Chicago.

RS Means. 2008. *Cost Works.*

Schoenau, G.J., and Kehrig, R.A. 1990. A Method for Calculating Degree-Days to Any Base Temperature. *Energy and Buildings*, 14, 299–302.

Sekine, K., Ooka, R. Yokoi, M., Shiba, Y., and Hwang, S.-H. 2007. Development of a Ground-Source Heat Pump System with Ground Heat Exchanger Utilizing the Cast-in-Place Concrete Pile Foundations of Buildings. *ASHRAE Transactions*, 113(1), 558–56.

Seo, D., and Krarti, M. 2010. *Energy Analysis of Cecil Residential Complex*. Final Report. Building Systems Program, University of Colorado, Boulder.

Shapiro, I. 2011. 10 Common Problems in Energy Audits. *ASHRAE Journal*, 53(2), 26–32.

Sherman, M., and Dickerhoff, D. 1998. *Air-Tightness of U.S. Dwellings*. Energy Performance of Buildings Group, Energy and Environment Division, Lawrence Berkeley National Laboratory, Berkeley, CA.

Sherman, M.H., and Matson, N.E. 1993. Ventilation-Energy Liabilities in US Dwellings. In *Proceedings of 14th AIVC Conference*, Copenhagen, Denmark, pp. 23–41.

Shui, B., Evans, M., Lin, H., Jiang, W., Liu, B., Song, B., and Somasundaram, S. 2009. *Country Report on Building Energy Codes in China*. PNNL-17909. Pacific Northwest National Laboratory, Richland, WA.

Simmonds, P. 1993. Thermal Comfort and Optimal Energy Use. *ASHRAE Transactions*, 99(1), 1037–1048.

Slusher, D., *Automatic Building Energy Model Simulation and Calibration Tool*, MS Report, University of Colorado, Boulder, CO, 2010.

Sterling, R.L., Gupta, S., Shen, L.S., and Goldberg, L.F. 1993. *Assessment of Soil Thermal Conductivity for Use in Building Design and Analysis*. Final Report, ASHRAE Project 701-RP.

Slusher, D., and Krarti, M. 2011. Automatic Analysis Tool for Building Energy Simulation Modeling and Calibration. Submitted to *Building and Energy Journal*.

Sonderegger, R.C. 1985. Thermal Modeling of Buildings as a Design Tool. In *Proceedings of CHMA 2000*, Vol. 1.

SRCC. 1994. *Operating Guidelines and Minimum Standards for Certifying Solar Water Heating Systems*. Solar Rating and Certification Corporation, Cocoa, FL.

Studer, D., and Krarti, M. 2010. Evaluation of Ground Source Heat Pump Energy, Demand, and Greenhouse Potential in Colorado Residential Buildings. Proceedings of ASME Energy Sustainability Conference, Phoenix, AZ.

Surles, W.A. 2011. Development and Application of a Control Analytic Tool for Evaluating Automated Residential Smart Grid Control Strategies. MS thesis, University of Colorado, Boulder.

Taha, H. 1997. Urban Climates and Heat Islands: Albedo, Evapotranspiration, and Anthropogenic Heat. *Energy and Buildings*, 25, 99.

TEMA. 2010. *Elaboration d'un Plan d'Action pour la Renovation Thermique et Energetique des Batiments Existants en Tunisie—Synthese*. Report by TEMA Consulting and CEESEN. Tunis, Tunisia.

Tuhus-Dubrow, D., and Krarti, M. 2009. Comparative Analysis of Optimization Approaches to Design Building Envelope for Residential Buildings. *ASHRAE Transactions*, 115(2), 205–219.

Tuhus-Dubrow, D., and Krarti, M. 2010. Genetic Algorithm Based Approach to Optimize Building Envelope Design for Residential Buildings. *Building and Environment*, 45(7), 1574–1584.

Tuluca A., and Steven Winter Associates. 1997. *Energy Efficient Design and Construction for Commercial Buildings.* New York: McGraw Hill.

Tumura, G.T., and Shaw, C.Y. 1976. Studies on Exterior Wall Air-Tightness and Air Infiltration of Tall Buildings. *ASHRAE Transactions,* 82(1), 122–129.

Turiel, I. 1997. Present Status of Residential Appliance Energy Efficiency Standards, an International Review. *Energy and Buildings,* 26(1), 5.

Verderber, R., Morse, C., and Alling, R. 1993. Harmonics from Compact Fluorescent Lamps. *IEEE Transactions on Industry Applications,* 29(3), 670.

Waide, P., Lebot, B., and Hinnells, M. 1997. Appliance Energy Standards in Europe. *Energy and Buildings,* 26(1), 45.

Wanga, W., Zmeureanu, R., and Rivard, H. 2005. Applying Multi-Objective Genetic Algorithms in Green Building Design Optimization. *Building and Environment,* 40, 1512–1525.

Watson, R.D., and Chapman, K.S. 2002. *Radiant Heating and Cooling Handbook.* McGraw-Hill, New York.

WBCSD. 2009. *Energy Efficiency in Buildings: Transforming the Market.* Report by the World Business Council for Sustainable Development, Washington, DC.

Weiss, W. 2003. *Solar Heating Systems for Houses: A Design Handbook for Solar Combisystem.* James & James (Science Publishers) Ltd., London.

Weiss, W., Bergmann, I., and Faninger, G. 2008. *Solar Heat Worldwide, Markets and Contributions to the Energy Supply.* AEE—Institute for Sustainable Technologies and International Energy Agency, Solar Heating and Cooling Programme, Gleisdorf, p. 47.

Wetter, M. 2004. GenOpt®, Generic Optimization Program. Lawrence Berkeley National Laboratory, Berkeley, CA. http://gundog.lbl.gov/GO/download/documentation.pdf.

Wilson, E.J.H. 2011. The Energy Implications of Air-Side Fouling in Constant Air Volume HVAC Systems. MS thesis, University of Colorado, Boulder.

Wright, J.A., Loosemore, H.A., and Farmani, R. 2002. Optimization of Building Thermal Design and Control by Multi-Criterion Genetic Algorithm. *Energy and Buildings,* 43, 959–972.

WSEO. 2009. MotorMaster Electric Motor Selection Software. Washington State Energy Office, Olympia, WA.

Xcel Energy. 2009. Xcel Energy Announces Energy Star Programs for New and Existing Homes. http://www.xcelenergy.com/Company/Newsroom/Pages/NewsRelease2009-03-24-XcelEnergyannouncesENERGYSTAR%C2%AEprograms fornewandexistinghomes.aspx.

Yoon, J., Lee, E., and Claridge, D. 2003. Calibration Procedure for Energy Performance Simulation of a Commercial Building. *Journal of Solar Energy Engineering,* 125, 251–257.

Zhang, J., Wang, R.Z., and Wu, J.Y. 2007. System Optimization and Experimental Research on Air Source Heat Pump Water Heater. *Applied Thermal Engineering,* 27, 1029–1035.

Zhu, Y. 2006. Applying Computer-Based Simulation to Energy Auditing: A Case Study. *Energy and Buildings,* 38, 421–428.

Index

C

Conversion Factors (Metric to English)

Area	1 m²	= 1550.0 in²
		=10.764 ft²
Energy	1 J	= 9.4787×10^{-4} Btu
Heat transfer rate	1 W	= 3.4123 Btu/h
Heat flux	1 W/m²	= 0.3171 Btu/hr·ft²
Heat generation rate	1 W/m³	= 0.09665 Btu/hr·ft³
Heat transfer coefficient	1 W/m² ·K	= 0.17612 Btu/hr·ft²·°F
Latent heat	1 J/kg	= 4.2995×10^{-4} Btu/lb$_m$
Length	1 m	= 3.2808 ft
	1 km	= 0.62137 mile
Mass	1 kg	= 2.2046 lb$_m$
Mass density	1 kg/m3	= 0.062428 lb$_m$/ft³
Mass flow rate	1 kg/s	= 7936.6 lb$_m$/h
Mass transfer coefficient	1 m/s	= 1.1811×10^4 ft/h
Pressure	1 N/m²	= 0.020886 lb$_f$/ft²
		= 1.4504×10^{-4} lb$_f$/in.²
		= 4.015×10^{-3} in. water
Power	1 kW	= 1.340 HP
		= 3,412 Btu/hr
	1×10^5 N/m²	= 1 bar
Refrigeration capacity	1 kJ/hr	= 94,782 Btu/hr
		= 7.898×10^{-5} ton
	1 kW	= 0.2844 ton
Specific heat	1 J/kg·K	= 2.3886×10^{-4} Btu/lb$_m$·°F
Temperature	K	= (5/9) °R
		= (5/9)(°F + 459.67)
		= °C + 273.15
Temperature difference	1 K	= 1 °C
		= (9/5) °R = (9/5) °F
Thermal conductivity	1 W/m·K	= 0.57782 Btu/hr·ft·°F
Thermal resistance	1 K/W	= 0.52750 °F/hr·Btu
Volume	1 m³	= 6.1023×10^4 in³
		= 35.314 ft³
		= 264.17 gal
Volume flow rate	1 m³/s	= 1.2713×10^5 ft³/hr
	1 L/s	= 127.13 ft³/hr
		= 2.119 ft³/min

Printed and bound by CPI Group (UK) Ltd, Croydon, CR0 4YY

18/10/2024

01776266-0011